T0305735

MAGNETIC ACTUATORS AND SENSORS

IEEE Press
445 Hoes Lane
Piscataway, NJ 08854

IEEE Press Editorial Board
John Anderson, *Editor in Chief*

Linda Shafer	Saeid Nahavandi	George Zobrist
George W. Arnold	Om P. Malik	Tariq Samad
Ekram Hossain	Mary Lanzerotti	Dmitry Goldgof

Kenneth Moore, *Director of IEEE Book and Information Services (BIS)*

Technical Reviewers
Mark Solveson, *ANSYS Corporation*
Mark A. Juds, *Eaton Corporation*
Yogeshwarsing Calleecharan, Ph.D.

MAGNETIC ACTUATORS AND SENSORS

SECOND EDITION

John R. Brauer

IEEE PRESS

WILEY

Copyright © 2014 by The Institute of Electrical and Electronics Engineers, Inc.

Published by John Wiley & Sons, Inc., Hoboken, New Jersey. All rights reserved.
Published simultaneously in Canada.

No part of this publication may be reproduced, stored in a retrieval system, or transmitted in any form or
by any means, electronic, mechanical, photocopying, recording, scanning, or otherwise, except as
permitted under Section 107 or 108 of the 1976 United States Copyright Act, without either the prior
written permission of the Publisher, or authorization through payment of the appropriate per-copy fee to
the Copyright Clearance Center, Inc., 222 Rosewood Drive, Danvers, MA 01923, (978) 750-8400,
fax (978) 750-4470, or on the web at www.copyright.com. Requests to the Publisher for permission
should be addressed to the Permissions Department, John Wiley & Sons, Inc., 111 River Street, Hoboken,
NJ 07030, (201) 748-6011, fax (201) 748-6008, or online at http://www.wiley.com/go/permission.

Limit of Liability/Disclaimer of Warranty: While the publisher and author have used their best efforts in
preparing this book, they make no representations or warranties with respect to the accuracy or
completeness of the contents of this book and specifically disclaim any implied warranties of
merchantability or fitness for a particular purpose. No warranty may be created or extended by sales
representatives or written sales materials. The advice and strategies contained herein may not be suitable
for your situation. You should consult with a professional where appropriate. Neither the publisher nor
author shall be liable for any loss of profit or any other commercial damages, including but not limited to
special, incidental, consequential, or other damages.

For general information on our other products and services or for technical support, please contact our
Customer Care Department within the United States at (800) 762-2974, outside the United States at
(317) 572-3993 or fax (317) 572-4002.

Wiley also publishes its books in a variety of electronic formats. Some content that appears in print may
not be available in electronic formats. For more information about Wiley products, visit our web site at
www.wiley.com.

Library of Congress Cataloging-in-Publication Data:

Brauer, John R., 1943–
  Magnetic actuators and sensors / John R. Brauer. – Second edition.
    pages cm
  ISBN 978-1-118-50525-0 (hardback)
  1. Actuators. 2. Detectors. I. Title.
  TJ223.A25B73 2013
  621.34–dc23

                                          2013018187

Printed in the United States of America

10 9 8 7 6 5 4 3 2 1

■ CONTENTS

Since the publication of this book 7 years ago, I have received hundreds of emails from dozens of readers. I would like to thank all of you for your feedback. Your thoughtful questions, helpful suggestions, and many minor corrections have encouraged me to undertake an improved second edition.

Another important reason for a second edition is that considerable progress has been made over the past 7 years in the analysis and design of magnetic actuators and sensors. Accordingly, this edition has a number of added sections to cover new material. Other changes over the past 7 years include the availability of software products. While the free version of Maxwell software (Maxwell SV, a subset of Maxwell version 9) can no longer be downloaded, the Maxwell files used in both the editions still function in the commercial Maxwell version 16 now sold by ANSYS, Inc. Two free magnetic finite-element software products currently available are mentioned in Chapter 4.

Additions for the second edition are summarized here. In Part I, Chapter 1 has a new section on mechatronics and added figures. Chapter 2 adds magnetization and magnetization curve material. Chapter 3 has a new large figure comparing various types of circuits. Chapter 4 updates available finite-element software. Chapter 5 has new material on Halbach magnets, and a new section on magnetic volume forces on permeable particles.

In Part II, Chapter 7 has two new sections, one on magnetic bearings and one on magnetic separators. Chapter 9 has its Section 9.3 greatly expanded to begin with Maxwell's equations. It also has a new large section added on magnetic infusion and effusion.

In Part III, Chapter 10 has a new figure on the range of magnetic field magnitudes, new material on encoders and current sensors, and a new section on GMR spin valve sensors. Chapter 11 has four new sections: Chattock coils, SQUID magnetometers, magnetoimpedance and miniature sensors, and MEMS sensors.

In Part IV, Chapter 14 clarifies both its electromagnetic and mechanical sections, and it ends with new material on reciprocating linear actuators. Chapter 15 has a new frequency domain analysis, a new section on actuators for 2D planar motion, and a new final section containing two detailed examples of optimization of magnetic actuators and systems. Chapter 16 adds new sections on optimizing an electrohydraulic system and on digital hydraulic valves. Finally, Appendix A has been expanded and new Appendices B and C have been added. Files for all examples in this book now appear at http://booksupport.wiley.com.

I thank all of my colleagues over many decades for their friendship and help, and I would like to especially thank the following for their contributions to this second edition. Mark Juds has contributed material including the B–H data of Appendix B. Mark Solveson is the first author of the design optimization studies added to Chapters 15 and 16. The reviewers of both editions made many helpful suggestions which I have endeavored to fulfill. Also I would like to thank my wife, Susan, for again reading aloud every word of this edition and suggesting changes for clarity. Finally, this year marks the 100th anniversary of the birth of my mother, Elizabeth, who taught me many good things including her love of books. I hope this book helps the next generation of engineers.

JOHN R. BRAUER

jbrauer@ieee.org; http://johnrbrauer.com
Fish Creek, Wisconsin

▬▬ PREFACE TO THE FIRST EDITION

This book is written for practicing engineers and engineering students involved with the design or application of magnetic actuators and sensors. The reader should have completed at least one basic course in electrical engineering and/or mechanical engineering. This book is suitable for engineering college juniors, seniors, and graduate students.

IEEE societies whose members will be interested in this book include the Magnetics Society, Computer Society, Power Engineering Society, Industry Applications Society, and Control System Society. Readers of the *IEEE/ASME Transactions on Mechatronics,* sponsored by the IEEE Industrial Electronics Society, may also want to read this book. Many members of the Society of Automotive Engineers (SAE) might also be very interested in this book because the magnetic devices discussed here are commonly used in automobiles and aircraft.

This book is a suitable text for upper-level engineering undergraduates or graduate students in courses with titles such as "Actuators and Sensors" or "Mechatronics." It can also serve as a supplementary text for courses such as "Electromagnetic Fields," "Electromechanical Energy Conversion," or "Feedback Control Systems." It is also appropriate as a reference book for "Senior Projects" in electrical and mechanical engineering. Its basic material has been used in a 16-hour seminar for industry that I have taught many times at the Milwaukee School of Engineering. More than twice as many class hours, however, will be required to thoroughly cover the contents of this book.

The chapters on magnetic actuators are intended to replace a venerable book by Herbert C. Roters, *Electromagnetic Devices,* published by John Wiley & Sons in 1941. Over the decades since 1941, many technological revolutions have occurred. Perhaps, the most wide-ranging revolution has been the rise of the modern computer. The computer not only uses magnetic actuators and sensors in its disk drives and external interfaces but also enables new ways of analyzing and designing magnetic devices. Hence this book includes the latest computer-aided engineering methods from the most recently published technical papers. The latest software tools are used, especially the electromagnetic finite-element software package Maxwell SV, which are available to students at no charge from Ansoft Corporation, for which I am a part-time consultant. Other software tools used include SPICE, MATLAB, and Simplorer. Simplorer SV, the student version, is also available to students free of charge from Ansoft Corporation. If desired, the reader can work the computational examples and problems with other available software packages, which should yield similar results.

To download Maxwell SV and Simplorer SV along with their example files, please visit the website for this book:

ftp://ftp.wiley.com/public/sci_tech_med/magnetic_actuators/

This book is divided into four parts, each containing several chapters. Part I, on *magnetics,* begins with an introductory chapter defining magnetic actuators and sensors and why they are important. The second chapter is a review of basic electromagnetics, needed because magnetic fields are the key to understanding magnetic actuators and sensors. Chapter 3 is on the reluctance method, a way to approximately calculate magnetic fields by hand. Chapter 4 covers the finite-element method, which calculates magnetic fields very accurately via the computer. Magnetic force is a required output of magnetic actuators and is discussed in Chapter 5, and other magnetic performance parameters are the subject of Chapter 6.

Part II is on *actuators.* Chapter 7 discusses direct current (DC) actuators, while Chapter 8 deals with alternating current (AC) actuators. The last chapter devoted strictly to magnetic actuators is Chapter 9, on their transient operation.

Part III of the book is on *sensors.* Chapter 10 describes in detail the Hall effect and magnetoresistance, and applies these principles to sensing position. Chapter 11 covers many other types of magnetic sensors. However, types of sensors involving quantum effects are not included, because quantum theory is beyond the scope of this book.

Part IV of the book, on *systems,* covers many system aspects common to both magnetic actuators and sensors. Chapter 12 presents coil design and temperature calculations. Electromagnetic compatibility issues common to sensors and actuators are discussed in Chapter 13. Electromechanical performance is analyzed in Chapter 14 using coupled finite elements, while Chapter 15 uses electromechanical system software. Finally, Chapter 16 shows the advantages of electrohydraulic systems that incorporate magnetic actuators and/or sensors. Many examples are presented throughout the book because my teaching experience has shown that they are vital to learning. The examples that are numbered are simple enough to be fully described, solved, and repeated by the reader. In addition, problems at the end of the chapters enable the reader to progress beyond the solved examples.

I would like to thank the many engineers whom I have known for making this book possible. Starting with my father, Robert C. Brauer PE, it has been my great pleasure to work with you for many decades. I thank my wife, Susan McCord Brauer, for her encouragement and advice on writing. Thanks also go to the reviewers of this book for their many excellent suggestions. All of you have taught me many things. This book is my attempt to summarize some of what I have learned and to pass it on.

JOHN R. BRAUER

jbrauer@ieee.org
Fish Creek, Wisconsin
January 2006

LIST OF 66 EXAMPLES

PART I

MAGNETICS

CHAPTER 1

Introduction

Magnetic actuators and sensors use magnetic fields to produce and sense motion. Magnetic actuators enable applied electric voltage or current signals to move objects. To sense the motion with an electric signal produced by magnetic fields, magnetic sensors are often used.

Since computers have inputs and outputs that are electrical signals, magnetic actuators and sensors are ideal for computer control of motion. Hence magnetic actuators and sensors are increasing in popularity. Motion control that was in the past accomplished by manual command is now increasingly carried out by computers with magnetic sensors as their input interface and magnetic actuators as their output interface.

Both magnetic actuators and magnetic sensors are energy conversion devices, using the energy stored in static, transient, or low frequency magnetic fields. This book is focused on these magnetic devices, not on devices using electric fields or high frequency electromagnetic fields.

1.1 OVERVIEW OF MAGNETIC ACTUATORS

Figure 1.1 is a block diagram of a magnetic actuator. Input electrical energy in the form of voltage and current is converted to magnetic energy. The magnetic energy creates a magnetic force, which produces mechanical motion over a limited range, typically along a straight line but sometimes rotating over an arc. Thus magnetic actuators convert input electrical energy into output mechanical energy. As mentioned in the caption of Figure 1.1, the blocks are often nonlinear (output not proportional to input), as will be discussed later in this book.

Typical magnetic actuators include the following.

- Contactors, circuit breakers, and relays to control electric motors and circuits.
- Switchgear and relays for electric power transmission and distribution.
- Head positioners for computer disk drives.

Magnetic Actuators and Sensors, Second Edition. John R. Brauer.
© 2014 The Institute of Electrical and Electronics Engineers, Inc. Published 2014 by John Wiley & Sons, Inc.

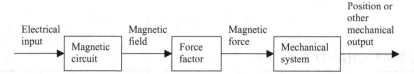

FIGURE 1.1   Block diagram of a magnetic actuator. The blocks are not necessarily linear. Both the magnetic circuit block and the force factor block are often nonlinear. The force factor block often produces a force proportional to the square of the magnetic field.

- Loudspeakers.
- Fuel injectors in engines of automobiles, trucks, and locomotives.
- Electrohydraulic valves in airplanes, tractors, robots, automobiles, and other mobile or stationary equipment.
- Biomedical prosthetic devices for artificial hearts, limbs, ears, and other organs.
- Magnetic separators that produce forces on magnetic objects large and small, including particles smaller than a micron targeted within the human body for tumors, etc.

Since magnetic actuators produce motion over a limited range, other electrome-chanical energy converters with large ranges of motion are not discussed in this book. Thus electric motors that produce multiple 360° rotations are not covered here. However, "step motors" which produce only a few degrees of rotary motion are classified as magnetic actuators and are included in this book.

1.2   OVERVIEW OF MAGNETIC SENSORS

A magnetic sensor has the block diagram shown in Figure 1.2. Compared to a magnetic actuator, the energy flow is different, and the amount of energy is often much smaller. The main input is now a mechanical parameter such as position or velocity, although electrical and/or magnetic input energy is usually needed as well. Input energy is converted to magnetic field energy. The output of a magnetic sensor

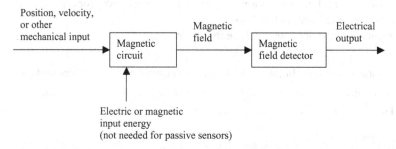

FIGURE 1.2   Block diagram of a magnetic sensor. The blocks are not necessarily linear.

is an electrical signal. In many cases the signal is a voltage with very little current, and thus the output electrical energy is often very small.

Magnetic devices that output large amounts of electrical energy are not normally classified as sensors. Hence typical generators and alternators are not discussed in this book.

Typical magnetic sensors include the following.

- Proximity sensors to determine presence and location of conducting objects for factory automation, bomb or weapon detection, and petroleum exploration.
- Microphones that sense air motion (sound waves).
- Linear variable differential transformers to determine object position.
- Velocity sensors for antilock brakes and stability control in automobiles.
- Hall effect position or velocity sensors.
- Magnetoresistive position or field sensors.

Design of magnetic actuators and sensors involves analysis of their magnetic fields. The actuator or sensor should have geometry and materials that utilize magnetic fields to produce maximum output for minimum size and cost.

1.3   ACTUATORS AND SENSORS IN MOTION CONTROL SYSTEMS

Motion control systems can use nonmagnetic actuators and/or nonmagnetic sensors. For example, electric field devices called *piezoelectrics* are sometimes used as sensors instead of magnetic sensors. Other nonmagnetic sensors include global positioning system (GPS) sensors that use high frequency electromagnetic fields, radio frequency identification (RFID) tags, and optical sensors such as television cameras. Nonmagnetic actuators and sensors are not discussed in this book.

An example of a motion control system that uses both a magnetic actuator and a magnetic sensor is the computer disk drive head assembly shown in Figure 1.3. The head assembly is a magnetic sensor that senses ("reads") not only the computer data magnetically recorded on the hard disk, but also the position (track) on the disk. To position the heads at various radii on the disk, a magnetic actuator called a *voice coil actuator* is used.

Often the best way to control motion is to use a feedback control system as shown in Figure 1.4. Its block diagram contains both an actuator and a sensor. The sensor may be a magnetic sensor measuring position or velocity, while the actuator may be a magnetic actuator producing a magnetic force. It is found that accurate control requires an accurate sensor. Control systems books widely used by electrical and mechanical engineers describe how to analyze and design such control systems [1–4]. The system design requires mathematical models of both actuators and sensors, which will be discussed throughout this book.

Another example of a magnetic actuator and a magnetic sensor is shown in Figure 1.5. It shows a tubular magnetic actuator and a magnetic Hall effect sensor

FIGURE 1.3   Typical computer disk drive head assembly. The actuator coil is the rounded triangle in the upper left. The four heads are all moved inward and outward toward the spindle hub by the magnetic force and torque on the actuator coil. Portions of the actuator and all magnetic disks are removed to allow the coil and heads to be seen.

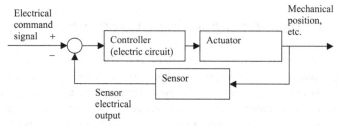

FIGURE 1.4   Basic feedback control system which may use both a magnetic actuator and a magnetic sensor.

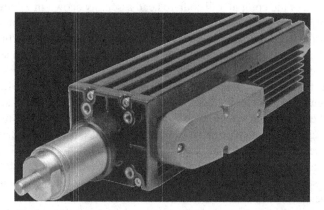

FIGURE 1.5   Magnetic actuator with built-in magnetic sensor, producing straight-line motion along its axis. Figure courtesy of Dunkermotoren Linear Systems.

packaged together to produce and sense motion along a straight line. This linear motion is accomplished without any gears or chains, thus enabling long maintenance-free life with low friction. Associated electronic controls enable precise motion control.

1.4 MAGNETIC ACTUATORS AND SENSORS IN MECHATRONICS

The word "mechatronics" is a blend of the words mechanics and electronics [5]. Mechatronic systems contain both mechanical components and electronics with controlling software. To enable the electronics to control mechanical motion, electromechanical devices are used, often containing magnetic actuators and sensors, as shown in Figures 1.1–1.5.

Figure 1.6 depicts mechatronics as made up of four major overlapping systems [6]. The mechanical systems are controlled by electrical/electronic systems, computer systems, and control systems—all working together. Note all four major systems have overlaps; one overlap area is called electromechanical systems. Magnetic actuators and sensors are important components of electromechanical systems.

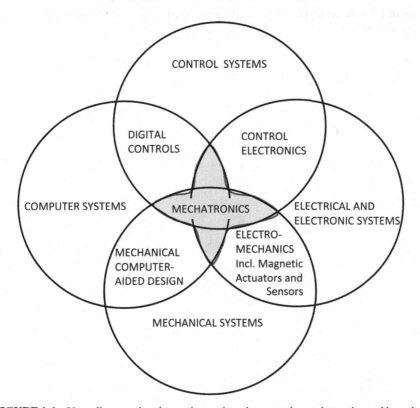

FIGURE 1.6 Venn diagram showing major engineering areas in mechatronics and how they relate to magnetic actuators and sensors.

Figure 1.6 is actually a simplified picture of the overlapping and multidisciplinary or "multiphysics" nature of mechatronics. This book also deals with additional overlaps not explicitly indicated in Figure 1.6, for example, the use of computer software to analyze and design magnetic actuators and sensors. To understand mechatronic systems containing magnetic actuators and sensors, this book is ordered in parts devoted to Magnetics, then Actuators, then Sensors, and finally to the resulting Systems.

REFERENCES

1. Dorf RC, Bishop RH. *Modern Control Systems*, 9th ed. Upper Saddle River, NJ: Prentice-Hall Inc.; 2001.

2. Dorsey J. *Continuous and Discrete Control Systems*, New York: McGraw-Hill; 2002.

3. Phillips CL, Harbor RD. *Feedback Control Systems*, 4th ed. Upper Saddle River, NJ: Prentice-Hall Inc.; 2000.

4. Lumkes JH Jr. *Control Strategies for Dynamic Systems*, New York: Marcel Dekker, Inc.; 2002.

5. Cetinkunt S. *Mechatronics*, Hoboken, NJ: John Wiley & Sons, Inc.; 2007.

6. Kevan T. Mechatronics primer: Reinventing machine design, *Desktop Engineering*, February, 2009. pp. 14–16.

CHAPTER 2

Basic Electromagnetics

Study of magnetic fields provides an explanation of how magnetic actuators and sensors work. Hence this chapter presents the basic principles of *electromagnetics*, a subject that includes magnetic fields.

In reviewing electromagnetic theory, this chapter also introduces various parameters and their symbols. The symbols and notations used in this chapter will be used throughout the book, and most are also listed in Appendix A along with their units.

2.1 VECTORS

Magnetic fields are vectors, and thus it is useful to review mathematical operations involving vectors. A *vector* is defined here as a parameter having both magnitude and direction. Thus it differs from a *scalar*, which has only magnitude (and no direction). In this book, vectors are indicated by **bold** type, and scalars are indicated by italic non-bold type.

To define *direction*, rectangular coordinates are often used. Also called *Cartesian coordinates*, the position and direction are specified in terms of x, y, and z. This book denotes the three rectangular direction unit vectors as $\mathbf{u}_x$, $\mathbf{u}_y$, and $\mathbf{u}_z$; they all have magnitude equal to one.

Common to several vector operations is the "del" operator (also termed "nabla"). It is denoted by an upside down (inverted) delta symbol, and in rectangular coordinates is given by:

$$\nabla = \frac{\partial}{\partial x}\mathbf{u}_x + \frac{\partial}{\partial y}\mathbf{u}_y + \frac{\partial}{\partial z}\mathbf{u}_z \tag{2.1}$$

2.1.1 Gradient

A basic vector operation is *gradient*, also called "grad" for short. It involves the del operator operating on a scalar quantity, for example, temperature T. In rectangular coordinates the gradient of T is expressed as:

$$\nabla T = \frac{\partial T}{\partial x}\mathbf{u}_x + \frac{\partial T}{\partial y}\mathbf{u}_y + \frac{\partial T}{\partial z}\mathbf{u}_z \tag{2.2}$$

Magnetic Actuators and Sensors, Second Edition. John R. Brauer.
© 2014 The Institute of Electrical and Electronics Engineers, Inc. Published 2014 by John Wiley & Sons, Inc.

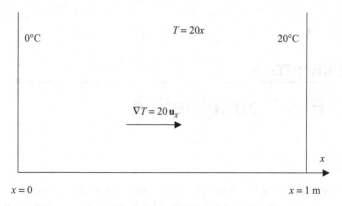

FIGURE 2.1 Temperature distribution and gradient versus position x.

An example of temperature gradient is shown in Figure 2.1. A block of ice is placed to the left of $x = 0$, for position x values less than zero. At $x = 1$ m, a wall of room temperature 20°C is located. Assuming the temperature varies linearly from $x = 0$ to $x = 1$ m, then:

$$T = 20x \qquad (2.3)$$

To find the temperature gradient, substitute (2.3) into (2.2), obtaining:

$$\nabla T = 20\,\mathbf{u}_x\,°C/m \qquad (2.4)$$

The direction of the gradient is the direction of maximum rate of change of the scalar (here temperature). The magnitude of the gradient equals the maximum rate of change per unit length. Since this book uses the SI (Système International) or metric system of units, all gradients here are per meter.

Two other vector operations involve multiplication with the del operator. Another word for multiplication is product, and there are two types of vector products.

Example 2.1 Gradient Calculations Find the gradient of the following temperature distribution at locations $(x,y,z) = (1,2,3)$ and $(4,-2,5)$:

$$T = 5x + 8y^2 + 3z \qquad (E2.1.1)$$

Solution You must be careful in taking the partial derivatives in the gradient equation (2.2), and you must first find the expression for the gradient before evaluating it at any location. Thus the first step is to find the gradient expression:

$$\nabla T = \frac{\partial(5x + 8y^2 + 3z)}{\partial x}\mathbf{u}_x + \frac{\partial(5x + 8y^2 + 3z)}{\partial y}\mathbf{u}_y + \frac{\partial(5x + 8y^2 + 3z)}{\partial z}\mathbf{u}_z$$

$$(E2.1.2)$$

The partial derivative of y with respect to x is zero, and so are all other partial derivatives of non-alike variables, and thus we obtain:

$$\nabla T = \frac{\partial(5x)}{\partial x}\mathbf{u}_x + \frac{\partial(8y^2)}{\partial y}\mathbf{u}_y + \frac{\partial(3z)}{\partial z}\mathbf{u}_z \qquad \text{(E2.1.3)}$$

Carrying out the derivatives gives:

$$\nabla T = 5\mathbf{u}_x + 16y\mathbf{u}_y + 3\mathbf{u}_z \qquad \text{(E2.1.4)}$$

Finally, the gradient can be evaluated at the two specified locations:

$$\nabla T(1, 2, 3) = 5\mathbf{u}_x + 16(2)\mathbf{u}_y + 3\mathbf{u}_z = 5\mathbf{u}_x + 32\mathbf{u}_y + 3\mathbf{u}_z \qquad \text{(E2.1.5)}$$

$$\nabla T(4, -2, 5) = 5\mathbf{u}_x + 16(-2)\mathbf{u}_y + 3\mathbf{u}_z = 5\mathbf{u}_x - 32\mathbf{u}_y + 3\mathbf{u}_z \qquad \text{(E2.1.6)}$$

Recall that the gradient must always be a vector. Its magnitude is always the square root of the sum of the squares of its x, y, and z components.

2.1.2 Divergence

The *scalar product* or *dot product* obtains a scalar and is denoted by a "dot" symbol. Applying it to the del operator and a typical vector, here called **J**, obtains "del dot **J**," called the *divergence* of the vector:

$$\nabla \cdot \mathbf{J} = \frac{\partial J_x}{\partial x} + \frac{\partial J_y}{\partial y} + \frac{\partial J_z}{\partial z} \qquad \text{(2.5)}$$

The divergence of a vector is its net outflow per unit volume, which is a scalar. In some cases, the divergence is zero, that is, the vector is *divergenceless*. For example, if **J** is current density (to be defined later), then Kirchhoff's law which shows that total current at a point is zero ($\sum I = 0$) can be expressed as a divergenceless **J**:

$$\nabla \cdot \mathbf{J} = 0 \qquad \text{(2.6)}$$

Figure 2.2 shows typical fields with and without divergence.

2.1.3 Curl

The other type of vector product obtains a vector and is called the *vector product* or *cross product*. It is expressed using a cross or x sign. If it is the product of the del

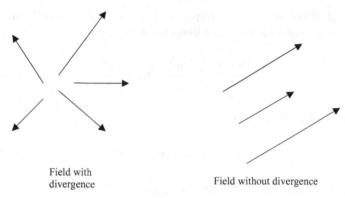

Field with
divergence   Field without divergence

FIGURE 2.2   Field with and without divergence.

operator and a typical vector, here called **A**, one obtains a vector "del cross **A**," called the *curl* of a vector. It can be expressed as a 3 by 3 determinant:

$$
\nabla \times \mathbf{A} = \begin{pmatrix} \dfrac{\partial}{\partial x} & \dfrac{\partial}{\partial y} & \dfrac{\partial}{\partial z} \\ A_x & A_y & A_z \\ \mathbf{u}_x & \mathbf{u}_y & \mathbf{u}_z \end{pmatrix}
\tag{2.7}
$$

A 3 by 3 determinant is evaluated by the "basket-weave" method. The reader is recommended to write out such a 3 by 3 determinant and connect its elements with diagonal lines. Row 1 column 1 (the (1,1) or top left entry) is multiplied by row 2 column 2 and then by row 3 column 3, resulting in one of six terms of the cross product. The next term is found by multiplying the (1,2) entry by the (2,3) entry and the (3,1) entry. The next term multiplies the (1,3), (2,1), and (3,2) entries. The next three terms must be subtracted, and consist of (3,1) times (2,2) times (1,3), then (3,2) times (2,3) times (1,1), and finally (3,3) times (2,1) times (1,2). Thus (2.7) can be rewritten as:

$$
\nabla \times \mathbf{A} = \left(\frac{\partial A_z}{\partial y} - \frac{\partial A_y}{\partial z} \right) \mathbf{u}_x + \left(\frac{\partial A_x}{\partial z} - \frac{\partial A_z}{\partial x} \right) \mathbf{u}_y + \left(\frac{\partial A_y}{\partial x} - \frac{\partial A_x}{\partial y} \right) \mathbf{u}_z
\tag{2.8}
$$

Besides the rectangular (x,y,z) coordinates analyzed above, engineers often use cylindrical coordinates or spherical coordinates. In cylindrical and spherical coordinates there are differences in the gradient, divergence, and curl equations [1].

In general, curl is analogous to a wheel that rotates, and thus curl is sometimes called *rot*. Figure 2.3 shows a water wheel and its curl. The wheel has paddles and is rotated by a stream of water. The stream may either be a river (mill stream) or may be a diversion channel inside a dam. Note that the wheel has curl (rotation) about its z axis because its velocities vary with position. Plots of the x and y components of velocity **v** are shown as functions of y and x, respectively. The partial derivatives of (2.8) produce a z-component of curl. The partial of velocity component v_y with

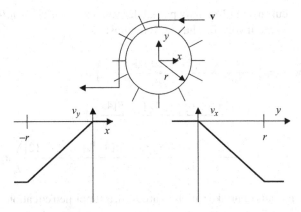

FIGURE 2.3   A rotating wheel is analogous to curl.

respect to x gives a positive contribution, and the negative partial of v_x with respect to y also gives a positive contribution, resulting in a curl of **v** with a large positive z component. Thus the curl is directed along the axis of rotation.

Figure 2.3 also indicates a relation between curl and the integral all around the circular path on the surface of the wheel. The velocity follows the outer circular path. The integral is called the *circulation*. In the next section, Stokes' law will mathematically define the relation between circulation and curl.

Example 2.2   Divergence and Curl of a Vector   Find the divergence and curl at location $(3,2,-1)$ of the vector:

$$\mathbf{A} = [8\,x^4 + 6(y^2 - 2)]\mathbf{u_x} + [9x + 10y + 11z]\mathbf{u_y} + [4x]\,\mathbf{u_z} \qquad \text{(E2.2.1)}$$

Solution   You must first find the expressions for divergence and curl, and then evaluate them at the desired location.

In finding the divergence using (2.5), only partial derivatives of *like variables* (such as x with respect to x) are involved. Thus we obtain:

$$\nabla \cdot \mathbf{A} = \frac{\partial A_x}{\partial x} + \frac{\partial A_y}{\partial y} + \frac{\partial A_z}{\partial z}$$

$$= \frac{\partial[8x^4 + 6(y^2 - 2)]}{\partial x} + \frac{\partial[9x + 10y + 11z]}{\partial y} + \frac{\partial[4x]}{\partial z} \qquad \text{(E2.2.2)}$$

Again, a partial derivative with respect to x treats y and z as constants, the partial with respect to y treats x and z as constants, and the partial with respect to z treats x and y as constants. Thus we obtain for the divergence expression:

$$\nabla \cdot \mathbf{A} = \frac{\partial[8x^4]}{\partial x} + \frac{\partial[10y]}{\partial y} + \frac{\partial[0]}{\partial z} = 32x^3 + 10 \qquad \text{(E2.2.3)}$$

which evaluated at $(3,2,-1)$ gives 874. Recall that the divergence is always a scalar.

In finding the curl using (2.8), only partial derivatives of *unlike variables* (such as y with respect to x) are involved. Thus (2.8) obtains:

$$\nabla \times \mathbf{A} = \left(\frac{\partial [4x]}{\partial y} - \frac{\partial [9x + 10y + 11z]}{\partial z} \right) \mathbf{u}_x$$

$$+ \left(\frac{\partial [8x^4 + 6y^2 - 12]}{\partial z} - \frac{\partial [4x]}{\partial x} \right) \mathbf{u}_y$$

$$+ \left(\frac{\partial [9x + 10y + 11z]}{\partial x} - \frac{\partial [8x^4 + 6y^2 - 12]}{\partial y} \right) \mathbf{u}_z \quad \text{(E2.2.4)}$$

Again, since partials of unlike variables are zero, a simplified equation is obtained:

$$\nabla \times \mathbf{A} = \left(\frac{\partial [0]}{\partial y} - \frac{\partial [11z]}{\partial z} \right) \mathbf{u}_x + \left(\frac{\partial [0]}{\partial z} - \frac{\partial [4x]}{\partial x} \right) \mathbf{u}_y + \left(\frac{\partial [9x]}{\partial x} - \frac{\partial [6y^2]}{\partial y} \right) \mathbf{u}_z$$

$$\text{(E2.2.5)}$$

which yields the curl expression:

$$\nabla \times \mathbf{A} = (-11)\mathbf{u}_x + (-4)\mathbf{u}_y + (9 - 12y)\mathbf{u}_z \quad \text{(E2.2.6)}$$

Substituting the point $(3,2,-1)$ yields the final answer:

$$\nabla \times \mathbf{A} = (-11)\mathbf{u}_x + (-4)\mathbf{u}_y + (-15)\mathbf{u}_z \quad \text{(E2.2.7)}$$

2.2 AMPERE'S LAW

With the background in vector operations presented above, the fundamental source of magnetic fields can now be presented. The origin of magnetic fields is expressed by Ampere's law, named after André-Marie Ampere of France.

Ampere's law at any point in space states that the curl of static magnetic field intensity $\mathbf{H}$ equals current density $\mathbf{J}$:

$$\nabla \times \mathbf{H} = \mathbf{J} \quad (2.9)$$

where $\mathbf{H}$ is magnetic field intensity in A/m, and current density $\mathbf{J}$ is in A/m^2. These units are SI, and will be used throughout this book as listed in Appendix A.

In air, related to $\mathbf{H}$ is magnetic flux density $\mathbf{B}$ by:

$$\mathbf{B} = \mu_o \mathbf{H} \quad (2.10)$$

where $\mathbf{B}$ is magnetic flux density in teslas and μ_o is the permeability of free space (vacuum) or air. The unit, tesla, has the symbol T and is named after the renowned American inventor Nikola Tesla. One tesla (1 T) equals one weber per square meter (1 Wb/m^2). The value $\mu_o = 12.57 \times 10^{-7}$ H/m. Thus in air, vector $\mathbf{B}$ equals vector $\mathbf{H}$ multiplied by a very small number. Often in this book and other books, $\mathbf{B}$ is simply called the magnetic field. A field in this context is any quantity that can vary over space, and $\mathbf{B}$ is a vector field.

Because the curl of $\mathbf{H}$ equals $\mathbf{J}$, $\mathbf{H}$ "circles" around an axis consisting of current much as the wheel circles around its axis in Figure 2.3. Another way to express Ampere's law of (2.9) is to integrate it over a surface $\mathbf{S}$ to obtain:

$$\int (\nabla \times \mathbf{H}) \cdot \mathbf{dS} = \int \mathbf{J} \cdot \mathbf{dS} \qquad (2.11)$$

The units of both sides of this equation are amperes. The surface $\mathbf{S}$ is a vector with direction normal (perpendicular) and magnitude equal to the surface area.

There is a purely mathematical vector identity that can be used to replace the surface integral of the curl in the left side of (2.11). As mentioned at the end of the preceding section, Stokes' law replaces the surface integral by a *closed* path (or line) integral, giving the most common expression for Ampere's law in integral form:

$$\oint \mathbf{H} \cdot \mathbf{dl} = \int \mathbf{J} \cdot \mathbf{dS} = NI \qquad (2.12)$$

where $\mathbf{l}$ is the vector path length in meters, and the path being closed is indicated by the circle on the integral sign. Note that the total current, the surface integral of current density, can also be written as the product of current I times the number of conductors N carrying that current. To create large $\mathbf{H}$ and $\mathbf{B}$ with reasonably small current I values, often a coil winding with many conductors (or turns) N is used in magnetic devices.

While $\mathbf{B}$ is magnetic flux density, its integral over any surface $\mathbf{S}$ is called magnetic flux. Flux is the Latin word for the English word flow, and magnetic flux flows around much like other fluids such as water. Since the SI units of $\mathbf{B}$ are teslas or webers per square meter, magnetic flux has units of webers. Using ϕ for flux, the surface integral can be written as:

$$\phi = \int \mathbf{B} \cdot \mathbf{dS} \qquad (2.13)$$

An important property of magnetic flux is that if $\mathbf{S}$ is a *closed* surface, such as a spherical surface or any surface that completely encloses a volume, then the total magnetic flux through the closed surface is zero. A closed surface integral is indicated by a circle on the integral sign, and thus magnetic flux through a closed surface obeys:

$$\phi_c = \oint \mathbf{B} \cdot \mathbf{dS} = 0 \qquad (2.14)$$

Since divergence has been previously defined as the net output flux per unit volume, the zero flux of (2.14) applied to a tiny volume and its closed surface means that:

$$\nabla \cdot \mathbf{B} = 0 \qquad (2.15)$$

Thus magnetic flux density is always divergenceless. Since the total magnetic flux through any closed surface is zero, magnetic flux flows in a manner similar to that for an incompressible fluid. In air, since $\mathbf{B} = \mu_o\mathbf{H}$, both $\mathbf{B}$ and $\mathbf{H}$ must both "circle around" the "axis" of a current-carrying wire.

Example 2.3 Ampere's Law at a Point and Along a Closed Path Apply Ampere's law to two situations:

(a) Given the magnetic field intensity expression (in A/m):

$$\mathbf{H} = [8x^4 + 6(y^2 - 2)]\,\mathbf{u_x} + [9x + 10y + 11z]\,\mathbf{u_y} + [4x]\mathbf{u_z} \qquad (E2.3.1)$$

Find the expression for current density $\mathbf{J}$ at location (2,4,6).

(b) Given a region of four conductors, each carrying current $I = 5$ A outward. As shown in Figure E2.3.1, two closed paths are defined, l_1 and l_2. Find the integral of $\mathbf{H}$ along each of the closed paths.

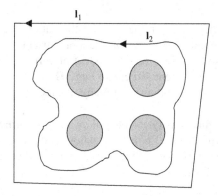

FIGURE E2.3.1 Two closed paths adjacent to a current-carrying region.

Solution

(a) The current density $\mathbf{J}$ at any point is the curl of $\mathbf{H}$ at that point. The $\mathbf{H}$ expression (E2.3.1) is recognized as identical to (E2.2.1) for $\mathbf{A}$, for which the curl has already been found. Thus:

$$\mathbf{J} = \nabla \times \mathbf{H} = (-11)\mathbf{u_x} + (-4)\mathbf{u_y} + (9 - 12y)\mathbf{u_z} \qquad (E2.3.2)$$

which evaluated at (2,4,6) gives:

$$\mathbf{J} = (-11)\mathbf{u}_x + (-4)\mathbf{u}_y + (-39)\mathbf{u}_z \qquad \text{(E2.3.3)}$$

in A/m^2.

(b) Both the number of conductors and their individual currents are known. Their product $NI = 20$ A, since number N is dimensionless. Since both closed paths enclose all conductors, the integral of $\mathbf{H}$ along each of the closed paths equals 20 A.

2.3 MAGNETIC MATERIALS

The main magnetic property that can vary among materials is permeability. The permeability of free space (vacuum), μ_o, is also applicable in air. Also, many other materials, including copper and aluminum, have free space permeability μ_o.

The general symbol for permeability is μ, and materials with permeability other than μ_o are often expressed in terms of relative permeability μ_r defined using:

$$\mu = \mu_r \mu_o \qquad (2.16)$$

where we recall that $\mu_o = 12.57E-7$ H/m, where throughout this book yEz means $y \times 10^z$, and thus $\mu_o = 12.57 \times 10^{-7}$ H/m. Several materials have $\mu_r \gg 1$ and thus are much more permeable than air. Permeability is another word common to both magnetics and fluid flow. Highly fluid-permeable earth, such as sand, conducts water flow much better than does low fluid-permeable rock. Thus magnetic flux, like fluids, prefers to flow in materials of high permeability.

Magnetic materials with high relative permeability μ_r are said to be magnetically *soft*. The most common soft magnetic materials are *ferromagnetics,* including iron (Fe or ferrous material), cobalt (Co), and nickel (Ni). These elements are neighbors on the periodic table, with atomic numbers 26, 27, and 28, respectively. The Curie temperatures of these elements are 770, 1131, and 358°C, respectively, and the high permeability exists as long as the elements are kept below the Curie temperature. Besides these three pure elements, alloys containing these elements are usually also ferromagnetic with high permeability. The most common ferromagnetic alloys are steels, which contain both iron and carbon. The relative permeability of typical steel is often on the order of 2000. Steel is often used as the main inner or *core* material of magnetic devices.

The reason for the high permeability of certain materials is that they contain many *magnetic domains*. Each domain has $\mathbf{B}$ in a particular direction created by its atomic electron motion (to be discussed further in Chapter 10). As $\mathbf{H}$ is applied and increased, more domains rotate and/or expand in the direction of $\mathbf{H}$, causing magnetization $\mathbf{M}$ that increases $\mathbf{B}$ in accordance with:

$$\mathbf{B} = \mu_0(\mathbf{H} + \mathbf{M}) \qquad (2.17)$$

where sometimes **M** is proportional to the applied **H**:

$$\mathbf{M} = \chi\mathbf{H} \tag{2.18}$$

where the dimensionless proportionality constant χ is called magnetic susceptibility. Thus in magnetic materials:

$$\mathbf{B} = \mu\mathbf{H} = \mu_o(1 + \chi)\mathbf{H} \tag{2.19}$$

Each magnetic domain is of size as large as 0.1 mm = 100 μm [2]. Since permeability μ is a macroscopic or average number over all domains, it usually applies only for materials with at least one dimension exceeding a few micrometers. The macroscopic concept of permeability therefore does not usually apply for material samples of nanometer dimensions in all three directions. Often nanotechnology relies instead on microscopic effects such as those of quantum mechanics, which is not covered in this book except briefly in Section 11.8.

As an example of the effect of permeability on magnetic field **B**, Figure 2.4 shows the application of Ampere's law to a single circular wire carrying current I. Ampere's law and material permeability are here used to find **B** at any location of radius $r > r_w$, the wire radius. First, Ampere's law over the closed path of radius r completely enclosing the current I gives:

$$\oint \mathbf{H} \cdot \mathbf{dl} = 2\pi rH = I \tag{2.20}$$

Note that the closed path integral can be replaced by the path length $2\pi r$ times a scalar constant H, the magnitude of the magnetic field intensity, because symmetry requires the field to be independent of angular position, that is to be a constant magnitude at any given radius. The direction of **H** must circle around the current as shown in Figure 2.4. Such a peripheral direction is usually called a circumferential polar direction $\mathbf{u}_\phi$. The direction follows the *right-hand rule:* For your right-hand

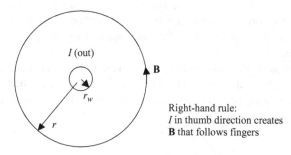

FIGURE 2.4   Magnetic field of a wire found using Ampere's law.

thumb pointing in the direction of the current, the direction of **H** circles around in the direction of the right-hand fingers. Then solving (2.20) gives the magnitude:

$$H = I/(2\pi r) \tag{2.21}$$

Finally, the magnetic flux density **B** is found by multiplying **H** by permeability μ of the material surrounding the wire. If the material is air, then the flux density magnitude is:

$$B = \mu_o H = \mu_o I/(2\pi r) \tag{2.22}$$

Note that the magnetic field is inversely proportional to radius from the current, assuming that the radius is no smaller than the wire radius.

If the material surrounding the wire is steel of relative permeability 2000, then the flux density magnitude is increased to:

$$B = \mu_r \mu_o H = 2000\mu_o I/(2\pi r) \tag{2.23}$$

Note that the direction of **B** is the same as the direction of **H**, which is true for all materials except *anisotropic* (directionally dependent) magnetic materials. Air and many ferromagnetic materials are magnetically *isotropic* with a scalar permeability. Anisotropic magnetic materials require a tensor permeability which is not discussed in this book but has been investigated elsewhere [3,4]. Tensor material properties are used in Chapter 10.

An important property of magnetic permeability to be discussed next is the *nonlinear B–H curve*. Such nonlinearity is exhibited by all ferromagnetic materials. Their high relative permeability, such as the 2000 assumed in (2.23), is only applicable for low values of flux density. For high values of flux density, *saturation* is said to occur. Flux no longer flows as easily, similar to a towel being saturated with water and no longer picking up as much liquid. The ratio *B/H* gradually decreases and thus the permeability is no longer a constant. Figure 2.5 shows a typical nonlinear *B–H* curve for steel. It is customarily plotted with *H* along the horizontal axis. Note that in the neighborhood of 1.5 T or so, the curve has a "knee" that transitions to a much flatter slope. The *incremental permeability,* defined as the slope, gradually decreases to the permeability of free space. Usually the slope equals μ_o at a *saturation flux density* somewhat above 2 T.

To obtain *B–H* curves, steel suppliers and other sources can be consulted. While their curves are often in the SI units of T and A/m, sometimes they are instead in CGS units. For flux density *B* values given in gauss (G), multiply by 1.E−4 to obtain teslas. For field intensity *H* values in oersteds (Oe), multiply by 79.577 to obtain amperes per meter. Steel suppliers can provide better *B–H* curves (with higher permeabilities) with more expensive heat treatment or by adding alloy materials such as Co.

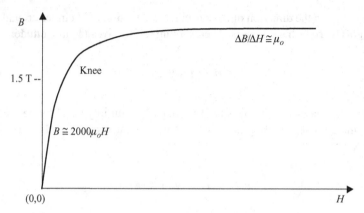

FIGURE 2.5 Typical nonlinear magnetization curve of steel.

A useful approximate expression for steel B–H curves uses three constants and was published in 1975 [5]:

$$H = (k_1 e^{k_2 B^2} + k_3)B \tag{2.24}$$

where values of the three constants are listed in Table 2.1 for typical types of steel. Cast steel is cast or forged into its desired shape. Cold rolled steel has been formed into a thin sheet in a steel rolling mill at typical room temperature. Annealed steel has been heat treated, where annealing involves high temperature heating often in low oxygen atmospheres, followed by slow cooling. The annealing temperature, length of time, atmosphere, cooling process, and other details all can significantly affect the B–H curve. The B–H relation also may actually be a set of multiple curves depending on history of applied H; such *hard magnetic* materials will be discussed in Chapter 5. Mechanical hardness also tends to follow the magnetic hardness (or softness) of materials. However, hard magnetic materials often are brittle, not nearly as mechanically strong as steel.

An improvement on the nonlinear B–H relation of (2.24) has been proposed recently by Mark A. Juds, who notes that (2.24) gives permeability:

$$\mu = B/H = 1/(k_1 e^{k_2 B^2} + k_3) \tag{2.25}$$

TABLE 2.1 Constants for (2.24) uncorrected B–H Curves of Three Typical Types of Steel

	k_1	k_2	k_3
Cast	49.4	1.46	520.6
Cold-rolled	3.8	2.17	396.3
Annealed	2.6	2.72	154.4

But since the slope (incremental μ) should approach μ_o at high B and high H, Juds adds one term to give [6]:

$$\mu = B/H = [1/(k_1 e^{k_2 B^2} + k_3)] + \mu_o \qquad (2.26)$$

This new equation improves the B–H curve fit at high B and significantly enhances the numerical convergence of nonlinear iterative solutions. Appendix B lists the three constants k_1, k_2, and k_3 determined by Juds for a large number of different ferromagnetic materials for the B–H relation (2.26) with the corrected final permeability.

Another important property of magnetic materials is their electrical conductivity. Air has zero electrical conductivity (except in special cases such as lightning), but ferromagnetic materials usually have high electrical conductivities. To see why electrical conductivity is important, Faraday's law must be presented.

Example 2.4 Magnetic Flux Density in Various Materials Surrounding a Wire

(a) A copper wire of radius 1 mm is placed at location $(1,0,0)$ m, carrying current of 100 kA in the $+z$ direction. Find the vector **B** at location $(3,0,0)$ for the wire embedded in the following materials that extend infinitely far in all three directions: (1) air, (2) steel with constant (assuming no saturation) relative permeability $= 2500$, (3) steel with a B–H curve with the following (B,H) values: $(0,0)$, $(1.5,1000)$, $(1.8,7958)$, ...

(b) Repeat the above if the current direction is reversed to the $-z$ direction.

Solution

(a) The radius from $(1,0,0)$ to $(3,0,0)$ is 2 m. From the right-hand rule, the direction of **B** at point $(3,0,0)$ is in the $+y$ direction. From (2.23), the magnitude is:

$$B = \mu_r \mu_o I/(2\pi r) = \mu_r (12.57\text{E}{-}7)(100\text{E}3)/(4\pi) = \mu_r(1.\text{E}{-}2)$$

$$(\text{E2.4.1})$$

(1) Thus for air, $\mathbf{B} = 1.\text{E}{-}2$ T $\mathbf{u}_y$

(2) For steel with relative permeability $= 2500$, $\mathbf{B} = 25$ T $\mathbf{u}_y$

(3) For steel with the nonlinear B–H curve,

$$H = I/(2\pi r) = 1.\text{E}5/(4\pi) = 7958 \text{ A/m} \qquad (\text{E2.4.2})$$

From the (B,H) curve, the corresponding flux density $\mathbf{B} = 1.8$ T $\mathbf{u}_y$

(b) When the current direction is reversed in the above three cases, the right-hand rule shows that the direction of **B** changes to $-\mathbf{u}_y$

2.4 FARADAY'S LAW

Following Ampere's law in importance for magnetic devices is Faraday's law. It was discovered by Michael Faraday of England. While Ampere's law deals primarily with current, Faraday's law deals primarily with voltage.

Faraday's law at any point in space equates the negative partial time derivative of **B** with the curl of the electric field intensity **E**:

$$\nabla \times \mathbf{E} = -\frac{\partial \mathbf{B}}{\partial t} \tag{2.27}$$

where **E** is in volts per meter. The line or path integral of **E** is the negative of voltage V_1:

$$V_1 = -\int \mathbf{E} \cdot d\mathbf{l} \tag{2.28}$$

Taking the integral of both sides of (2.27) over any unchanging surface **S** (a surface changed by motion will be discussed later in this section) gives:

$$\int (\nabla \times \mathbf{E}) \cdot d\mathbf{S} = -\frac{\partial}{\partial t} \int \mathbf{B} \cdot d\mathbf{S} \tag{2.29}$$

As for Ampere's law in integral form, Stokes' law can be used to replace the surface integral of curl by a closed line integral. Thus (2.24) becomes:

$$\oint \mathbf{E} \cdot d\mathbf{l} = -\frac{\partial}{\partial t} \mathbf{B} \cdot d\mathbf{S} \tag{2.30}$$

which is Faraday's law in integral form. The electric field (intensity) is *induced* by a time-varying magnetic field (flux density).

As previously mentioned, often magnetic devices have N conductors, usually arranged in coils of wire with N turns. The voltage induced in such a coil, using the usual sign convention, is $-N$ times that of one turn, and thus from (2.28) the coil voltage V is:

$$V = N \oint \mathbf{E} \cdot d\mathbf{l} \tag{2.31}$$

where the line integral is over the closed path of one turn. Finally, multiplying both sides of (2.30) by N and using (2.31) gives the expression for induced coil voltage:

$$V = -N \frac{\partial}{\partial t} \int \mathbf{B} \cdot d\mathbf{S} \tag{2.32}$$

Recall from (2.13) that the surface integral of flux density is flux, giving:

$$V = -N\frac{\partial \varphi}{\partial t} \tag{2.33}$$

This form of Faraday's law states that voltage is $(-N)$ times the time rate of change of magnetic flux. Note that the flux in a stationary coil must change with time in order for voltage to be induced.

Assuming that the number of turns does not change with time, another expression of Faraday's law is:

$$V = -\frac{\partial \lambda}{\partial t} \tag{2.34}$$

where λ is called *flux linkage* and is defined by:

$$\lambda = N\phi \tag{2.35}$$

Since N is dimensionless, flux linkage has units of webers.

If a conductor moves with a velocity **v** through a magnetic flux density **B**, Faraday's law (for a changing surface) shows that a *motional electric field* is induced given by:

$$\mathbf{E}_{\text{motion}} = \mathbf{v} \times \mathbf{B} \tag{2.36}$$

The velocity **v** is assumed in this book to be much less than the speed of light, so that relativistic effects can be ignored. The motionally induced voltage is found by the line integral of (2.28). Note that a motional voltage can be produced even by a constant (DC) magnetic flux density.

Since a voltage is induced by a time-varying magnetic field, current I may also flow according to Ohm's law of electric circuits:

$$I = V/R \tag{2.37}$$

where R is resistance in ohms, named after Georg Ohm of Germany. Another form of Ohm's law is for fields:

$$\mathbf{J} = \sigma \mathbf{E} \tag{2.38}$$

where σ is electrical conductivity in the reciprocal of ohm-meters, also called siemens per meter (S/m). The reciprocal of conductivity is resistivity ρ in ohm-meters.

If **E** is induced by magnetic fields, then the current density **J** is also said to be induced, and in the same direction for isotropic materials. Induced current densities and induced currents occur in many magnetic devices with time-varying magnetic fields. These induced currents can be either desirable or undesirable.

A prime example of desirable induced current is the current in the secondary (or output) coil of a transformer. Transformers produce secondary voltage and current from Faraday's law as long as the primary (input) coil voltage and current are time varying. Usually the primary voltage and current are AC sinusoids, producing from Ampere's law a time-varying magnetic field and flux. The flux *links* or passes through both the primary and secondary windings. Faraday's law states that this flux induces a secondary voltage and current which depend in magnitude on the number of secondary turns. Thus transformers produce useful induced currents. The induced currents produce their own flux (from Ampere's law) which can be shown to oppose (by on opposite direction to) the original primary magnetic flux according to Lenz's law, a corollary of Faraday's law.

Induced currents may also be undesirable. For example, the transformer described above with its desirable secondary induced current may also have undesirable induced currents in other parts. The transformer is customarily surrounded by a steel case to confine its magnetic fields, and steel has high electrical conductivity. Typical steel conductivities range from about 5.E5 S/m to 1.E7 S/m. The lower values are for steels with high silicon content. Because of the high steel conductivity, the transformer housing often contains induced *eddy* currents, the name coming from their flow patterns which somewhat resemble the eddy patterns of turbulent rivers. The housing eddy currents are of no use and consume power (the product of voltage and current). They also produce heating. Thus eddy currents are usually undesirable.

A common way to reduce eddy currents in steel (or other conductive materials) is to *laminate.* Figure 2.6 shows typical steel *laminations,* or thin sheets. Due to the thin air spaces, surface oxidation, and/or surface treatments, electric current cannot flow from one lamination to another. Instead, any eddy currents are confined to flow within each lamination. Thus steel eddy current loss, often given in watts per cubic meter, is usually greatly reduced by using laminations.

Alternatives to steel laminations are *ferrites* and *composites.* Both have much lower conductivities than does steel, on the order of 1 S/m. However, their *B–H* curves are poorer than most steels. Ferrites saturate at only about 0.4 T. Composites, made of iron powder and insulating binders, have relative permeability of only a few hundred [7].

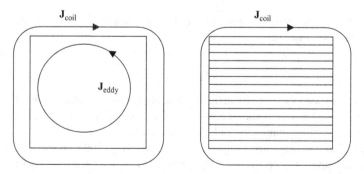

FIGURE 2.6   Comparison of eddy current flow patterns in conductive steel without (left) and with (right) laminations. Due to Lenz' law, the eddy currents tend to flow in the direction opposing the applied coil current.

Eddy currents and their effects will be considered in detail in Chapters 6–16.

Example 2.5   Induced Voltage and Current

(a) A coil called the primary coil establishes the magnetic flux density $\mathbf{B} = 1.1\sin(2\pi ft)\,\mathbf{u}_z$ T. The frequency $f = 60$ Hz. A secondary coil is of thin uniform copper (conductivity 5.8E6 S/m) wire and has resistance $R = 2\ \Omega$. Assume $I = V/R$ and that the magnetic field is not changed by the secondary current. The secondary coil is square in shape, connecting the points $(x,y,z) = (0,0,0)$, $(0.5\text{ m},0,0)$, $(0.5\text{ m},0.6\text{ m},0)$, $(0,0.6\text{ m},0)$ and back to the origin, with a total of 50 turns.

　　Find the voltage and the current induced in the secondary, including their polarities (directions). Also find both $\mathbf{E}$ and $\mathbf{J}$ in the wire.

(b) Repeat for frequency $f = 50$ Hz.

Solution

(a) The coil has turns $N = 50$ and area $A = (0.5\text{ m})(0.6\text{ m}) = 0.3\text{ m}^2$. Faraday's law gives:

$$V = -NA\frac{\partial B}{\partial t} = -50(0.3)\frac{\partial}{\partial t}(1.1\sin 377t) = -15(1.1)(377)\cos 377t$$

$$= -6221\cos(2\pi 60t) \tag{E2.5.1}$$

　　This negative voltage has polarity following the right-hand rule, which is counterclockwise when viewed from the $+z$ axis.

　　The secondary current is the voltage divided by $2\ \Omega$, or $-3111\cos(2\pi 60t)$ A. The electric field is the voltage divided by the wire length; the wire length is $(50)(1\text{ m} + 1.2\text{ m}) = 110$ m. Thus the electric field is $-56.55\cos(2\pi 60t)$ V/m, directed around the loop in the same direction as the current. To find the current density, multiply by the conductivity 5.8E7 S/m, giving 3.28E9 A/m^2 in the same direction as the current.

(b) The lower frequency changes both frequency and amplitude of the secondary voltage and current. The voltage becomes $-5184\cos(2\pi 50t)$ V, and the current is $-2592\cos(\pi 50t)$ A. The electric field is $-47.13\cos(2\pi 50t)$ V/m and the current density is 2.73E9 A/m^2. Directions and polarities do not change.

2.5   POTENTIALS

Potentials are very useful in electrical engineering. The most common example is voltage, which is a scalar, not a vector.

A DC or static electric field is the negative gradient of electrostatic scalar potential, here denoted as ϕ_v:

$$\mathbf{E} = -\nabla \phi_v \qquad (2.39)$$

where ϕ_v has units of volts. For time-varying fields, the induced electric field of Faraday's law must also be added.

Similar to magnetic flux density, there is an electric flux density $\mathbf{D}$ (in units of coulombs per square meter) defined by:

$$\mathbf{D} = \varepsilon \, \mathbf{E} \qquad (2.40)$$

where ε is permittivity in farads per meter, a material property. Air and vacuum have $\varepsilon = \varepsilon_o = 8.854\text{E}{-}12$ F/m.

Materials with permittivity other than ε_o are often expressed in terms of relative permittivity, ε_r defined using:

$$\varepsilon = \varepsilon_r \, \varepsilon_o \qquad (2.41)$$

Another common term for relative permittivity is dielectric constant.

There is an important relation between $\mathbf{D}$ and electric charge density:

$$\nabla \cdot \mathbf{D} = \rho_v \qquad (2.42)$$

where ρ_v is electric charge density per unit volume in units of coulombs per cubic meter.

The relation between electric scalar potential, electric field, and charge is illustrated in the capacitor shown in Figure 2.7. The capacitor consists of two metal plates spaced 1 m apart in air. Because they are made of high conductivity copper or aluminum, there is essentially zero voltage drop along either plate. The lower plate at $y = 0$ is

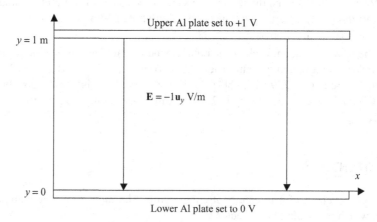

FIGURE 2.7   Two plates of a capacitor spaced 1 m apart and with 1 V DC applied.

grounded, that is, set to 0 V. The upper plate at $y = 1$ is attached to a 1 V DC battery. Thus the expression for the electric scalar potential in volts in Figure 2.6 is $\phi_v = y$ volts. Substituting into (2.39) gives:

$$E = -\nabla y = -\frac{\partial y}{\partial x}\mathbf{u}_x - \frac{\partial y}{\partial y}\mathbf{u}_y - \frac{\partial y}{\partial z}\mathbf{u}_z = -1\mathbf{u}_y \text{ V/m} \qquad (2.43)$$

Note that the electric field in Figure 2.7 points from high scalar potential in volts to low scalar potential. Since **D** is **E** times the permittivity of air, and from (2.42) is *not* divergenceless, both **D** and **E** terminate on the plates, which contain electric charge. Unlike magnetic flux lines, which must always "circle" due to their divergencelessness, the electric flux lines of Figure 2.7 can terminate or stop.

For magnetic fields, the potential required is a vector, not a scalar. Magnetic vector potential **A** is defined using:

$$\mathbf{B} = \nabla \times \mathbf{A} \qquad (2.44)$$

Substituting into Faraday's law at a point (2.27) gives:

$$\nabla \times \mathbf{E} = -\frac{\partial}{\partial t}(\nabla \times \mathbf{A}) = -\left(\nabla \times \frac{\partial \mathbf{A}}{\partial t}\right) \qquad (2.45)$$

where the above **E** is the induced electric field intensity:

$$\mathbf{E}_{\text{ind}} = -\frac{\partial \mathbf{A}}{\partial t} \qquad (2.46)$$

Adding this induced electric field to the electric field due to electric scalar potential of (2.39) gives the total electric field expression:

$$\mathbf{E} = -\nabla \phi_v - \frac{\partial \mathbf{A}}{\partial t} \qquad (2.47)$$

Example 2.6   Fields from Potentials   Given the following potentials in a region, find its fields **B** and **E**:

$$\phi_v = 2x, \qquad \mathbf{A} = 0.2y \, \sin(2\pi 60t) \, \mathbf{u}_z \qquad (E2.6.1)$$

Solution

$$\mathbf{B} = \nabla \times \mathbf{A} = (\nabla \times 0.2y\mathbf{u}_z) \sin(2\pi 60t) = \frac{\partial(0.2y)}{\partial y}\mathbf{u}_x(\sin 2\pi 60t)$$

$$= 0.2 \sin(2\pi 60t)\mathbf{u}_x \qquad (E2.6.2)$$

$$\mathbf{E} = -\nabla \phi_v - \frac{\partial \mathbf{A}}{\partial t} = -2\mathbf{u}_x - (0.2)(2\pi 60) \cos(2\pi 60t)\mathbf{u}_z \, y \qquad (E2.6.3)$$

2.6 MAXWELL'S EQUATIONS

James Clerk Maxwell of Scotland summarized all of electromagnetics in four equations. Expressed in point form, they are:

$$\nabla \cdot \mathbf{B} = 0 \tag{2.48}$$

$$\nabla \cdot \mathbf{D} = \rho_v \tag{2.49}$$

$$\nabla \times \mathbf{E} = -\frac{\partial \mathbf{B}}{\partial t} \tag{2.50}$$

$$\nabla \times \mathbf{H} = \mathbf{J} + \frac{\partial \mathbf{D}}{\partial t} \tag{2.51}$$

These four equations need not be in any particular order. As listed above, the first three have been previously given in this chapter: (2.48) expresses the continuity of magnetic flux, (2.49) expresses that electric flux terminates on charge, and (2.50) expresses Faraday's law. Note that the final equation, (2.51), is Ampere's law that has been enhanced by an additional term on its right-hand side. The additional term is called *displacement current density*:

$$\mathbf{J}_{\text{disp}} = \frac{\partial \mathbf{D}}{\partial t} \tag{2.52}$$

Displacement current density and its surface integral, displacement current, only exist for time-varying fields. Examples of displacement current include current through capacitors and coupled high frequency electromagnetic fields. Coupled electromagnetic fields [1, 8] are often of frequencies from 1 MHz to over 1 GHz, and are further discussed in Chapter 13.

Maxwell's equations can also be written in integral form by integrating the preceding four differential Maxwell equations, obtaining respectively:

$$\oint \mathbf{B} \cdot d\mathbf{S} = 0 \tag{2.53}$$

$$\oint \mathbf{D} \cdot d\mathbf{S} = \rho_v dV \tag{2.54}$$

$$\oint \mathbf{E} \cdot d\mathbf{l} = -\frac{\partial}{\partial t} \int \mathbf{B} \cdot d\mathbf{S} \tag{2.55}$$

$$\oint \mathbf{H} \cdot d\mathbf{l} = \int \left(\mathbf{J} + \frac{\partial \mathbf{D}}{\partial t} \right) \cdot d\mathbf{S} \tag{2.56}$$

This entire chapter is summarized by the above Maxwell's equations. Now that they are known and understood, their application to real world problems can begin in the next chapter.

Example 2.7   Displacement Current in a Capacitor   The parallel-plate capacitor of Figure E2.7.1 has the following electric field **E** between its plates:

$$\mathbf{E} = 2500 \sin(2\pi 60 t)\,\mathbf{u}_y \text{ V/m}$$

where y is directed between the plates, which are separated by 50 mm of air. Each plate is of area 40 mm by 30 mm. Find the displacement current density and the total current in the capacitor.

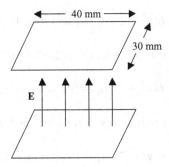

FIGURE E2.7.1   A capacitor made of two 30 by 40 mm plates spaced by 10 mm and with a known AC electric field intensity.

Solution

$$\mathbf{J}_{\text{disp}} = \frac{\partial \mathbf{D}}{\partial t} = \varepsilon_o \frac{\partial \mathbf{E}}{\partial t} = (8.854\text{E}{-}12)(2\pi 60)(2500)\cos(2\pi 60 t)\mathbf{u}_y$$

$$= 8.34\text{E}{-}6\cos(2\pi 60 t)\mathbf{u}_y \text{ A/m}^2 \qquad (\text{E2.7.1})$$

$$I_C = \int \mathbf{J}_{\text{disp}} \cdot d\mathbf{S} = (8.34\text{E}{-}6\cos 377t)(0.04)(0.03)$$

$$= 1.\text{E}{-}8\cos 377t \text{ A} \qquad (\text{E2.7.2})$$

PROBLEMS

2.1   A temperature distribution in Cartesian coordinates follows the equation:

$$T(x, y, z) = 10x + 20y^3 + 30z$$

Find the expression for the temperature gradient.

2.2 A mosquito flies in the direction of maximum rate of change of temperature, thereby seeking warm-blooded animals. If the temperature distribution obeys:

$$T(x, y, z) = 40x + 10y^2 + 30z$$

and the mosquito is sitting at location (1,2,3) m, find the direction it will fly initially.

2.3 Find the divergence and curl at location (1,2,−3) of the vector:

$$\mathbf{A} = [8x^4 + 6(y^2 - 2)]\,\mathbf{u}_x + [9x + 10y + 11z]\,\mathbf{u}_y + [4x]\,\mathbf{u}_z$$

2.4 Find the divergence and curl at location (2,−3,4) of the vector:

$$\mathbf{A} = [5x^4 + 6(y^3 - 2)]\,\mathbf{u}_x + [9x + 11y + 12z]\,\mathbf{u}_y + [4y]\mathbf{u}_z$$

2.5 Given the magnetic field intensity distribution (in A/m):

$$\mathbf{H} = [5x^4 + 6(y^3 - 2)]\,\mathbf{u}_x + [9x + 11y + 12z]\,\mathbf{u}_y + [4x]\,\mathbf{u}_z$$

Find the expression for current density **J** at location (2,4,3).

2.6 Given a region of 20 conductors, each carrying current $I = 10$ A. As shown in Figure E2.3.1, two closed paths are defined, $\mathbf{l}_1$ and $\mathbf{l}_2$. Find the integral of **H** along each of the closed paths.

2.7 A copper wire of radius 2 mm is placed at location (0,3,0) m, carrying current of 50 kA in the $+z$ direction. Find the vector **B** at location (4,0,0) for the wire embedded in the following materials that extend infinitely far in all the three directions: (1) air, (2) steel with constant relative permeability = 2500, (3) steel with a B–H curve with the following (B,H) values: (0,0), (1.5,1000), (1.55,1592), and (1.8,7958).

2.8 Same as the preceding problem, except the current direction is reversed to the $-z$ direction.

2.9 A coil called the primary coil establishes the magnetic flux density $\mathbf{B} = 1.0\sin(2\pi ft)\,\mathbf{u}_z$ T. The frequency $f = 60$ Hz. A secondary coil is of thin uniform aluminum (conductivity 3.54E7 S/m) wire and has resistance $R = 4$ Ω. Assume $I = V/R$ and that the magnetic field is not changed by the secondary current. The secondary coil is square in shape, connecting the points $(x,y,z) = (0,0,0)$, (0.5 m,0,0), (0.5 m,0.6 m,0), (0,0.6 m,0) and back to the origin, with a total of 40 turns.

 Find the voltage and the current induced in the secondary, including their polarities (directions). Also find both **E** and **J** in the wire.

2.10 Same as preceding problem, except the frequency is changed to 400 Hz.

2.11   Given the following potentials in a region, find its fields **B** and **E**:

$$\phi_v = 4y, \quad \mathbf{A} = 0.2x \ \sin(2\pi 60t)\,\mathbf{u}_z$$

2.12   Given the following potentials in a region, find its fields **B** and **E**:

$$\phi_v = 4xy, \quad \mathbf{A} = 0.3x \ \sin(2\pi 50t)\,\mathbf{u}_z$$

2.13   The parallel-plate capacitor of Figure 2.7 has the following electric field **E** between its plates:

$$\mathbf{E} = 4500 \ \sin(2\pi 50t)\,\mathbf{u}_y \ \text{V/m}$$

where y is directed between the plates, which are separated by 5 mm of air. Each plate is of area 70 mm by 50 mm. Find the displacement current density and the total current in the capacitor.

2.14   The parallel-plate capacitor of Figure 2.7 has the following electric field **E** between its plates:

$$\mathbf{E} = 3500 \ \sin(2\pi 400t)\,\mathbf{u}_y \ \text{V/m}$$

where y is directed between the plates, which are separated by 8 mm of air. Each plate is of area 40 mm by 90 mm. Find the displacement current density and the total current in the capacitor.

REFERENCES

1. Sadiku M. *Elements of Electromagnetics*, 3rd ed. New York: Oxford University Press; 2001.
2. Kittel C. *Introduction to Solid State Physics*, 3rd ed. New York: John Wiley & Sons; 1967.
3. Brauer JR, Nakamoto E. Finite element analysis of nonlinear anisotropic B-H Curves. *Int J Applied Electromagn Mater* 1992;3.
4. Brauer JR, Hirtenfelder F. Anisotropic materials in electromagnetic finite element analysis, *Int. Conf. on Computation in Electromagnetics*, London, November 1991.
5. Brauer JR. Simple equations for the magnetization and reluctivity curves of steel. *IEEE Trans Magn* 1975;11:81.
6. Juds MA. *Notes on Solenoid Design*, Milwaukee, WI: Eaton Corporate R&D; 2012.
7. Skibinski GL, Schram BG, Brauer JR, Badics Z. Finite element prediction of losses and temperatures of laminated and composite inductors for AC drives, *Proc. IEEE Int. Electric Machines and Drives Conference*, June 2003.
8. Morgenthaler FR. *The Power and Beauty of Electromagnetic Fields*, Hoboken, NJ: Wiley IEEE Press; 2011.

CHAPTER 3

Reluctance Method

The reluctance method is a way of using Ampere's law of the preceding chapter to solve for magnetic fluxes and magnetic fields. Simplifying assumptions must be made, and thus the reluctance method cannot always obtain accurate results. For very simple problems, its results are often reasonably accurate, and can be quickly obtained "on the back of an envelope." Thus it often serves as a first step in the process of designing magnetic actuators and sensors.

3.1 SIMPLIFYING AMPERE'S LAW

The reluctance method begins with Ampere's law in integral form from the preceding chapter:

$$\oint \mathbf{H} \cdot \mathbf{dl} = NI \tag{3.1}$$

where ampere-turns NI are assumed given, and magnetic field intensity $\mathbf{H}$ and magnetic flux density $\mathbf{B}$ are to be found. The closed line integral is replaced by a summation:

$$\sum_k H_k l_k = NI \tag{3.2}$$

where the closed path consists of line segments of subscript k, each of which is in the direction of the field intensity. Thus the direction of the field and flux is assumed known. Because flux density $\mathbf{B}$ prefers to flow through high permeability materials, the line segments (straight or curved) are assumed to follow a path through high permeability materials such as steel. If the steel has a gap made of air, the closed path follows the shortest part of the airgap.

Magnetic Actuators and Sensors, Second Edition. John R. Brauer.
© 2014 The Institute of Electrical and Electronics Engineers, Inc. Published 2014 by John Wiley & Sons, Inc.

To account for the permeability of the closed path of (3.2), recall that **B** is permeability μ times **H**, giving:

$$\sum_{k} (B_k/\mu_k)l_k = NI \tag{3.3}$$

where μ_k is the permeability of path segment k. Next, recall from the preceding chapter that flux is the surface integral of flux density:

$$\phi = \int \mathbf{B} \cdot \mathbf{dS} \tag{3.4}$$

Assuming that each path segment has cross-sectional surface area S_k normal to the segment direction carrying B_k, each segment carries the flux:

$$\varphi_k = B_k S_k \tag{3.5}$$

Substituting into (3.3) gives:

$$\sum_{k} [(\varphi_k/(\mu_k S_k)]l_k = NI \tag{3.6}$$

Recall from the preceding chapter that flux is continuous (because the divergence of flux density is zero). Thus the flux through all segments of (3.6) is the same value, giving:

$$\varphi \sum_{k} [(l_k/(\mu_k S_k)] = NI \tag{3.7}$$

The term being summed is called *reluctance*, symbolized by the script letter $\mathcal{R}$. Thus (3.7) becomes:

$$\varphi \sum_{k} \mathcal{R}_k = NI \tag{3.8}$$

Units of reluctance must be amperes per weber; see Appendix A for alternative expression of its units.

From the preceding two equations, *reluctance* is defined as:

$$\mathcal{R} = l/(\mu S) \tag{3.9}$$

If all path reluctances are known, then (3.8) can be used to find the unknown flux:

$$\varphi = (NI)/\left(\sum_{k} \mathcal{R}_k\right) \tag{3.10}$$

TABLE 3.1   Basic Analogous Parameters of Electric Circuits and Magnetic Circuits

Parameter	Electric Circuit	Magnetic Circuit
Flow	Current I in amperes (A)	Flux ϕ in webers (Wb)
Potential	EMF in volts (V)	MMF in ampere-turns
Potential/flow	Resistance R in ohms (Ω)	Reluctance $\mathcal{R}$ in A/Wb
Flow/potential	Conductance G in siemens (S)	Permeance $\mathcal{P}$ in Wb/A
Flow density	Current density J in A/m^2	Flux density B in teslas (T)

With flux known, individual flux densities can be found using (3.5) to give:

$$B_k = \varphi_k / S_k \qquad (3.11)$$

Thus reluctances can be used in the reluctance method to solve for flux and flux density everywhere along the closed flux path. The simplest equation for the reluctance method is a simplification of (3.8):

$$\varphi\mathcal{R} = NI \qquad (3.12)$$

which is analogous to the familiar Ohm's law of electric circuits:

$$IR = V \qquad (3.13)$$

Thus the reluctance method is also called *the magnetic circuit method*. In magnetic circuits, reluctance is analogous to resistance, and flux is analogous to current. The excitation of magnetic circuits is ampere-turns, analogous to voltage. Since voltage is sometimes called electromotive force (EMF), NI is sometimes called *magnetomotive force* (MMF). Also, just as any voltage drop in an electric circuit can be called an *EMF drop*, the corresponding drop in a magnetic circuit is an *MMF drop*. Table 3.1 summarizes the basic analogies between electric and magnetic circuits. Figure 3.1 shows the magnetic circuit parameters.

The reluctance method or magnetic circuit method is also sometimes called the *permeance method*. Permeance is defined as the reciprocal of reluctance:

$$\mathcal{P} = 1/\mathcal{R} \qquad (3.14)$$

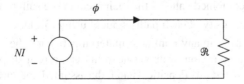

FIGURE 3.1   Basic magnetic circuit parameters.

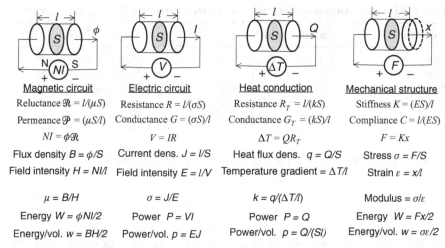

Magnetic circuit	Electric circuit	Heat conduction	Mechanical structure
Reluctance $\mathcal{R} = l/(\mu S)$	Resistance $R = l/(\sigma S)$	Resistance $R_T = l/(kS)$	Stiffness $K = (ES)/l$
Permeance $\mathcal{P} = (\mu S/l)$	Conductance $G = (\sigma S)/l$	Conductance $G_T = (kS)/l$	Compliance $C = l/(ES)$
$NI = \phi \mathcal{R}$	$V = IR$	$\Delta T = Q R_T$	$F = Kx$
Flux density $B = \phi/S$	Current dens. $J = I/S$	Heat flux dens. $q = Q/S$	Stress $\sigma = F/S$
Field intensity $H = NI/l$	Field intensity $E = l/V$	Temperature gradient $= \Delta T/l$	Strain $\varepsilon = x/l$
$\mu = B/H$	$\sigma = J/E$	$k = q/(\Delta T/l)$	Modulus $= \sigma/\varepsilon$
Energy $W = \phi NI/2$	Power $P = VI$	Power $P = Q$	Energy $W = Fx/2$
Energy/vol. $w = BH/2$	Power/vol. $p = EJ$	Power/vol. $p = Q/(Sl)$	Energy/vol. $w = \sigma\varepsilon/2$

FIGURE 3.2 Analogous static (DC) parameters for magnetic circuits, electric circuits (Section 2.4), heat conduction (Section 12.4.1), and mechanical structures (Section 14.3). For more information on units, see Appendix A.

and thus permeance is proportional to permeability. Figure 3.2 shows reluctance and permeance (and other parameters) in magnetic circuits and how they compare with analogous parameters in electric circuits, thermal circuits, and mechanical structures. Thermal and mechanical problems will be discussed further in Part IV of this book.

The reluctance method consists of the following steps.

(1) Find the closed flux path (or paths) that "circles" the given ampere-turns NI. This path is usually through high permeability materials such as steel, but may also contain air segments, which should be as short as possible. The path direction must follow the right-hand rule, where your thumb is in the direction of NI.

(2) Find the lengths and cross-sectional areas of all path segments, assuming average values of lengths and the inverses of areas. Also note the permeability of each path segment. For nonlinear materials, assume initially that their permeability is the constant value below their "knee."

(3) Find the reluctance of each path segment using (3.9).

(4) Combine the reluctances and use (3.10) or (3.8) to find the flux. Reluctances in series add directly. Reluctances in parallel combine just like resistances in parallel; the combined value is the reciprocal of the sum of reciprocals.

(5) Find the flux density of each path segment using (3.11).

(6) If the flux density in any nonlinear material is beyond the knee, then make a new (lower) assumption of its permeability and repeat steps (3), (4), and (5) until the calculated flux densities match the assumed nonlinear permeabilities. For details and various examples consult books such as that by Roters [1].

3.2   APPLICATIONS

Example 3.1   Reluctance Method for "C" Steel Path with Airgap   The first example of the reluctance method is shown in Figure E3.1.1. A "C"-shaped piece of steel of uniform thickness 0.1 m lies in the plane of the paper. The steel is of depth 0.1 m into the page. The opening of the "C" is an airgap of length 0.1 m between the steel poles. (A magnetic *pole* is a surface where magnetic flux leaves to form a North pole or enters to form a South pole.) The object is to find magnetic flux density **B** throughout the steel and airgap produced by the 10,000 ampere-turns directed out of the page in a coil inside the "C" and returning to its left. The steel is assumed to have a relative permeability of 2000.

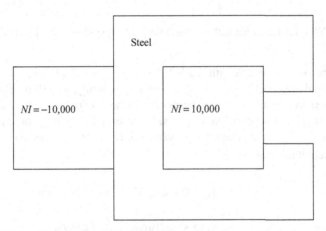

FIGURE E3.1.1   Steel magnetic structure with a single airgap. Each side of the steel "C" core is 0.4 m long and the steel thickness and depth are 0.1 m.

Solution   Follow the six steps at the end of Section 3.1.

(1) From the right-hand rule, the flux flows counterclockwise (CCW) around the steel "C" as shown in Figure E3.1.2. The flux path shown is in the middle of the steel and airgap, and thus is of average length. The longest flux path (not shown) follows the outer surface of the steel, while the shortest flux path follows the interface between the winding and the steel.

(2) The steel segment length along the flux path is:

$$\ell_{Fe} = (0.4 + 0.4 + 0.4 + 0.15 + 0.15)\,m = 1.5\,m$$

The steel segment cross-sectional area is $(0.1\ m)(0.1\ m) = 0.01\ m^2$. (At the four corners the area is bigger, but the corners are only a small portion of the steel.) The steel relative permeability is given as 2000.

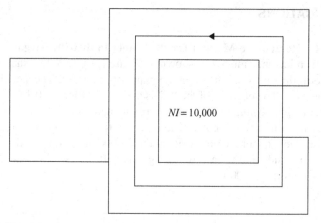

FIGURE E3.1.2   Flux path of Figure E3.1.1 following the right-hand rule.

The air segment length, called the airgap, is 0.1 m. The airgap cross-sectional area is $(0.1 \text{ m})(0.1 \text{ m}) = 0.01 \text{ m}^2$. Actually, as will be discussed in the next section, because air extends beyond the steel poles, the flux can expand or "fringe" to a larger area. For now, however, fringing is ignored and hence 0.01 m^2 is assumed. The relative permeability of air (or vacuum) is 1.

(3) The reluctances are

$$\mathcal{R}_{\text{Fe}} = 1.5/[(20{,}000)(12.57\text{E–}7)(0.01)] = 5966 \qquad \text{(E3.1.1)}$$

$$\mathcal{R}_{\text{air}} = 0.1/[(12.57\text{E–}7)(0.01)] = 7.955\text{E6} \qquad \text{(E3.1.2)}$$

(4) The flux obeys:

$$\phi(59{,}666 + 7.955\text{E6}) = 10{,}000 \qquad \text{(E3.1.3)}$$

$$\varphi = 12.477\text{E–}4 \text{ Wb} \qquad \text{(E3.1.4)}$$

(5) Since the cross-sectional area is approximately the same everywhere, everywhere the flux density is the same:

$$B = \frac{\varphi}{S} = \frac{12.477\text{E–}4}{1.\text{E–}2} = 0.125 \text{ T} \qquad \text{(E3.1.5)}$$

(6) Since the above flux density is less than 1.5 T, which is the approximate knee of steel B–H curves, the above calculated values are valid. However, one must remember that all reluctance method calculations are approximate.

Example 3.2   Reluctance Method for Sensor with Variable Airgap   The second example of the reluctance method is shown in Figure E3.2.1. It shows a simplified magnetic sensor with stationary stator and movable armature. Both stator and armature are made of steel, with the stator also having a coil made of copper or aluminum. The armature is shown in two positions. In the first position the armature and stator teeth are aligned. In the second position the armature teeth are moved halfway between the stator teeth, so they are as misaligned as possible.

The stator coil in Figure E3.2.1 has 1000 ampere-turns. The stator and armature teeth are shown to have width equal to 1 cm. The airgap between the stator and rotor teeth is 1 mm when they are aligned. In the misaligned position, the airgap is assumed to increase to 1 cm. The depth into the page of the sensor of Figure E3.2.1 is given as 10 cm.

The steel permeability is given as infinitely high, and thus only airgap reluctances are needed. Hence dimensions of the steel are not given.

The airgap flux densities are to be found for both armature positions in Figure E3.2.1.

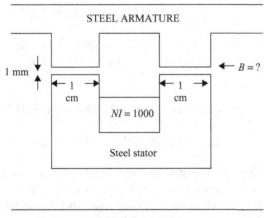

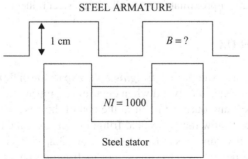

FIGURE E3.2.1   Portion of simple magnetic sensor.

Solution   Follow the six steps at the end of the preceding Section 3.1.

(1) From the right-hand rule, the flux flows CCW around the steel stator "U" shape. It must cross the airgap twice, once entering the armature on its right and then leaving the armature on its left.
(2) The air segment length is total air length over the entire closed path. Thus it equals the right airgap plus the left airgap. Hence in the aligned position the air length is 2 mm, while in the misaligned position the air length is 2 cm = 20 mm.
(3) The reluctances are airgap reluctances only. For the two positions, aligned and misaligned, they are proportional to total air lengths:

$$\mathcal{R}_{\text{align}} = 0.002/[(12.57\text{E}{-}7)(0.01)(0.10)] = 1.591\text{E}6 \quad \text{(E3.2.1)}$$

$$\mathcal{R}_{\text{misalign}} = 0.02/[(12.57\text{E}{-}7)(0.01)(0.10)] = 15.91\text{E}6 \quad \text{(E3.2.2)}$$

(4) The flux obeys:

$$\phi_{\text{align}} = (1000/\mathcal{R}_{\text{align}}) = 628.5\text{E}{-}6 \text{ Wb} \quad \text{(E3.2.3)}$$

$$\phi_{\text{misalign}} = (1000/\mathcal{R}_{\text{misalign}}) = 62.85\text{E}{-}6 \text{ Wb} \quad \text{(E3.2.4)}$$

(5) Assuming the cross-sectional area for the airgap flux is always 1 cm by 10 cm, the airgap flux density is the same on both sides. For the two positions, B varies as:

$$B_{\text{align}} = \phi_{\text{align}}/[(0.01)(0.10)] = 1000\phi_{\text{align}} = 0.6285 \text{ T} \quad \text{(E3.2.5)}$$

$$B_{\text{misalign}} = \phi_{\text{misalign}}/[(0.01)(0.10)] = 1000\phi_{\text{misalign}} = 0.06285 \text{ T} \quad \text{(E3.2.6)}$$

(6) Since the above flux densities are less than 1.5 T, which is the approximate knee of steel B–H curves, the assumption of infinite steel permeability might be approximately true. However, one must remember that all reluctance method calculations are approximate, especially if the steel reluctance is ignored.

3.3 FRINGING FLUX

As mentioned in solving Example 3.1, *fringing* is the expansion of flux in air. Magnetic flux paths in steel can expand outward when entering air, much like water flow paths through a hose can expand outward in air at the end of the hose or nozzle. Thus the area seen by the flux or flow increases, and fringing can cause the airgap reluctance to decrease significantly. Since, as in Examples 3.1 and 3.2, airgap reluctance is often the largest reluctance determining the flux, ignoring fringing can reduce the accuracy of the reluctance method.

In general, fringing flux can be ignored when the airgap path length is small compared with the airgap width. One can define a *fringing factor* as:

$$K_{\text{fringe}} = \mathcal{R}_{\text{nofringe}}/\mathcal{R}_{\text{fringe}} \qquad (3.15)$$

where the factor is usually greater than 1, because the reluctance with fringing is less than that without fringing. For example, if a steel leg is cylindrical with diameter D and the airgap length is g, then K_{fringe} is a function of the ratio g/D that increases with the ratio. For g/D less than approximately 0.04, K_{fringe} is close to 1 and fringing can usually be ignored. For larger g/D ratios, however, fringing can cause greater than 5% changes in magnetic parameters such as reluctance and force.

Reluctances or permeances due to fringing have been derived for several common geometries. Examples include quarter and half cylinders and shells [1–3]. Example 5.2 will display fringing flux. In many geometries, parallel flux paths mean that reluctances or permeances in magnetic circuits must be placed in parallel as well as in series [4], analogous to parallel and series electric circuits.

3.4 COMPLEX RELUCTANCE

The reluctance derived above is a real number and is based upon Ampere's law. To include AC eddy losses produced by Faraday's law currents in conducting materials such as steel, the reluctance becomes a complex number. Complex numbers are customarily used in AC electrical engineering to account for phase shifts between AC sinusoids. Complex permeability, which produces complex permeance and complex reluctance, is often specified for semiconducting ferrites [5, 6].

In general, the reluctance $\mathcal{R}$ is the sum of ferromagnetic (Fe) and airgap reluctances, thus:

$$\phi = NI/(\mathcal{R}_{\text{Fe}} + \mathcal{R}_{\text{gap}}) \qquad (3.16)$$

The airgap reluctance is real (lossless), but the steel reluctance may be complex in order to include eddy current losses and/or other power losses [7]. Hence $\mathcal{R}_{\text{Fe}}$ has both a real part and a positive imaginary part:

$$\mathcal{R}_{\text{Fe}} = \mathcal{R}_{\text{RE}} + j\mathcal{R}_{\text{IM}} = |\mathcal{R}_{\text{Fe}}|\angle\theta \qquad (3.17)$$

where θ is the phase angle. Chapter 8 will discuss and apply complex reluctance to AC devices.

3.5 LIMITATIONS

The reluctance method described above has several limitations. The limitations and ways to address them are the following.

(1) Fringing flux in air is either ignored or approximated with fringing factors, or derived assuming the shape of the flux path.

(2) The path area of each part is often assumed to be average path area. A more accurate method is to use the reciprocal of the average reciprocal area, since reluctance is proportional to the reciprocal of area.

(3) Since reluctance is inversely proportional to permeability, predicting nonlinear B–H effects in steel is difficult. However, B–H curves can be used in the reluctance method [1, 3].

(4) As discussed in the preceding section, losses in steel can be represented by the use of complex reluctance. However, obtaining values of complex reluctance is often very difficult.

(5) Most real world devices have complicated geometries with multiple flux paths that are difficult to analyze. However, magnetic circuits can be constructed with series and parallel reluctances to more accurately model such devices.

(6) For many magnetic devices, however, the necessity of assuming flux paths makes the reluctance method inaccurate. Thus the reluctance method cannot be accurately applied to many devices.

PROBLEMS

3.1   Redo Example 3.1 with steel relative permeability equal to 100.

3.2   Change both the tooth width and the airgap in Figure E3.2.1 of Example 3.2. The tooth width on both sides is reduced to 8 mm. The airgap is 2 mm in the aligned position and 11 mm in the misaligned position.

3.3   Apply Ampere's law directly to Example 3.2 to obtain flux density in both airgaps for both positions. Ampere's law can be applied directly because H and B are the same on both left and right sides due to their exact symmetry, and the steel is assumed to have zero reluctance.

3.4   Apply Ampere's law directly to Problem 3.2.

REFERENCES

1. Roters HC. *Electromagnetic Devices*. New York: John Wiley & Sons; 1941.
2. Kallenbach E, Eick R, Quendt P, Ströhla T, Feindt K, Kallenbach M. *Elektromagnete*, 2nd ed. Weisbaden, Germany: B. G. Teubner Verlag/Springer; 2003 (in German).
3. Juds MA, *Notes on Solenoid Design*. Milwaukee, WI: Eaton Corporate R&D; 2012.
4. Batdorff MA, Lumkes JH. High-fidelity magnetic equivalent circuit model for an axisymmetric electromagnetic actuator. *IEEE Trans Magn* 2009;45:3064–3072.
5. Brauer JR, Hirtenfelder F. Anisotropic materials in electromagnetic finite element analysis, *Conference on Computation in Electromagnetics*, London, November 1991.
6. Valstyn EP, Hadiono T. The transfer function of a ferrite head. *IEEE Trans Magn* 1990;26:2424–2426.
7. Payne A, Cain W, Hempstead R, McCrea G. Measuring core losses in thin film heads. *IEEE Trans Magn* 1996; 32:3518–3520.

CHAPTER 4

Finite-Element Method

Because of the limitations of the reluctance method of the preceding chapter, engineers have long sought more accurate ways to calculate magnetic fields. This chapter is devoted to the finite-element method, the most general and accurate method yet found [1–3].

Even before digital computers existed, a few scientists and engineers employed *relaxation techniques* [4] to iteratively solve governing differential equations of phenomena including heat and magnetics. For magnetic fields, for example, Ampere's law at a point can be solved using a method commonly called the *finite-difference method*. Whether solved on a computer or not, the finite-difference method involves replacing differentials such as ∂x with differences such as Δx. Usually the geometric spacing, such as Δx and Δy in two dimensions, must be uniform everywhere, although some finite-difference software allows a nonuniform "grid" of spacing. The process of subdividing the problem region into such a gridwork is called *discretization*, and a major problem with the finite-difference method is that a uniform discretization is not appropriate for many engineering devices. Most devices have regions with intricate geometry requiring fine discretization, along with regions without intricate geometry not requiring such detailed modeling.

4.1 ENERGY CONSERVATION AND FUNCTIONAL MINIMIZATION

The *finite-element method* can be derived using the basic principle of conservation of energy. Assuming for now that the region analyzed has no power loss, then the energy input must equal the energy stored:

$$W_{in} = W_{stored} \tag{4.1}$$

The finite-element method has been applied to structural analysis using mechanical energies, to thermal analysis using thermal energies, and to fluid dynamics using fluidic energies [1].

Magnetic Actuators and Sensors, Second Edition. John R. Brauer.
© 2014 The Institute of Electrical and Electronics Engineers, Inc. Published 2014 by John Wiley & Sons, Inc.

The finite-element equation (4.1) is here applied to magnetic field problems. Energy is stored in magnetic fields and is input using a current density **J**. Thus expressions for the energies over the problem volume v can be substituted to obtain:

$$\frac{1}{2} \int \mathbf{J} \cdot \mathbf{A} dv = \int \frac{B^2}{2\mu} dv \qquad (4.2)$$

where constant permeability μ (linear B–H) is assumed for now, and integrations are taken on both sides over the problem volume. Recall from Chapter 2 that **A** is magnetic vector potential, here unknown. Recall also that magnetic field (flux density) **B** is the curl of **A**:

$$\mathbf{B} = \nabla \times \mathbf{A} \qquad (4.3)$$

The next step in finite-element analysis is to define an *energy functional* as the difference between stored energy and input energy [1]:

$$F = W_{\text{stored}} - W_{\text{input}} \qquad (4.4)$$

For linear magnetic fields, substituting (4.2) gives the functional:

$$F = \int \left[\frac{B^2}{2\mu} - \frac{1}{2} \int \mathbf{J} \cdot \mathbf{A} \right] dv \qquad (4.5)$$

The Law of Energy Conservation requires the functional to be zero. However, rather than setting (4.5) directly to zero, the finite-element method minimizes the functional. That is, the correct solution for unknown fields **B** and **A** is found by setting the partial derivative of the functional to zero:

$$\frac{\partial F}{\partial A} = 0 \qquad (4.6)$$

As shown in most calculus texts, the derivative is a slope. At a minimum of a function, the slope is zero, as shown for example in Figure 4.1.

Substituting the functional of (4.5) into (4.6) obtains:

$$\frac{\partial}{\partial A} \int \frac{B^2}{2\mu} dv = \int J \, dv \qquad (4.7)$$

The above equation is the basis for finite-element analysis of linear magnetostatic (DC magnetic) fields.

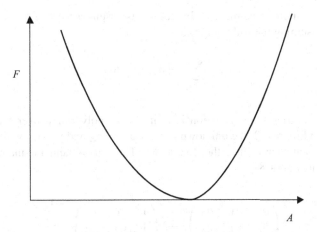

FIGURE 4.1   Functional F with a minimum having zero slope (derivative).

4.2   TRIANGULAR ELEMENTS FOR MAGNETOSTATICS

Besides the energy functional, the second main requirement of finite-element analysis is to discretize (break up) the volume analyzed into small pieces called finite elements. The most basic finite element is the triangle.

Figure 4.2 shows the simplest triangular finite element for magnetostatic fields. It lies in the xy plane and its magnetic flux density **B** is assumed to lie in the same plane. As in the reluctance examples of the preceding chapter, the planar **B** is produced by a current coming out of the plane. Thus the current density **J** is in the z direction normal to the plane of the page. Since **J** lies only in the z direction (plus or minus), the unknown magnetic vector potential **A** lies only in the z direction. In many magnetic devices, especially those made of steel laminations, **B** lies in the plane and the devices have geometry independent of the direction out of the plane as shown in Figure 4.2.

Note that the triangle of Figure 4.2 has three vertices called *nodes* labeled as L, M, N. The unknown $A_z = A$ must be found at the three nodes, and thus the triangle of Figure 4.2 is called a *nodal* finite element.

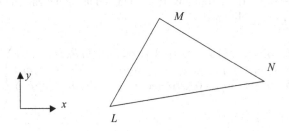

FIGURE 4.2   Triangular finite element lying in the xy plane.

The triangular finite element of Figure 4.2 is assumed to be a *first-order finite element* by assuming the following *shape function*:

$$A(x, y) = \sum_{k=L,M,N} [A_k(a_k + b_k x + c_k y)] \tag{4.8}$$

where the first-order shape function is a first-order polynomial over the plane of interest. From Figure 4.2, the unknown $A = A_K$ at $x = x_K$ and $y = y_K$, with two more such equalities at each of the other two nodes. Thus one obtains the matrix relation for the constants of (4.8):

$$\begin{pmatrix} a_L & a_M & a_N \\ b_L & b_M & b_N \\ c_L & c_M & c_N \end{pmatrix} = \begin{pmatrix} 1 & x_L & y_L \\ 1 & x_M & y_M \\ 1 & x_N & y_N \end{pmatrix}^{-1} \tag{4.9}$$

Hence for any triangle, the polynomial constants of (4.8) are known.

Substituting the shape function of (4.8) into the functional minimization Equation 4.7 yields:

$$\int \frac{\partial}{\partial A_k} \left(\frac{B^2}{2\mu} - \frac{1}{2} JA \right) dv = 0 \tag{4.10}$$

where evaluating the curl of **A** for the triangle gives:

$$B^2 = \left(\frac{\partial A}{\partial x} \right)^2 + \left(\frac{\partial A}{\partial y} \right)^2 \tag{4.11}$$

4.3 MATRIX EQUATION

The final step of finding the unknown vector potentials at the three nodes of the triangle of Figure 4.2 is to integrate over its volume, which is actually its area times the depth into the page. If the triangle area is denoted as S, one can show [1] that a 3 by 3 matrix equation is obtained:

$$(S/\mu) \begin{pmatrix} (b_L b_L + c_L c_L) & (b_L b_M + c_L c_M) & (b_L b_N + c_L c_N) \\ (b_M b_L + c_M c_L) & (b_M b_M + c_M c_M) & (b_M b_N + c_M c_N) \\ (b_N b_L + c_N c_L) & (b_N b_M + c_N c_{ML}) & (b_N b_N + c_N c_N) \end{pmatrix} \begin{bmatrix} A_L \\ A_M \\ A_N \end{bmatrix}$$

$$= (S/3) \begin{bmatrix} J \\ J \\ J \end{bmatrix} \tag{4.12}$$

which is commonly denoted as:

$$[K]\{A\} = \{J\} \tag{4.13}$$

The matrix K is commonly called the *stiffness matrix* from its origin in structural finite-element analysis. The column vector J is usually known, and then (4.13) enables the unknown column vector A to be found.

When several triangles are used to discretize the region analyzed, then the 3 by 3 matrix equation of (4.12) and (4.13) becomes an n by n matrix equation, where n is the total number of nodes. Usually many triangles are required, each of which may have a different constant permeability and applied current density (zero or nonzero). With modern software and hardware, matrices with thousands of rows and columns (and millions of matrix entries) can be solved within a few minutes. In general, both storage requirements and computer solution times are proportional to the number of unknowns n to a power between 2 and 3.

After all nodal vector potentials are found using (4.13), the magnetic flux density in each triangle can be found using (4.11). Substituting the first-order shape function into (4.11), it is seen that the **B** in first-order triangles is a constant value. Since each first-order triangle has constant flux density, more and smaller triangular finite elements are needed wherever the magnetic field varies the most. Since the field is expected to vary the most near fine steel geometries such as corners, which must naturally be modeled with many triangles, the finite-element method is practical and can be very accurate.

To increase accuracy over each finite element, shape functions with higher polynomial order (called p) can be used. For example, a second-order or *quadratic* shape function for magnetic vector potential over a triangle is:

$$A(x, y) = \sum_{k=L,M,N,O,P,Q} [A_k(a_k + b_k x + c_k y + d_k x^2 + e_k y^2 + f_k xy)] \tag{4.14}$$

In most cases the three extra variables create three extra nodes placed at the middle of each triangle edge and thus called *midnode* variables.

Extensions of the above equations enable better analysis of real-world magnetic problems. For saturable nonlinear materials, Newton's iterative method is used [1]. For conducting materials with eddy current power losses, an additional energy term can be added to the energy functional. For true 3D problems, often the triangular finite element is replaced by a tetrahedron. With only corner nodes, the tetrahedron has 4 nodes, whereas with midedge nodes, it has 10 nodes. However, to allow discontinuity in the normal component of 3D magnetic vector potential, for highest accuracy the variables preferred in tetrahedrons are the tangential components of **A** along its edges. Such finite elements are called *edge elements* or tangential vector elements [5]. In some finite-element matrix derivations, as an alternative to the energy methods described here, the differential equation is directly used in Galerkin's method [3].

Example 4.1   Matrix Equation for Two Finite Elements   Two first-order triangular finite elements are used to model a region made of copper wire and containing current density of 40,000 A/m^2 in the direction normal to the plane of the elements. The first element connects node numbers 1, 2, and 3. It has coefficient values:

$$b_1 = 1, \ b_2 = 2, \ b_3 = 3, \ c_1 = 4, \ c_2 = 5, \ c_3 = 6, \ \text{and area} \ S = 1.\text{E--}4\,\text{m}^2$$

The second element connects node numbers 3, 4, and 5. It has coefficient values (for those nodes in that order):

$$b_1 = 1, \ b_2 = 2, \ b_3 = 3, \ c_1 = 4, \ c_2 = 5, \ c_3 = 6, \ \text{and area} \ S = 2.\text{E--}4\,\text{m}^2$$

The matrix equation for all unknown vector potentials is to be written in form ready to be solved.

Solution   Set up (4.12) for each element, and then combine the matrices into a single 4 by 4 matrix equation. For the first element, (4.12) is:

$$(1.\text{E--}4/12.57\text{E--}7) \begin{pmatrix} (1+16) & (2+20) & (3+24) \\ (2+20) & (4+25) & (6+30) \\ (3+24) & (6+30) & (9+36) \end{pmatrix} \begin{bmatrix} A_1 \\ A_2 \\ A_3 \end{bmatrix}$$

$$= (1.\text{E--}4/3) \begin{bmatrix} 40.\text{E}3 \\ 40.\text{E}3 \\ 40.\text{E}3 \end{bmatrix} \tag{E4.1.1}$$

$$\begin{pmatrix} (1352) & (1750) & (2148) \\ (1750) & (2307) & (2864) \\ (2148) & (2864) & (3580) \end{pmatrix} \begin{bmatrix} A_1 \\ A_2 \\ A_3 \end{bmatrix} = \begin{bmatrix} 1.333 \\ 1.333 \\ 1.333 \end{bmatrix} \tag{E4.1.2}$$

where the rows and columns correspond to node numbers 1, 2, and 3, respectively. For the second element, the rows and columns correspond to node numbers 3, 4, and 5, respectively, with the matrix equation:

$$(2.\text{E--}4/12.57\text{E--}7) \begin{pmatrix} (1+16) & (2+20) & (3+24) \\ (2+20) & (4+25) & (6+30) \\ (3+24) & (6+30) & (9+36) \end{pmatrix} \begin{bmatrix} A_1 \\ A_2 \\ A_3 \end{bmatrix}$$

$$= (2.\text{E--}4/3) \begin{bmatrix} 40.\text{E}3 \\ 40.\text{E}3 \\ 40.\text{E}3 \end{bmatrix} \tag{E4.1.3}$$

$$
\begin{pmatrix}
(2704) & (3500) & (4296) \\
(3500) & (4614) & (5728) \\
(4296) & (5728) & (7160)
\end{pmatrix}
=
\begin{bmatrix}
2.667 \\
2.667 \\
2.667
\end{bmatrix}
\tag{E4.1.4}
$$

Combining (E4.1.2) for rows and columns 1, 2, 3 with (E4.1.4) for rows and columns 3, 4, 5 gives the final 5 by 5 matrix equation:

$$
\begin{pmatrix}
1352 & 1750 & 2148 & 0 & 0 \\
1750 & 2307 & 2864 & 0 & 0 \\
2148 & 2864 & (3580+2704) & 3500 & 4296 \\
0 & 0 & 3500 & 4614 & 5728 \\
0 & 0 & 4296 & 5728 & 7160
\end{pmatrix}
\begin{bmatrix}
A_1 \\
A_2 \\
A_3 \\
A_4 \\
A_5
\end{bmatrix}
=
\begin{bmatrix}
1.333 \\
1.333 \\
4 \\
2.667 \\
2.667
\end{bmatrix}
\tag{E4.1.5}
$$

Note that zero entries exist for nodes that are not directly connected by a finite element. As more elements are added, more zeros occur, and the matrix becomes dominated by zeros. A zero-dominated matrix is said to be *sparse*. Modern matrix solution software takes advantage of the *sparsity* to reduce storage requirements and solution time [6].

4.4 FINITE-ELEMENT MODELS

For the magnetic fields to be found in any magnetic device, a *finite-element model* of that device must be created. Making a finite-element model consists of the following four steps.

(1) **Geometry, subdivided into finite elements.** All 2D or 3D geometry must be specified, and the user and/or the software must subdivide it into finite elements. Geometry specification is difficult for many real-world devices of complicated shape. For example, Figure 4.3 shows a magnetic actuator with 2D geometry. If the geometry already exists in drawing software, then some finite-element software will accept it. However, either the finite-element software or the user must subdivide all geometric regions, including air for magnetic fields, into finite elements, such as those shown in Figure 4.4.

(2) **Materials.** For electromagnetic fields, the three material properties of Chapter 2 may be required: permeability, conductivity, and permittivity. Recall that for nonlinear magnetic materials, the *B–H* curve is required. For magnetostatics, (4.12) shows that conductivity and permittivity are not required.

(3) **Excitations.** For magnetic fields, the excitations or sources input are current densities and/or currents according to (4.12). As will be seen in later chapters, voltages and/or permanent magnets can be used instead if desired.

(4) **Boundary conditions.** On the outer boundary of the region analyzed, boundary conditions are required. For example, since Figure 4.4 is a one-half model

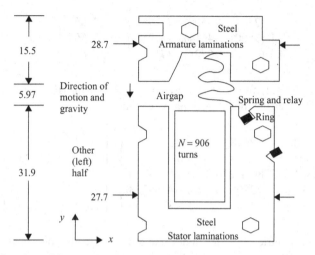

FIGURE 4.3 Magnetic actuator manufactured by Eaton Corporation [7], showing the geometry and materials of the right half. Dimensions are in millimeters. The steel laminations are stacked to a depth of 28.5 mm into the page.

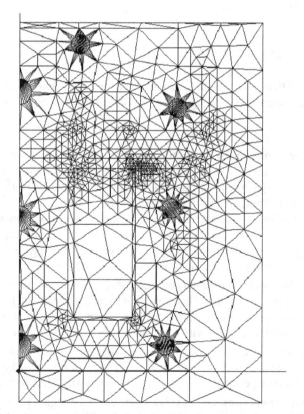

FIGURE 4.4 Computer display of triangular finite elements used to model Figure 4.3.

of the right side of an actuator, the left (symmetry) boundary must be specified. On the other three sides of the outer boundary, one must specify whether magnetic flux will enter or not. If flux does not enter at the boundary of a 2D model, then the next chapter shows that boundary has constant A_z, usually set to zero. Modeling only one-half reduces the computer time substantially. In some magnetic devices with multiple poles, periodic boundary conditions enable even smaller fractions to be modeled. In many cases, the magnetic field extends beyond the region modeled into surrounding air space, and special *open boundaries* are appropriate.

In some finite-element software, the four steps listed above appear in a task menu. The four steps are often called *preprocessing*, because they must be accomplished in order for the matrix equation to be set up and solved for the magnetic vector potential distribution throughout the device analyzed. Most magnetic finite-element software available today, both paid and free, has preprocessing steps similar to the above four steps. At this writing, free magnetic finite-element software is available from FEMM and Quickfield™.

Detailed examples of software use are described at many websites, such as at www.ansys.com, where examples explain the use of the magnetic finite-element software Maxwell®. This book uses Maxwell for many examples in several chapters. The version of Maxwell used for solutions in this book is version 9 (formerly incorporating student version SV), but all Maxwell files in this book can presently be converted to current Maxwell version 16 software files. The website http://booksupport.wiley.com contains example input files for both versions 9 and 16 of Maxwell. The input procedures in this book are given for Maxwell version 9 except when stated that another version of Maxwell is used.

Example 4.2   Finite-Element Analysis of the "C" Steel Path with Airgap of Example 3.1   The problem shown in Figure E3.1.1 and solved by the reluctance method in Example 3.1 is to be solved here using Maxwell.

Solution   Follow the above four steps to make a finite-element model of the problem in Maxwell SV.

(1) Enter the geometry. Maxwell requires a *background* or *region* which must be at least as big as the device modeled. The overall background chosen is the same as in Figure E3.1.1, 0.8 m horizontally and 0.6 m vertically. The geometry is entered as straight lines that are connected to form the various parts. Make sure that the *xy* plane is selected on the upper left tab on the main menu just below the tab selecting the *magnetostatic* problem type.

(2) Enter the materials. Since the steel is to have a relative permeability of 2000, you must add a special linear steel material.

(3) Enter the *sources* (version SV) or *excitations* (version 16) in the menu. The coil region in the center of the "C" is set to 10,000 ampere-turns. Its return on the left side of the "C" must then have $-10,000$ ampere-turns.

(4) Enter the boundary conditions. Here the flux is assumed confined to the 0.8 m by 0.6 m region analyzed, and thus the vector potential is set to zero on the entire outer boundary.

You should now note on the Maxwell SV main menu that the above steps all have check marks next to them, indicating that you can now click on its "Solve" command. (In version 16, you can click on "Validate" to ensure that all the steps have been completed, and then click on "Analyze" to initiate the solution.) Maxwell then automatically places triangular finite elements in the geometry you have entered. It adds elements in passes as shown in Table E4.2.1. For each pass it computes both the total stored energy and the estimated energy error in percent. Its default energy error is 1%, but in Table E4.2.1 it has been reduced to 0.1%, requiring 11 passes. The corresponding finite-element model or *mesh* is shown in Figure E4.2.1.

TABLE E4.2.1   Convergence of "C" core example finite-element analysis in Maxwell

Pass	Triangles	Total Energy (J)	Energy Error (%)
1	46	1.331176E+002	8.8755
2	124	1.38127 E+002	2.6007
3	160	1.38824 E+002	1.2903
4	206	1.39575 E+002	0.8061
5	262	1.40271 E+002	0.7050
6	340	1.40619 E+002	0.4578
7	440	1.41122 E+002	0.3302
8	572	1.41274 E+002	0.2237
9	737	1.41454 E+002	0.1833
10	965	1.41588 E+002	1.1220
11	1241	1.41651 E+002	0.0851

The energy listed in Table E4.2.1 is for a depth of 1 m. To compare its energy of 141.65 J with that of the reluctance solution of Example 3.1, use the right hand side of (4.2):

$$W = \int \frac{B^2}{2\mu} dv \tag{E4.2.1}$$

Since the reluctance method of Example 3.1 obtained $B = 0.125$ T, and most of the energy is in the airgap volume of size 0.1 m by 0.1 m by a depth of 1 m (to agree with Maxwell), (E4.2.1) becomes:

$$W = (0.125)^2/(12.57E{-}7)(0.1)(0.1)(1) = 124.3 \text{ J} \tag{E4.2.2}$$

This energy of the reluctance method is about 12% lower than the 141.65 J computed by Maxwell. Since it accurately accounts for fringing flux, the finite-element solution is thus considerably more accurate than the reluctance method solution.

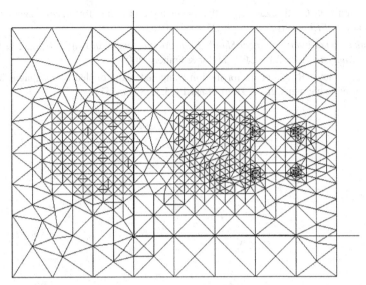

FIGURE E4.2.1   Finite-element model of steel path with airgap of Examples 3.1 and 4.2 produced by 11 adaptive passes.

PROBLEMS

4.1   Repeat Example 4.1 with the second finite element (of area 2.E-4 m^2) changed to steel with relative permeability 1000 and containing no current. Assuming the first element remains the same, find the final matrix equation.

4.2   Substitute the first-order triangular shape function of (4.8) into $\mathbf{B} = \nabla \times \mathbf{A}$ to show that $\mathbf{B}$ is constant in each such finite element.

4.3   Use Maxwell to solve Example 4.2 with the steel having the nonlinear $B\text{–}H$ curve called "steel_1010" in the Maxwell material list.

4.4   Use Maxwell to solve Example 4.2 with the steel having the nonlinear $B\text{–}H$ curve called "steel_1008" in the Maxwell material list.

REFERENCES

1. Brauer JR (ed.). *What Every Engineer Should Know About Finite Element Analysis*, 2nd ed. New York: Marcel Dekker; 1993.
2. Hastings JK, Juds MA, Brauer JR, Accuracy and economy of finite element magnetic analysis, *National Relay Conference*, Stillwater, OK, April 1985.
3. Ida N, Bastos JPA. *Electromagnetics and Calculation of Fields*, 2nd ed. New York: Springer-Verlag; 1997.
4. Holman JP. *Heat Transfer*, New York: McGraw-Hill Book Co., Inc., 1963.

5. Lee JF, Sun DK, Cendes ZJ. Tangential vector finite elements for electromagnetic field computation. *IEEE Trans Magn* 1991;27:4032–4035.

6. Komzsik L. *What Every Engineer Should Know About Computational Techniques of Finite Element Analysis*. Boca Raton, FL: Taylor & Francis; 2005.

7. Juds MA, Brauer JR. AC contactor motion computed with coupled electromagnetic and structural finite elements. *IEEE Trans Magn* 1995;31:3575–3577.

CHAPTER 5

Magnetic Force

The magnetic vector potential **A** and the magnetic flux density **B** computed by the finite-element method of the preceding chapter can be used to find magnetic forces of magnetic actuators and sensors. Related to magnetic force are magnetic flux plots, magnetic energy, magnetic pressure, permanent magnets, and magnetic torque. All are discussed in this chapter, and all are examples of *postprocessing* of finite-element solutions.

5.1 MAGNETIC FLUX LINE PLOTS

Magnetic vector potential **A** in the 2D planar triangular finite elements of the preceding chapter has been shown to consist of only one component, A_z. This component is directed perpendicular to the xy plane of the finite elements.

Plots of contours of constant A_z can be shown to be very useful plots for 2D planar models. They are called *magnetic flux line plots*, or *magnetic flux plots* for short. Figure 5.1 shows a magnetic flux line plot for the Eaton actuator of the preceding chapter. It is obtained by drawing contour lines connecting constant values of A_z, similar to plots of constant temperature appearing in weather maps. One can choose as many contour lines as one likes, but useful plots have all contours spaced by equal increments.

The 2D plot of contours of constant A_z is called a magnetic flux line plot because it shows the same pattern as obtained by sprinkling iron filings on a sheet of paper. Each iron filing is long and thin, resembling a compass needle. Like a compass needle, each filing will align itself in the direction of the magnetic flux density **B**, and hence flux lines are sometimes called *lines of force*. Also, the filings will concentrate in the regions of highest flux density magnitude. Thus the closer flux lines in Figure 5.1 represent higher flux density regions. A key advantage of finite-element flux plots over iron filing flux plots is that finite-element flux plots enable the engineer to visualize the flux flow within steel and other materials, where iron filings cannot be sprinkled.

Magnetic Actuators and Sensors, Second Edition. John R. Brauer.
© 2014 The Institute of Electrical and Electronics Engineers, Inc. Published 2014 by John Wiley & Sons, Inc.

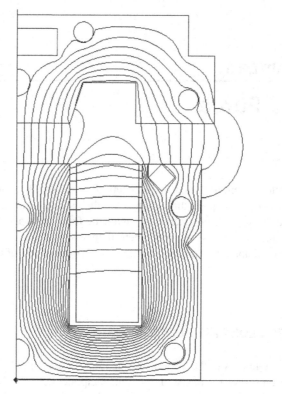

FIGURE 5.1 Computer display in black and white showing flux line plot for 2D planar Eaton actuator model in the preceding chapter.

Since most magnetic devices contain parts made of steel with nonlinear $B-H$ curves, the flux densities in the steel are of particular interest. If the flux line plots show steel regions with densities less than about 1.5 T, the approximate knee of most steel $B-H$ curves, then some of the steel may be wasted. Some of the steel in those regions can be removed, thereby saving the cost and weight of that steel and/or the copper coil which surrounds it. On the other hand, if the flux plot shows steel with flux densities much greater than the knee at approximately 1.5 T, then that steel is saturated, and it is likely that more steel should be added to reduce the current required and the associated I^2R power loss in the coil. Hence flux plots and flux density distributions are very helpful in the design of magnetic devices.

Besides problems modeled with planar triangular finite elements, another type of 2D problem is not planar but is *axisymmetric*. Figure 5.2 shows a magnetic actuator originally developed by Bessho et al. in Japan [1]. Note that its parts are all revolved around the central axis of symmetry, usually called the z axis. Thus its finite-element model, shown in Figure 5.3, lies in a plane (usually called the rz plane) but consists of axisymmetric finite elements. Instead of each element having the same depth into the page, the depths are proportional to radius from the axis of symmetry, and thus

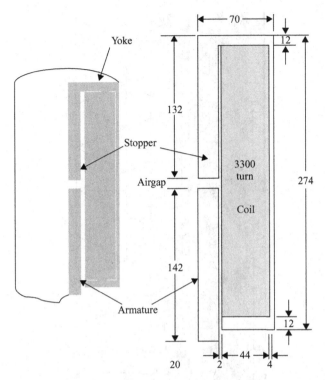

FIGURE 5.2   Axisymmetric magnetic actuator developed by Bessho et al. Dimensions shown are in millimeters.

the volume integrals for the finite-element matrix equation differ from those of the 2D planar finite elements of the preceding chapter. Similarly, the flux line plots must take into account the effects of the radius on volume. Accordingly, flux line plots for axisymmetric models are contours of constant radius times vector potential, where the vector potential only has a component in the direction around the axis of symmetry.

For 3D models, flux line plots are not as meaningful, because no piece of paper with iron filings can represent 3D flux flow. Thus contours of constant magnitude of vector potential are not very meaningful for 3D problems. Instead, color plots of the magnitude of flux density are very useful in 3D (and 2D) problems; usually red is used for high flux density and blue or black for low flux density.

For both 2D and 3D finite-element models, it is vital that flux line plots and/or flux density plots be obtained and examined. The engineer should never report results without doing so, because the plots can reveal mistakes in the model. Typical mistakes might include the following.

- Wrong material selection
- Wrong boundary condition
- Wrong units (mm instead of m, etc.)

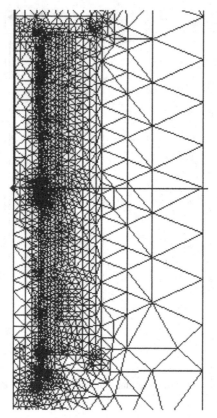

FIGURE 5.3 Computer display of 2D axisymmetric finite-element model of actuator of Figure 5.2.

- Wrong input excitation (current or voltage)
- Wrong number of turns

Any mistake must be corrected and the model rerun until the flux display appears reasonable, which is required for finite-element model validation. An additional check would be to compare the finite-element flux with the flux estimated by the reluctance method of Chapter 3, if such an estimate can be made.

Example 5.1   Relation Between A and B for 2D Planar Problem   Given a 2D planar finite-element solution of $A_z = y$, find the corresponding magnetic flux density **B** and describe the flux line plot.

Solution   To find **B**, use the curl of **A** of (2.8), which for A_z only becomes:

$$\mathbf{B} = \frac{\partial A_z}{\partial y}\mathbf{u}_x - \frac{\partial A_z}{\partial x}\mathbf{u}_y \qquad (E5.1.1)$$

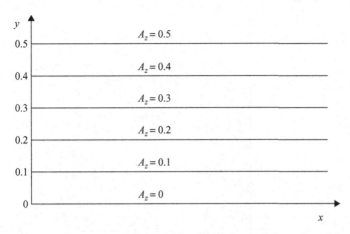

FIGURE E5.1.1   Flux line plot of Example 5.1, shown over a finite region of the xy plane.

Substituting the given expression for A_z, the partial with respect to x is zero, and thus the flux density has only an x component:

$$\mathbf{B} = \frac{\partial y}{\partial y}\mathbf{u}_x = 1\mathbf{u}_x \qquad (E5.1.2)$$

Note that the flux density is a uniform 1 T in the x direction. The corresponding flux line plot connects constant vector potential values and also shows uniformly spaced lines directed in the x direction. Figure E5.1.1 shows the flux line plot for a typical number of lines.

Example 5.2   Flux Line Plot of the "C" Steel Path with Airgap of Examples 3.1 and 4.2   Given the "C"-shaped steel path with airgap and winding of Examples 3.1 and 4.2, obtain the flux line plot using Maxwell.

Solution   When using Maxwell SV, after clicking on "Solve" in the main menu and obtaining the convergence Table 4.2.1 of Example 4.2, click on the next main menu item "Post Process." On the top of the screen, choose "Plot." On the pop-up that appears, choose "Flux Lines" on "Surface—all" in area "All." On the next pop-up, choose all defaults except make 21 divisions (and thus 20 flux lines) with spectrum color type "magenta."

   If Maxwell version 16 is used, after clicking "Analyze" and obtaining a convergence table similar to Table 4.2.1 of Example 4.2, create a flux line plot as follows. Select all the geometry, select Field Overlays, Fields, and then A. Choose all defaults and adjust the color scale to make 21 divisions (and thus 20 flux lines) with spectrum color type "magenta."

   The plot you obtain should be that of Figure E5.2.1. Note that it has four flux lines that pass through the center coil region without reaching the airgap of the "C"

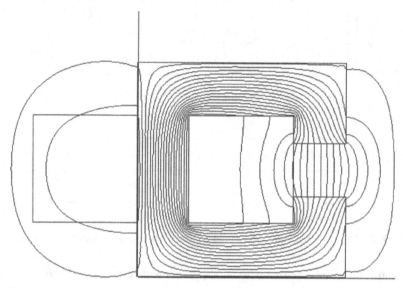

FIGURE E5.2.1 Computer display in black and white showing flux line plot of Examples 5.2, 4.2, and 3.1.

core. These four flux lines to the left of the airgap region are thus called *leakage flux*. They are a type of leakage flux sometimes called "partial" flux since it does not fully surround the coil. Note that Figure E5.2.1 also has four flux lines on the outside (to the right of the "C" airgap) that flow through the surrounding air without passing through the airgap region directly between the steel poles. These four lines represent *fringing flux* since their pattern extends outward much like fringe, as mentioned in Chapter 3.

5.2 MAGNETIC ENERGY

The expression for energy stored in a magnetic field has been presented in (4.2) of the preceding chapter as:

$$W_{\text{mag}} = \int \frac{B^2}{2\mu} dv \tag{5.1}$$

where constant permeability μ is assumed. In general, all types of energy W can be found by integrating energy density w over a volume, that is:

$$W = \int w \, dv \tag{5.2}$$

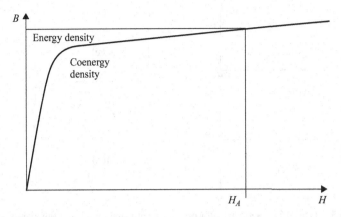

FIGURE 5.4 Energy density and coenergy density for a typical operating point on a typical nonlinear *B–H* curve.

Hence the energy density, in joules per cubic meter, for constant permeability material of (5.2) is easily seen to be:

$$w_{\text{mag}} = \frac{B^2}{2\mu} = \frac{1}{2}BH \tag{5.3}$$

For materials with nonlinear *B–H* curves, the energy density can be shown to be:

$$w_{\text{mag}} = \int H \cdot dB \tag{5.4}$$

Figure 5.4 shows that this energy density is the area to the left of the *B–H* curve. Note that Figure 5.4 also shows an area under the *B–H* curve called the *coenergy* density w_{co}. The sum of coenergy density and energy density is seen in Figure 5.4 to obey:

$$w_{\text{mag}} + w_{\text{co}} = HB \tag{5.5}$$

For constant permeability materials, energy density and coenergy density become equal and are given by (5.3).

5.3 MAGNETIC FORCE ON STEEL

Force is related to energy. Indeed, energy is in units of force times distance. One of the most basic methods to determine force is to use the method of virtual work. It

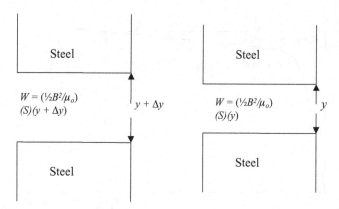

FIGURE 5.5 Typical magnetic actuator or sensor with steel poles and the energy stored in the airgap between them for two vertical positions.

states that force in a given direction, say the vertical y direction, equals the partial derivative of stored energy with respect to that direction:

$$F_y = \frac{\partial W}{\partial y} \tag{5.6}$$

An example of virtual work determining the force of a typical magnetic actuator or sensor is shown in Figure 5.5. Assuming for now that the steel (or other ferromagnetic material) has permeability that is thousands of times that of air, then the energy density in the air is thousands of times that in the steel, and is given by:

$$w_{\text{mag}} = B^2/(2\mu_o) \tag{5.7}$$

To determine the magnetic force acting in the vertical y direction in Figure 5.5, one may replace the derivative of (5.6) by its approximation:

$$F_y = \frac{\Delta W}{\Delta y} \tag{5.8}$$

where W is as always the volume integral of w. Assuming the flux density **B** in the airgap of Figure 5.5 is uniform, then magnetic energy W_{MAG} is simply w_{mag} times volume:

$$W_{\text{mag}} = [B^2/(2\mu_o)]v \tag{5.9}$$

As shown in Figure 5.5, during a *virtual displacement* Δy of a steel pole in the y direction, the volume of the airgap changes from its surface area S times $(y + \Delta y)$ to S times y. Hence from (5.9) we obtain:

$$F_y = \frac{\Delta W}{\Delta y} = \frac{S(y + \Delta y)B^2/(2\mu_o) - SyB^2/(2\mu_o)}{\Delta y} \tag{5.10}$$

and thus the magnetic force is:

$$F_{\text{mag}} = SB^2/(2\mu_o) \tag{5.11}$$

This magnetic force equation is very simple and useful, even though it assumes that the steel has infinitely high permeability. Note that the force is proportional to the square of flux density. The direction of the force causes the steel pole to move toward the airgap, that is, the steel poles of Figure 5.5 are attracted to each other independent of the direction of **B**. Because this force tends to reduce the reluctance of the magnetic circuit, it is sometimes called a *reluctance force*. Note that F_{mag} in (5.11) is always positive; some other authors insert a negative sign.

If the steel poles have finite permeability μ_s, then from Figure 5.5 the virtual displacement will produce an additional force term, because the steel volume changes with a virtual displacement. Thus the force of (5.11) becomes [2]:

$$F_{\text{mag}} = S[B^2/(2\mu_o) - B^2/(2\mu_s)] \tag{5.12}$$

Since $\mu_s > \mu_o$, the force direction remains such that the steel tends to move toward the air. As mentioned in Chapter 2, most steels have $\mu_s > 1000\mu_o$, in which case the simpler Equation 5.11 is sufficiently accurate.

When materials with nonlinear B–H curves are present, the magnetic force can be derived with the aid of Figure 5.6. It shows two B–H relations for an entire device with variable airgap. Curve 1 in Figure 5.6 is for a high reluctance position with large airgap. Curve 2 is for a lower reluctance position with the airgap reduced, thereby increasing B compared to Curve 1 for any $H > 0$. Since current is a common input for the magnetostatic finite-element analyses of Chapter 4, it is assumed to be constant in Figure 5.6, obtaining a constant applied field intensity H_A. The area between the two curves is the energy difference needed to obtain the force by the virtual work method, but is seen to actually be the difference in coenergies. Thus when nonlinear materials are present, the force equation becomes:

$$F_y = \frac{\partial W_{\text{co}}}{\partial y}\Big|_{I=\text{const}} \tag{5.13}$$

where W_{co} is the total magnetic coenergy and the applied current is assumed held constant.

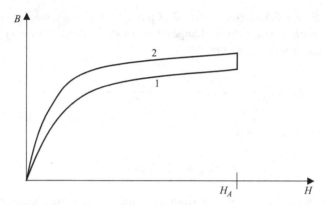

FIGURE 5.6 Nonlinear B–H relation of a typical magnetic device at positions 1 and 2. The applied current and thus applied field intensity H_A are constant.

Example 5.3 Force on One Pole of the "C" Steel Path with Airgap of Examples 3.1, 4.2, and 5.2 Given the "C"-shaped steel path with airgap and winding of Examples 3.1, 4.2, and 5.2, obtain the magnetic force on one of its two steel poles using Maxwell SV and compare it with the reluctance method result and the simple force formula (5.11).

Solution Maxwell uses the virtual work method of (5.6) in an algorithm designed to obtain total magnetic force on any geometric part. The entire "C" of steel of Figure E5.2.1 is expected to have zero total force, because the upper steel pole (above the airgap) will experience a downward force canceled by the equal upward force on the lower steel pole. Thus to determine the force on one pole, a one-half model must be constructed.

The upper-half is modeled by either altering the geometry of the model of Example 4.2 or by starting a new Maxwell project. Since only the upper-half is modeled, the half coil must contain half the ampere-turns, or 5000. Also, the boundary condition along the symmetry boundary (the horizontal plane at the bottom of the modeled upper-half) must be specified as *even symmetry* or left unconstrained to allow flux to cross it.

In Maxwell SV, before clicking on the "Solve" button, the button "Setup Executive Parameters" must be clicked. Within this setup, the force on the upper-half of the "C" must be requested. If Maxwell version 16 is used, before clicking on the "Analyze" button, select the upper steel pole and add a "Parameter" for force.

After solution, the resulting flux line plot is shown in Figure E5.3.1. In the "Convergence Data" table the energy stored is 70.82 J, agreeing closely with one-half the 141.65 J obtained by Maxwell for the full model of Example 4.2. The magnetic force in the table is approximately 993 N. Under the tab "Solutions," the "Force/torque" is seen to contain a small F_x and $F_y = -992.5$ N. Note that the force is downward, because steel pole faces always experience outward magnetic force. The force output

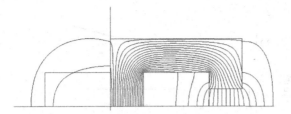

FIGURE E5.3.1 Computer display of flux lines in model of upper-half of steel "C" with airgap.

by Maxwell is for 1 m depth, and for the actual 0.1 m depth the force computation would be 99.3 N.

To compare the Maxwell computation with the reluctance method and (5.11), recall from Example 3.1 that its predicted $B = 0.125$ T. Substituting in (5.11) gives:

$$F_{mag} = [(0.125)^2/(2 \times 12.57E–7)](0.1)(0.1) = 62.15 \text{ N} \qquad \text{(E5.3.1)}$$

This approximate force differs considerably from the accurate 99.3 N force computed by Maxwell.

5.4 MAGNETIC PRESSURE ON STEEL

Pressure is defined as force per unit area. Thus from (5.11) for infinitely high permeability steel poles, the magnetic pressure acting on the pole area is:

$$P_{mag} = B^2/(2\mu_o) \qquad (5.14)$$

This pressure is normal to the pole area and is one component of the *Maxwell stress tensor*. For steel of finite permeability, (5.12) gives the pressure:

$$P_{mag} = B^2/(2\mu_o) - B^2/(2\mu_s) \qquad (5.15)$$

Since B is typically limited by most steel B–H curves to a practical maximum of approximately 2 T, the magnetic pressure is accordingly limited. However, instead of using (5.15) in the nonlinear case, the less well-known *nonlinear* Maxwell stress tensor should be used [3], which gives:

$$P_{mag} = B^2/(2\mu_o) - (\mathbf{B} \cdot \mathbf{H} - w_{mag}) \qquad (5.16)$$

where w_{mag} is magnetic energy density in the steel given in (5.4). Applying (5.5) obtains an even simpler expression:

$$P_{mag} = B^2/(2\mu_o) - w_{co} \qquad (5.17)$$

where coenergy appears as in (5.13).

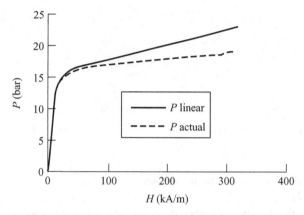

FIGURE 5.7 Magnetic pressures (bar) versus H (kA/m) in SAE 1010 steel. The upper curve graphs the pressure using the linear formula (5.15), whereas the lower (dashed) curve is from the actual nonlinear Maxwell stress tensor (5.17). At $H = 320$ kA/m, $B = 2.4$ T for this steel.

Using (5.17) instead of (5.15) further limits the maximum magnetic pressure. Figure 5.7 is a graph of the magnetic pressures in bar ($=10^5$ N/m^2 $= 10^5$ Pa) computed using (5.14) and (5.15) for SAE 1010 steel, using its B–H curve, which is assumed to be isotropic. This commonly used steel has $B = 2.4$ T at $H = 320$ kA/m. Note that the nonlinear pressure (5.17) gives pressures considerably lower than (5.15). At 2.4 T, (5.17) gives a pressure of 19.09 bar, which is 16.75% less than that given by (5.15). Figure 5.7 shows that magnetic pressures on typical steel have a practical limit of 20 bar or less. Recall that 1 bar is close to the standard pressure of air at sea level; the pressure varies with weather conditions but 1 bar is an approximate average. Discussion of all components of the Maxwell stress tensor will be given in Chapter 14.

5.5 LORENTZ FORCE

In addition to the magnetic force and pressure on steel described in the preceding two sections and called a *reluctance force*, a magnetic force called *Lorentz force* exists on currents. On any part, such as a wire, of length l and cross-sectional area s carrying current density $\mathbf{J}$, the Lorentz force is:

$$\mathbf{F}_L = \int (\mathbf{J}ds) \times (\mathbf{B}dl) \qquad (5.18)$$

where $\mathbf{B}$ is the magnetic flux density acting on the current density. Because (5.18) contains the cross product, the force direction is normal to both $\mathbf{J}$ and $\mathbf{B}$. The force

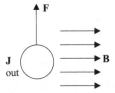

FIGURE 5.8   Lorentz force vectors. If the right hand is laid flat with its fingers in the direction of flux density **B** and the thumb in the direction of current density **J**, then the force is in the direction of the palm of the right hand.

direction follows the right-hand rule as shown in Figure 5.8. Note that an alternative expression of (5.18) is:

$$\mathbf{F}_L = \int (\mathbf{J}) \times (\mathbf{B}) dv \tag{5.19}$$

where the integration is over the volume of the current density.

If **J** and **B** are perpendicular to each other, then the current I is directed perpendicular to **B** as well. In this perpendicular case, the Lorentz force magnitude obeys the simple expression:

$$F = BIl \tag{5.20}$$

where l is the length of the wire. In Figure 5.8, the wire is coming out of the page.

Lorentz force can be computed by the method of virtual work [4], but is more easily and accurately computed using (5.19). Maxwell can be used to compute Lorentz force. Chapter 7 contains a further discussion of Lorentz force in magnetic actuators.

5.6 PERMANENT MAGNETS

A common magnetic actuator that uses Lorentz force is a voice coil actuator in loudspeakers. Such actuators obtain their magnetic flux density **B** using *permanent magnets*.

The key to understanding permanent magnet behavior is the *B–H* curve. Figure 5.9 shows a typical *B–H* relationship for a material that can behave as a permanent magnet. After starting at the origin (0,0) and increasing *H* by applying an increasing DC current in a nearby coil, the arrow to the right indicates the portion of the relation called the magnetization curve. Next, after *H* reaches a maximum, it is decreased and follows the arrow to the left. The curve going to the left is called the *demagnetization curve*. Since the demagnetization curve depends on the magnetization curve, demagnetization depends on the material history, a phenomenon called *hysteresis*. In contrast, the "soft" steel *B–H* relationship of Chapter 2 exhibited negligible hysteresis and was essentially a single-valued curve, unlike the multiple curves of Figure 5.9 for

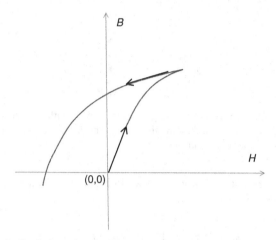

FIGURE 5.9 *B–H* relation of typical material that can exhibit permanent magnetization.

so-called magnetically "hard" materials. Both hard and soft magnetic materials exhibit nonzero magnetization **M** defined in Chapter 2.

Figure 5.10 is a labeled view of the second quadrant of Figure 5.9. Note that the intercept point of the demagnetization curve with the *B* axis occurs at B_r, usually called the *residual* flux density (sometimes called the *remanent* flux density). The intercept point of the demagnetization curve with the negative *H* axis occurs at $-H_c$, where H_c is usually called the *coercive* field intensity (sometimes misleadingly called the coercive force).

Table 5.1 lists some of the main types of permanent magnets and their typical properties. Also listed are the approximate years of first development. Table 5.1 shows that the earliest permanent magnets were made of steel that had been hardened [5] by adding elements other than iron and by quenching (heating followed by very rapid cooling). Such hardened steels are in contrast to the soft steels of Chapter 2, which have much smaller values of B_r and H_c. Besides listing B_r and H_c, Table 5.1

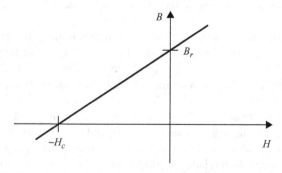

FIGURE 5.10 Linearized second quadrant of Figure 5.9 with definitions of B_r and H_c.

TABLE 5.1   Permanent Magnet Types and Typical Properties

Type	Approximate Year	B_r [T (teslas)]	H_c (A/m)	$(BH)_{max}$ (T-A/m)
Carbon steel (quenched)	1600	0.9	4.E3	1.6E3
Cobalt steel (quenched)	1916	0.96	18.E3	7.2E3
Alnico3	1931	0.68	39.E3	11.1E3
Alnico5	1948	1.03	49.E3	51.7E3
Ceramic5 (ferrite)	1952	0.38	192.E3	24.E3
Ceramic8 (ferrite)	1955	0.39	240.E3	28.E3
Samarium cobalt	1970	0.88	640.E3	150.E3
Neodymium iron (early)	1983	1.08	800.E3	223.E3
Neodymium iron (latest)	2004	1.43	1080.E3	400.E3

also lists the maximum second quadrant product of B times H; this energy product is the best indicator of the overall strength of a permanent magnet.

Table 5.1 shows that the first non-steel permanent magnets developed were Alnico magnets. Alnico stands for aluminum, nickel, and cobalt, the elements used in such magnets. Note that Alnico permanent magnet properties are definitely superior to those of steels.

Next developed were ceramic magnets made of barium ferrite [6]. Both barium and ferrite (iron oxide) are low cost and thus ceramic magnets are inexpensive yet provide good permanent magnet properties.

The next breakthrough was the samarium cobalt magnet [6]. More expensive than ceramics, its properties are clearly superior.

Perhaps the best modern permanent magnet is made of elements neodymium, iron, and boron. While neodymium is an expensive rare earth, less of it is required than of inexpensive iron. Note that the latest neodymium iron magnets have H_c exceeding 1,000,000 A/m, often with the addition of another rare earth element dysprosium.

Disadvantages of permanent magnets include the need to magnetize them and the fact that most of the properties of Table 5.1 often vary significantly with temperature. Recently the price of rare earth magnets has increased considerably due to supply and demand issues as well as geopolitical factors. Supply risks are especially high for neodymium and dysprosium [7].

A key advantage of permanent magnets over electromagnets (current-carrying coils) is that after magnetization they require no input energy and consume no power. This loss-free property is similar to that of superconductors, yet (unlike superconductors) permanent magnets usually function well at room temperatures and somewhat higher. Because extremely low temperatures are required for superconductors, they are not discussed further in this book except briefly in Section 11.8.

Permanent magnets provide magnetic flux by means of internal currents circulating around their domains. Usually these currents cancel inside a permanent magnet and exist only on magnet surfaces. Thus in finite-element and reluctance models, a permanent magnet can be replaced by an equivalent electromagnet with currents located on certain surfaces. Figure 5.11 shows a simple rectangular permanent magnet with two poles, north N and south S. North poles produce flux coming out, while

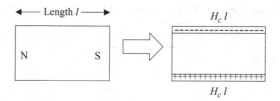

FIGURE 5.11　Simple permanent magnet and its equivalent coil.

south poles produce flux going into the magnet. Figure 5.11 also shows the equivalent electromagnet with a sheet current on two surfaces. Inside the sheet current, the equivalent electromagnet has the permeability equal to the slope of the demagnetization curve, which for most modern magnets is close to that of air.

The equivalent electromagnet of Figure 5.11 has ampere-turns that obey [2, 8]:

$$NI = H_c l \qquad (5.21)$$

where l is the length of the permanent magnet in the direction of magnetization. With the high H_c values of the modern magnets of Table 5.1, a typical modern permanent magnet has the equivalent of many thousands of ampere-turns.

By keeping the magnetization magnitude constant while properly varying the direction of magnetization (either continuously in one permanent magnet or with an array of smaller permanent magnets), a uniform Halbach field [9] is created. Figure 5.12 shows a Halbach field created inside a cylindrical permanent magnet with direction of magnetization θ_M obeying $\theta_M = 2\theta$, where θ is the circumferential angle

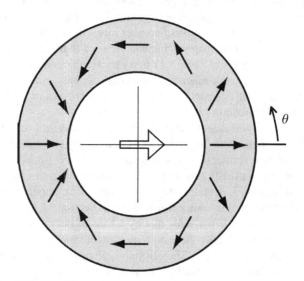

FIGURE 5.12　Halbach cylindrical permanent magnet with its magnetization direction equaling 2θ and the resulting uniform magnetic field inside the cylinder.

shown in Figure 5.12. Besides the cylindrical Halbach array shown in Figure 5.12, by rolling out the cylinder a linear Halbach array can be created that has a high unidirectional **B** on one side and essentially no magnetic field on the other side. A Halbach array will be used in a magnetic actuator to be discussed in Chapter 15.

Most finite-element software, including Maxwell, automatically analyzes permanent magnets using equivalent currents [8] such as those of (5.21). If the reluctance method is used, the equivalent current of permanent magnets is implemented by means of load lines, and results are usually much less accurate.

As experiments show, permanent magnets exhibit forces of attraction and repulsion between other magnets and steel. These forces can be very useful in magnetic actuators, and they can be calculated using the force equations and finite-element techniques presented and discussed earlier in this chapter.

Example 5.4   Force between Two Permanent Magnets   Given two identical cylindrical neodymium iron permanent magnets with a straight line $B–H$ curve with $B_r = 1.23$ T and $H_c = 8.9E5$ A/m. They both have radius 20 mm and height 10 mm, and are spaced 20 mm apart as shown in Figure E5.4.1. If both magnets have their north poles on their top, obtain the magnetic force on the lower magnet using Maxwell SV or version 16.

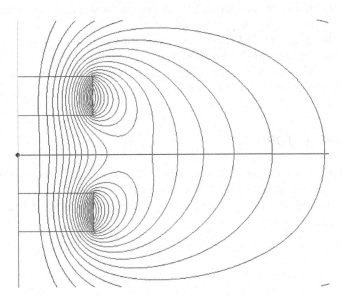

FIGURE E5.4.1   Computer display of two cylindrical permanent magnets and their flux lines.

Solution   If Maxwell SV is used, set "Solver" to "Magnetostatic" and its "Drawing" to "RZ Plane." In "Define Model" enter two boxes along the left (z) axis for the two magnets. In "Setup Materials" choose "NdFe35" for both magnets, and set their angle of magnetization as 90°, which is vertical with the north pole on top. Under "Setup

Boundaries/Sources" no sources need be specified, since the permanent magnets have been defined as materials. However, the boundary condition "balloon" should be specified on the outer boundary of the model, since there is no steel to confine the flux. A balloon boundary is a type of open boundary that allows flux to leave the finite-element model space. Next, use "Setup Executive Parameters" to specify force computation on the lower magnet. Under "Setup Solution Options" change the number of passes to 23. This high number is needed so that when "Solve" is clicked, the energy error will be less than 2%.

If Maxwell version 16 is used, set "Solution Type" to "Magnetostatic" and Geometry Mode to "Cylindrical about Z." Create two boxes along the left (z) axis for the two magnets. In "Assign Materials" choose "NdFe35" for both magnets, and set their direction of magnetization with "Z Component" equal to 1, which is vertical with the north pole on top. Under "Excitations" no sources need to be specified, since the permanent magnets have been defined as materials. However, the boundary condition "balloon" should be specified on the outer boundary of the model, since there is no steel to confine the flux. A balloon boundary is a type of open boundary that allows flux to leave the finite-element model space. Next, select the lower magnet and assign a "Force" under "Parameters" to specify force computation. Under "Analysis -> Setup1" change the number of passes to 23. This high number is needed so that when "Analyzed" is clicked, the energy error will be less than 2%.

Using either version of Maxwell, the computed force on the lower magnet is 30.92 N in the vertical upward direction. As expected, the north pole on the lower magnet is strongly attracted to the adjacent south pole on the upper magnet. The computed flux lines are shown in Figure E5.4.1.

5.7 MAGNETIC TORQUE

Important for rotary actuators is magnetic torque, which is closely related to magnetic force. As before, the virtual work method may be used.

The virtual work method for magnetic torque requires the partial derivative with respect to *angle in radians*:

$$T = \frac{\partial W}{\partial \theta} \tag{5.22}$$

The SI unit of both energy W and torque T is the newton-meter. For magnetic devices with constant ampere-turn excitation and materials with nonlinear B–H curves, the coenergy must be used as explained previously for magnetic force. Thus for magnetic devices, the torque is found using:

$$T = \frac{\partial W_{\text{co}}}{\partial \theta} \tag{5.23}$$

The finite-element software Maxwell uses (5.23) to compute torque on parts selected by the user. As mentioned before for the force expression (5.11), some other authors insert a negative sign in the torque expression (5.23). In any case magnetic torque tends to align (attract) steel parts for minimum reluctance.

5.8 MAGNETIC VOLUME FORCES ON PERMEABLE PARTICLES

In addition to the magnetic forces acting on steel structures such as described above in typical linear and rotary actuators, very small permeable particles made of steel or other soft magnetic materials can experience magnetic forces. These particles can be of size even smaller than a micrometer, and indeed may be only a few nanometers in diameter in the case of certain magnetic separators.

Instead of making tiny models of the very small permeable particles suspended in large volumes of air, water, or other nonpermeable solvents, a more practical approach is to find the force per unit volume of any permeable particle. For an infinitely permeable ($\mu \to \infty$) object in free space, the magnetic force per unit volume is the vector [10]:

$$\mathbf{f}_m = \nabla w_m = \nabla \left(\frac{1}{2} \mathbf{B} \cdot \mathbf{H} \right) = \nabla \left(\frac{B^2}{2\mu_o} \right) \tag{5.24}$$

Note that the force requires a gradient in magnetic energy density (here denoted as w_m), and thus iron particles in flux plots tend to concentrate in regions of high magnetic energy density, that is, higher B. In terms of vector components, the force density of (5.24) can be written for Cartesian xyz coordinates:

$$\mathbf{f}_m = \left(\frac{1}{2\mu_o} \right) \left(\frac{\partial}{\partial x} \mathbf{u}_x + \frac{\partial}{\partial y} \mathbf{u}_y + \frac{\partial}{\partial z} \mathbf{u}_z \right) \left(B_x^2 + B_y^2 + B_z^2 \right) \tag{5.25}$$

which can be expanded as:

$$\mathbf{f}_m = \left(\frac{1}{2\mu_o} \right) \left(\frac{\partial}{\partial x} \left(B_x^2 + B_y^2 + B_z^2 \right) \mathbf{u}_x + \frac{\partial}{\partial y} \left(B_x^2 + B_y^2 + B_z^2 \right) \mathbf{u}_y \right.$$
$$\left. + \frac{\partial}{\partial z} \left(B_x^2 + B_y^2 + B_z^2 \right) \mathbf{u}_z \right) \tag{5.26}$$

Carrying out the derivatives using the chain rule gives:

$$\mathbf{f}_m = \left(\frac{1}{\mu_o} \right) \left[\left(B_x \frac{\partial B_x}{\partial x} + B_y \frac{\partial B_y}{\partial x} + B_z \frac{\partial B_z}{\partial x} \right) \mathbf{u}_x \right.$$
$$\left. + \left(B_x \frac{\partial B_x}{\partial y} + B_y \frac{\partial B_y}{\partial y} + B_z \frac{\partial B_z}{\partial y} \right) \mathbf{u}_y + \left(B_x \frac{\partial B_x}{\partial z} + B_y \frac{\partial B_y}{\partial z} + B_z \frac{\partial B_z}{\partial z} \right) \mathbf{u}_z \right]$$
$$\tag{5.27}$$

For finite permeability of the particle, (5.27) is altered as in (5.15) to the following force per unit volume for a particle of permeability μ_p [11]:

$$
\mathbf{f}_m = \left(\frac{1}{\mu_o} - \frac{1}{\mu_p} \right) \left[\left(B_x \frac{\partial B_x}{\partial x} + B_y \frac{\partial B_y}{\partial x} + B_z \frac{\partial B_z}{\partial x} \right) \mathbf{u}_x \right.
$$
$$
\left. + \left(B_x \frac{\partial B_x}{\partial y} + B_y \frac{\partial B_y}{\partial y} + B_z \frac{\partial B_z}{\partial y} \right) \mathbf{u}_y + \left(B_x \frac{\partial B_x}{\partial z} + B_y \frac{\partial B_y}{\partial z} + B_z \frac{\partial B_z}{\partial z} \right) \mathbf{u}_z \right]
$$

$$(5.28)$$

For planar xy problems independent of z, the equation reduces to:

$$
\mathbf{f}_m = \left(\frac{1}{\mu_o} - \frac{1}{\mu_p} \right) \left[\left(B_x \frac{\partial B_x}{\partial x} + B_y \frac{\partial B_y}{\partial x} \right) \mathbf{u}_x + \left(B_y \frac{\partial B_y}{\partial y} + B_x \frac{\partial B_x}{\partial y} \right) \mathbf{u}_y \right]
$$

$$(5.29)$$

For fields expressed instead in axisymmetric cylindrical coordinates rz independent of angle $\emptyset$, the force density is:

$$
\mathbf{f}_m = \left(\frac{1}{\mu_o} - \frac{1}{\mu_p} \right) \left[\left(B_r \frac{\partial B_r}{\partial r} + B_z \frac{\partial B_z}{\partial r} \right) \mathbf{u}_r + \left(B_z \frac{\partial B_z}{\partial z} + B_r \frac{\partial B_r}{\partial z} \right) \mathbf{u}_z \right]
$$

$$(5.30)$$

where the particle is placed in a magnetic flux density with cylindrical components B_r, $B_\phi = 0$, and B_z. It should be noted that (5.29) and (5.30) have terms agreeing with others [12, 13] for 1D field variation as well as additional terms for 3D field variation.

The factor for finite permeability of (5.28)–(5.30) can be used to generalize (5.24) to:

$$
\mathbf{f}_m = \left(\frac{1}{\mu_o} - \frac{1}{\mu_p} \right) \nabla \left(\frac{1}{2} B^2 \right)
$$

$$(5.31)$$

Using $\mu_p = \mu_{rp}\mu_o$, where μ_{rp} is the relative permeability of the particle, the factor can be rewritten to give:

$$
\mathbf{f}_m = \left(1 - \frac{1}{\mu_{rp}} \right) \nabla \left(\frac{1}{2\mu_o} B^2 \right)
$$

$$(5.32)$$

which gives for axisymmetric force density in the z direction:

$$
f_{mz} = \left(1 - \frac{1}{\mu_{rp}} \right) \left[\frac{\partial}{\partial z} \left(\frac{1}{2\mu_o} B^2 \right) \right]
$$

$$(5.33)$$

Similarly, the Cartesian force density component f_{mx} is found by using the x derivative, and f_{my} is found by using the y derivative:

$$f_{my} = \left(1 - \frac{1}{\mu_{rp}}\right)\left[\frac{\partial}{\partial y}\left(\frac{1}{2\mu_o}B^2\right)\right] \qquad (5.34)$$

To verify (5.34), multiply both sides by its differential ∂y to obtain the magnetic surface pressure in the y direction:

$$P_{my} = \left(1 - \frac{1}{\mu_{rp}}\right)\left[\frac{1}{2\mu_o}B^2\right] \qquad (5.35)$$

which agrees with (5.15) for magnetic pressure.

To find the total force on a particle, one must integrate the above force density:

$$F = \int f dv \qquad (5.36)$$

which for 2D particles such as cylinders extending out of the page in the z direction becomes:

$$F = \int\int f(x, y) \cdot dx \cdot dy \int dz \qquad (5.37)$$

Assuming the length out of the page is L, the 2D force per unit length (in units of N/m) is simply:

$$F/L = \int\int f(x, y) \cdot dx \cdot dy \qquad (5.38)$$

The magnetic force densities and forces on permeable particles in typical magnetic separators will be discussed in Chapter 7.

PROBLEMS

5.1   Redo Example 5.1 but with $A_z = x$.

5.2   Redo Example 5.1 but with $A_z = x + 5$.

5.3   Redo Example 5.1 but with $A_z = y^2$.

5.4   Redo Example 5.2 but with 31 divisions to obtain 30 flux lines.

5.5   Find the total normal magnetic pressure in pascals (newtons per square meter) acting on steel surfaces in air with $B = 1.1$ T and steel relative permeability $\mu_r = 1500$.

5.6   Find the total normal magnetic pressure in pascals (newtons per square meter) acting on steel surfaces in air with $B = 1.2$ T and steel relative permeability $\mu_r = 500$.

5.7   (a) Redo Example 5.3 but with steel relative permeability $= 500$. (b) Also, use (5.12) as well as (5.11) to find the force using the reluctance method.

5.8   Redo the Maxwell computation of Example 5.3 using steel with the nonlinear curve of steel_1010.

5.9   A wire carrying current of 100 A in the x direction is subjected to a magnetic flux density $\mathbf{B} = 0.8$ T in the y direction. Find the Lorentz force, a vector.

5.10   Find the equivalent ampere-turns of a permanent magnet with $H_c = 1,000,000$ A/m. Its length in the direction of magnetization is 2 cm.

5.11   Redo Example 5.4 but with the permanent magnets changed to "Ceramic8D" ferrite magnets.

5.12   Redo Example 5.4 but with the permanent magnets changed to "SmCo28" samarium cobalt magnets.

5.13   Redo Example 5.4 but with the permanent magnet spacing reduced to 10 mm.

5.14   A rotating actuator is analyzed at two positions with constant ampere-turns applied. At $0°$, the total coenergy is 2.02 J. At $1°$, the total coenergy is 2.03 J. Find the torque and its direction. HINT: Be sure to use radians, not degrees.

5.15   Integrate the magnetic force density $f(r,z)$ for axisymmetric problems to show that the total force on an axisymmetric high permeable particle is of magnitude:

$$F = 2\pi \int\int f(r, z) \cdot r \cdot dr \cdot dz$$

REFERENCES

1. Bessho K, Yamada S, Kanamura Y. Analysis of transient characteristics of plunger type electromagnets. *Electrical Eng Japan* 1978;98:56–62.
2. Brauer JR (ed.). *What Every Engineer Should Know About Finite Element Analysis*, 2nd ed. New York: Marcel Dekker; 1993.
3. Brauer JR, Lumkes JH. Coupled model of a magnetic actuator controlling a hydraulic cylinder and load. *IEEE Trans Magn* 2002;38:917–920.
4. Brauer JR, Ruehl JJ, Juds MA, VanderHeiden MJ, Arkadan AA. Dynamic stress in magnetic actuator computed by coupled structural and electromagnetic finite elements. *IEEE Trans Magn* 1996;32:1046–1049.
5. Bozorth RM. *Ferromagnetism*. New York: John Wiley & Sons; 1951. Reprinted by Wiley IEEE Press.
6. Parker RJ. *Advances in Permanent Magnetism*. New York: John Wiley & Sons; 1990.
7. U.S. Department of Energy. *Critical Materials Strategy*. U.S. Department of Energy; 2011.

8. VanderHeiden RH, Brauer JR, Ruehl JJ, Zimmerlee GA. Utilizing permanent magnets in nonlinear three-dimensional finite element models. *IEEE Trans Magn* 1988;24:2931–2933.

9. Cullity BD, Graham CD. *Introduction to Magnetic Materials*, 2nd ed. Hoboken, NJ: Wiley IEEE Press; 2009. p 36.

10. Morgenthaler FR. *The Power and Beauty of Electromagnetic Fields*. Hoboken, NJ: Wiley IEEE Press; 2011. p. 42.

11. Brauer JR, Cook DL, Bray TE. Finite element computation of magnetic force densities on permeable particles in magnetic separators. *IEEE Trans Magn* 2007;43:3483–3487.

12. Gerber R, Birss R. *High Gradient Magnetic Separators*. New York: John Wiley & Sons; 1984.

13. Oberteuffer JA. Magnetic separation: a review of principles, devices, and applications. *IEEE Trans Magn* 1974;10:223–238.

CHAPTER 6

Other Magnetic Performance Parameters

The preceding chapter discussed force and related performance parameters of magnetic devices. This chapter discusses other magnetic performance parameters. After defining these key parameters, they are evaluated using reluctance and finite-element methods. Their relations with energy and force are also explained.

6.1 MAGNETIC FLUX AND FLUX LINKAGE

6.1.1 Definition and Evaluation

Magnetic flux through any surface **S** can be determined by integrating magnetic flux density **B**, using the surface integral from Chapter 2:

$$\phi = \int \mathbf{B} \cdot \mathbf{dS} \tag{6.1}$$

If magnetic vector potential **A** is used as in Chapter 4, then the flux can be expressed as:

$$\phi = \int (\nabla \times \mathbf{A}) \cdot \mathbf{dS} \tag{6.2}$$

Using Stokes' vector identity of Chapter 2, the surface integral can be changed to a closed line integral around the surface:

$$\phi = \oint A \cdot dl \tag{6.3}$$

The closed line integral of (6.3) is easily evaluated from finite-element solutions of **A**. For 2D planar solutions of depth d and vector potential component A_z obtained as described in Chapter 4, (6.3) becomes:

$$\phi_{12} = (A_{z1} - A_{z2})d \tag{6.4}$$

Magnetic Actuators and Sensors, Second Edition. John R. Brauer.
© 2014 The Institute of Electrical and Electronics Engineers, Inc. Published 2014 by John Wiley & Sons, Inc.

where the flux is found that flows between any points one and two on the planar model.

Recall from Chapter 2 [Eq. (2.35)] that flux linkage is defined as flux times the number of turns of a coil:

$$\lambda = N\phi \tag{6.5}$$

Thus if the flux of (6.1)–(6.4) is found through a coil, the flux linkage is easily evaluated.

Example 6.1   Finding Flux in Example 5.3 using Maxwell   Given the one-half model of the "C" steel path with airgap of Example 5.3, find the flux passing through the steel pole face using Maxwell finite-element software and compare it with the reluctance solution of Example 3.1.

Solution   If you are using Maxwell SV, in its main menu click on "Setup Executive Parameters" and then select "Flux Lines." Enter the points on the left corner and right corner of the pole face, separated 0.1 m apart in the x direction, thereby forming "Line1." Then return to the main menu and click on "Solve." You may have to request a smaller error under "Setup Solution Options" in order to execute a new finite-element solution. When the solution is completed, click on the upper "Solutions" tab and then on "Flux Lines." It displays "Flux Linkage" $= -0.014607$ Wb.

If instead you are using Maxwell version 16, its model has a Polyline object, Line1 created on the steel pole face. To create the flux calculation, open the Fields Calculator, enter Quantity -> B, Geometry -> Line, select Line1, Vector -> Normal, Scalar -> Integrate, Output -> Eval. This is also added as a Named Expression, "Flux," for later plotting. You may choose to request a smaller error in the Solve Setup dialog box. "Analyze" the model and find the Data Table result for Flux under Results. It displays -0.014611 Wb.

The Maxwell value is obtained using (6.4) for 1 m depth and for one turn. Thus the finite-element value for 0.1 m depth is -0.00146 Wb for one turn. One would have to multiply by the number of turns to obtain the flux linkage of (6.5).

The reluctance method solution of Example 3.1 obtains a flux density of 0.125 T. Multiplying by the pole area of 0.1 m times 0.1 m gives a flux of 0.00125 Wb, or 125/146th the Maxwell flux. Thus, as before, the reluctance method is considerably less accurate than the finite-element method.

6.1.2   Relation to Force and Other Parameters

Flux linkage is useful in several ways. One way, mentioned in Chapter 2, is to determine voltage induced by Faraday's law (2.34):

$$V = -\frac{\partial \lambda}{\partial t} \tag{6.6}$$

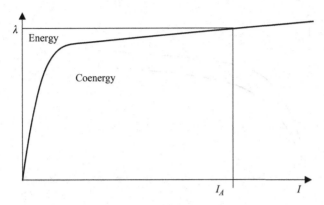

FIGURE 6.1   Magnetic energy and coenergy at a typical current value on a typical λ–I curve of a magnetic device.

Another use of flux linkage is to obtain alternative expressions for the energy, coenergy, and force of the preceding chapter. Energy input is power (voltage times current) integrated over time, giving [1]:

$$W = \int VI dt = \int I d\lambda \qquad (6.7)$$

where in general I versus λ is a nonlinear relation shown in Figure 6.1. Figure 6.1 has a shape similar to the nonlinear relation of B–H in a magnetic device graphed in Figure 5.4. From the relations ϕ equals B times area, λ equals N times ϕ, and H equals N times I, Figure 6.1 is a scaled version of Figure 5.4. From (6.7), the energy stored is the area to the left of the curve:

$$W_{\text{mag}} = \int I d\lambda \qquad (6.8)$$

Similarly to Figure 5.4, the coenergy is the area below the curve, which is [1,2]:

$$W_{\text{co}} = \int \lambda dI \qquad (6.9)$$

With the coenergy and energy known, the magnetic force may be obtained using techniques of the preceding chapter. As shown in Figure 6.2, and in agreement with (5.13), the force in the x direction is:

$$F_x = \frac{\partial W_{\text{co}}}{\partial x}\Big|_{I=\text{const}} \qquad (6.10)$$

Similarly, torque can be found by the derivative with respect to angle as described previously in Section 5.7.

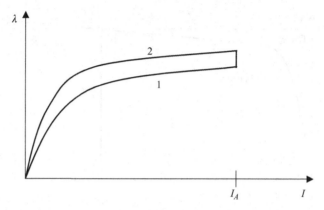

FIGURE 6.2 Nonlinear λ–I relation of a typical magnetic device at positions one and two. The applied current is assumed to be held constant.

Example 6.2 Finding Force Given Flux Linkage Versus Current and Position
Given the following relation for a magnetic actuator, find its magnetic force for $I = 2$ A and position $x = 0.05$ m:

$$\lambda = 0.15\sqrt{I}/x \qquad \text{(E6.2.1)}$$

Solution The coenergy of the actuator is found using (6.9):

$$W_{co} = \int \lambda dI = \int 0.15 I^{0.5} x^{-1} dI = (0.15/1.5) I^{1.5} x^{-1} = 0.1 I^{1.5} x^{-1} \qquad \text{(E6.2.2)}$$

The magnetic force is then found using (6.10):

$$F_{mag} = \frac{\partial W_{co}}{\partial x}\Big|_{I=\text{const}} = 0.1 I^{1.5} \frac{\partial}{\partial x} x^{-1} = -0.1 I^{1.5} x^{-2} \qquad \text{(E6.2.3)}$$

which, evaluated for the given current and position, gives:

$$F_{mag} = -0.1(2)^{1.5}/(0.05)^2 = -113\,\text{N} \qquad \text{(E6.2.4)}$$

6.2 INDUCTANCE

6.2.1 Definition and Evaluation

The basic definition of *inductance* is:

$$L = \lambda/I \qquad \text{(6.11)}$$

In the case of devices with purely linear *B–H* materials (of constant permeability), Ampere's law gives flux and flux linkage proportional to current. Hence in linear devices, inductance is a constant, independent of current. Inductance units are henrys (H). All coils have inductance and can be called *inductors*.

For a coil with the flux linkage and current of (6.11), the inductance obtained is the *self inductance*. For multiple coils, the ratios of different flux linkages and currents may be taken, expressed as:

$$L_{jk} = \lambda_j / I_k \tag{6.12}$$

If $j \neq k$, then L_{jk} is called *mutual inductance*. Self inductance is for $j = k$. An inductance matrix is a way to express (6.12), where diagonal elements are self inductances and off-diagonal elements are mutual inductances. The matrix is usually symmetric, that is,

$$L_{kj} = L_{jk}$$

For devices with nonlinear *B–H* materials, flux linkage is not proportional to current and thus inductance is not uniquely defined. Figure 6.3 shows two inductances equal to two slopes. *Secant inductance* (L_{sec}), sometimes also called *apparent inductance*, is the term often given to the ratio λ/I at a particular current. *Incremental inductance* (also called differential inductance) is the slope of the λ–I curve:

$$L_{\text{inc}} = \frac{\partial \lambda}{\partial I} \tag{6.13}$$

Both secant and incremental inductances vary with current. Only if the current is low enough that no saturation occurs does a unique inductance apply, for then secant and incremental inductances are the same constant values.

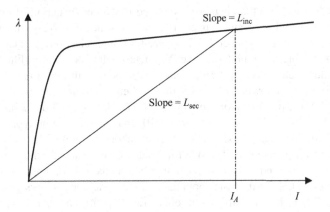

FIGURE 6.3 Inductances L_{sec} (secant) and L_{inc} (incremental) shown as slopes on a typical λ–I curve of a magnetic device.

Example 6.3   Finding an Inductance Matrix using Maxwell   Two coils are wound on a steel cylinder of length 20 mm and radius 10 mm. Both have cross section 4 mm by 4 mm and are placed flush with both ends of the cylinder as shown in Figure E6.3.1. The steel has relative permeability of 2000. Each coil has 10 turns and carries 1 A. Make an axisymmetric finite-element model in Maxwell, and use it to find the inductance matrix.

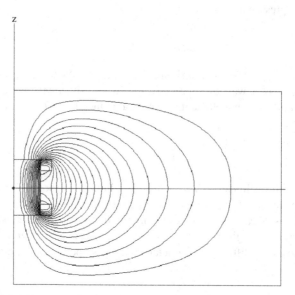

FIGURE E6.3.1   Computer display of two coils on a steel cylinder and the computed flux line plot for DC magnetostatics.

Solution   If you are using Maxwell SV, in its main menu click on "Solver" to make it "Magnetostatic" and click on "Drawing" to make it "RZ Plane." Then use "Define Model" to enter the geometry, followed by "Setup Materials" to enter the linear steel for the cylinder, and copper or vacuum for the coils. Next, under "Setup Boundaries/Sources" give both coils 10 ampere-turns and select zero potential for the outer boundary. Under "Setup Executive Parameters" click on "Matrix/Flux." Using its new window, select each coil region and (with the default return path) assign it to the matrix. After allowing 13 passes under "Setup Solution Options," click on "Solve." Obtain the flux line plot shown in Figure E6.3.1 using 21 divisions.

If instead you are using Maxwell version 10, change the solution type to "Magnetostatic," with Geometry Mode as "Cylindrical about Z." In the Maxwell drawing window, create the geometry and Assign Materials for linear steel for the cylinder and copper for the coils. Under Excitations, assign both coils with 10 ampere-turns, and zero Vector Potential for the outer boundary. Under "Parameters," assign a Matrix and "Include" both of the coils (with default return path). Optionally, specify the number of turns for each coil in the Post Processing tab. After allowing 13 passes in the Solve Setup, click on "Analyze."

Finally (for either version of Maxwell), click on the upper "Solutions" tab to obtain "Matrix." It shows an inductance matrix with $L_{11} = L_{22} = 6.063\text{E}{-}8$ H and with $L_{12} = L_{21} = 1.799\text{E}{-}8$ H. However, these inductances are for one turn in each coil. Each L_{jk} must be multiplied by $N_j N_k$, which is 100 in this problem, to obtain actual inductances. Note that the mutual inductances are much less than self inductances, which is often the case. Note also that due to the symmetry of this problem, both self inductances are the same, and both mutual inductances are the same.

6.2.2 Relation to Force and Other Parameters

The relation between inductance and reluctance is obtained as follows. Rewriting (6.11) as:

$$L = N\phi/I \tag{6.14}$$

and using the reluctance definition:

$$\mathscr{R} = NI/\phi \tag{6.15}$$

to solve for N and substitute it into (6.14), giving:

$$L = N^2/\mathscr{R} \tag{6.16}$$

The magnetic energy stored in a constant (linear) inductor is [3]:

$$W_{\text{mag}} = W_{\text{co}} = \int I\,d\lambda = \int \lambda\,dI = \int LI\,dI = \frac{1}{2}LI^2 \tag{6.17}$$

When inductance is constant over a current range but varies with position, it can be used to obtain magnetic force as follows. From (6.10), we obtain:

$$F_x = \frac{\partial W_{\text{co}}}{\partial x}\Big|_{I=\text{const}} = \frac{\partial}{\partial x}\left[\frac{1}{2}L(x)I^2\right]\Big|_{I=\text{const}} \tag{6.18}$$

$$F_x = \frac{1}{2}I^2\frac{dL(x)}{dx} \tag{6.19}$$

The above force equation is valid for a magnetic device with a single coil and single self inductance. When multiple coils are energized, energies and forces are added due to the other self inductances and mutual inductances. For example, for two coils coupled through mutual inductance L_{12}, magnetic energy is:

$$W_{\text{mag}} = \frac{1}{2}L_{11}I_1^2 + \frac{1}{2}L_{22}I_2^2 + L_{12}I_1I_2 \tag{6.20}$$

and the related force is:

$$F_x = \frac{1}{2}I_1^2\frac{dL_{11}(x)}{dx} + \frac{1}{2}I_2^2\frac{dL_{22}(x)}{dx} + I_1 I_2\frac{dL_{12}(x)}{dx} \qquad (6.21)$$

Similar expressions can be derived for torque as a function of angular position.

6.3 CAPACITANCE

6.3.1 Definition

Coils and other parts of magnetic devices possess *capacitance*. The basic definition of capacitance is:

$$C = Q/V \qquad (6.22)$$

where Q is the charge in coulombs and V is the voltage. The unit of capacitance is the farad (F), and a device possessing capacitance is called a capacitor.

A typical capacitor has been shown in Figure 2.7. As discussed in Chapter 2, capacitors are electric field devices, not magnetic field devices. Thus this book does not discuss capacitors in detail. However, because capacitance plays a role in the performance of magnetic actuators and sensors, especially at high frequencies, a few basics of capacitance are needed here.

The capacitance of (6.22) is called *self capacitance* if the voltage is across two conducting plates (as in Figure 2.7) and the charge is the magnitude of the charge on each plate. A capacitance matrix, analogous to an inductance matrix, is important if there are more than two plates.

Capacitance can be evaluated using the finite-element method for electrostatic fields, as will be shown in the next subsection.

6.3.2 Relation to Energy and Force

The electric field energy stored in a capacitor is:

$$W_{el} = \frac{1}{2}\int \mathbf{D}\cdot\mathbf{E}dv = \frac{1}{2}\int \varepsilon E^2 dv = \frac{1}{2}CV^2 \qquad (6.23)$$

where the various parameters, including permittivity ε, have been defined in Chapter 2.

When capacitance varies with position, an electric field force is created. The virtual work method can again be used to obtain the force as the derivative of energy:

$$F_x = \frac{\partial W_{el}}{\partial x}|_{V=\text{const}} = \frac{\partial}{\partial x}\left[\frac{1}{2}C(x)V^2\right]|_{V=\text{const}} \qquad (6.24)$$

$$F_x = \frac{1}{2}V^2\frac{dC(x)}{dx} \qquad (6.25)$$

If the force is desired in terms of fields rather than capacitance, then the change in stored electric field energy in Figure 2.7 can be found for a virtual displacement Δy of either plate. By observing the change in stored energy as was done for magnetic fields in Figure 5.5, the pressure due to an electric field E in air is found to be:

$$P_{\text{el}} = \frac{1}{2}\varepsilon_o E^2 \qquad (6.26)$$

Because electric field E is usually limited due to material breakdown (arcing and similar destructive behavior), electric field forces and pressures are usually much smaller than magnetic field forces and pressures. For example, in air the breakdown E is approximately 3 MV/m, varying somewhat with humidity, temperature, and pressure. Substituting that E into (6.26) gives an electric field pressure of 39.8 Pa, far below the maximum 20.E5 Pa produced by magnetic fields on steel poles in air.

Example 6.4   Finding Capacitance using Maxwell   Two aluminum plates 2 m wide are separated by 1 m as shown originally in Figure 2.7 and also in Figure E6.4.1. The lower plate is at 0-V DC and the upper at 1 V DC. The region between the two plates is assumed filled with polystyrene, which has a relative permittivity of 2.6. Find the voltage contours, electric field, energy stored, and capacitance using Maxwell. Validate the energy stored using (6.23).

FIGURE E6.4.1   Computer display of capacitor with computed voltage contours.

Solution   In the main Maxwell SV menu make sure that "Solver" is set to "Electrostatic" and "Drawing" to "XY Plane." If using Maxwell version 16, set "Solution Type" to "Electrostatic" and "Geometry Mode" to "Cartesian XY." Then for either version, the geometry of Figure E6.4.1 is entered using three "Box" commands. The material properties of aluminum and polystyrene are selected from the material list. The boundary conditions are then entered, consisting solely of setting the lower plate to 0 V and the upper plate to 1 V. The solution, obtained in one pass, has an energy stored of 2.41E−11 J. Postprocessing gives contours of constant phi (voltage) shown in Figure E6.4.1. A color plot of the magnitude of E shows it to be 1.0 V/m between the plates, as expected from Figure 2.7. The expected electric energy stored from (6.23) is:

$$W_{el} = \frac{1}{2} \int \varepsilon E^2 dv = (0.5)(2.6)(8.854E-12)(1)^2(1)(2)(1) = 2.3E-11 \text{ J}$$

$$\text{(E6.4.1)}$$

which is in reasonable agreement with the 2.41E−11 J output by Maxwell. Using this Maxwell energy value in (6.23) gives:

$$W_{el} = 2.41E-11 = \frac{1}{2}CV^2 = 0.5 \text{ C} \qquad \text{(E6.4.2)}$$

$$C = 2(2.41E-11) = 48.2\text{pF} \qquad \text{(E6.4.3)}$$

6.4 IMPEDANCE

Many magnetic devices carry sinusoidal alternating current, often of frequency 50 Hz or 60 Hz. For AC devices and systems, *impedance* is a complex number (in ohms) defined as:

$$Z = V/I \qquad (6.27)$$

where V is complex phasor voltage and I is complex phasor current.

In magnetic devices, voltage obeys Faraday's law (6.6), which becomes for AC frequency f (in Hz):

$$V = -j2\pi f \lambda \qquad (6.28)$$

where the time derivative of (6.6) has been replaced by $j2\pi f$, where j is the square root of minus one [4]. The imaginary number j causes the voltage V to be 90° out of phase from the flux linkage λ. In Chapter 2, such a 90° phase shift was shown in Example 2.7, in which a time derivative caused a sine to become a cosine.

In AC analyses, it is common to call $2\pi f$ the angular frequency ω (in radians/second), and thus (6.28) becomes:

$$V = -j\omega\lambda \qquad (6.29)$$

Substituting (6.3) and (6.5) gives:

$$V = -j\omega N \oint A \cdot dl \qquad (6.30)$$

and hence the impedance (with the sign reversed to follow normal sign conventions) of (6.26) is

$$Z = j\omega N \oint A \cdot dl / I = j\omega\lambda / I \qquad (6.31)$$

Z is in general a complex number [4], and can be split into real and imaginary components:

$$Z = R + jX \qquad (6.32)$$

X is called *reactance* and is shown using (6.14) and Faraday's law to be proportional to inductance:

$$X = j\omega L \qquad (6.33)$$

As discussed in Chapter 2, the resistance R can account for power losses due to induced eddy currents in conducting materials. Maxwell has a solution type "Eddy Current" which solves for AC fields and impedance. Similar to the inductance and capacitance matrices discussed above, impedance matrices can be obtained for multiple coils. Further discussions of AC fields will appear in Chapter 8.

Example 6.5   Finding an Impedance Matrix using Maxwell   As in Example 6.3, two coils are wound on a steel cylinder of length 20 mm and radius 10 mm. Both have cross section 4 mm by 4 mm and are placed flush with both ends of the cylinder as shown in Figures E6.3.1 and E6.5.1. The steel has relative permeability of 2000 and conductivity 2.E6 S/m. Each coil has 10 turns of stranded wire (the stranding eliminates eddy currents in the wire) and carries 1 A real of AC current of frequency 60 Hz. Make an axisymmetric finite-element model in Maxwell, and use it to find the impedance matrix.

Solution   If you are using Maxwell SV and have already solved Example 6.3, copy it to a new project. In the main Maxwell menu click on "Solver" to make it "Eddy Current" and click on "Drawing" to make it "RZ Plane." If you have not previously defined the geometry, use "Define Model" to enter the geometry. Next use "Setup Materials" to enter the linear steel for the cylinder, and copper or vacuum for the coils. Make sure that the steel has the conductivity 2.E6 S/m. Next, under "Setup Boundaries/Sources" give both coils 10 ampere-turns (be sure to select stranded and phase angle 0) and select zero potential for the outer boundary. Under "Setup Executive Parameters" click on "Matrix/Flux." Using its new window, select each coil region and (with the default return path) assign it to the matrix. After allowing 14 passes under "Setup Solution Options," click on "Solve."

If instead you are using Maxwell version 16 and have already solved Example 6.3, copy the Design to a new Project. Change the Solution Type to "Eddy Current" and select the Geometry Mode as "Cylindrical about Z." If you have not previously defined the geometry, create the objects and Assign Materials as linear steel for the

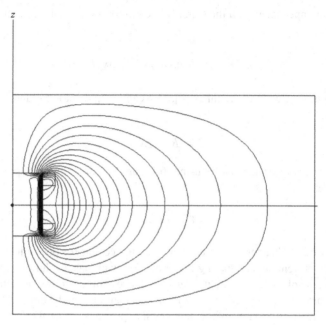

FIGURE E6.5.1   Computer display of two coils on a conducting steel cylinder and the computed 60-Hz flux line plot.

cylinder and copper for the coils. Make sure the steel has the conductivity 2.E6 S/m. Under Excitations, Assign both coils 10 ampere-turns (be sure to select stranded and phase angle 0) and select zero Vector Potential for the out boundary. Under "Parameters," Assign a Matrix and "Include" both coils. After allowing 14 passes under "Solve Setup" dialog box, click on "Analyze."

Either version of Maxwell obtains the flux line plot shown in Figure E6.5.1 using 21 divisions. Note that it is very different from the DC flux line plot of Figure E6.3.1, especially in the steel, evidently due to its eddy currents. Chapter 8 will explain that the eddy currents force the flux to flow near the skin of the conducting steel.

Finally, click on the upper "Solutions" tab to obtain "Matrix." It shows a matrix with $L_{11} = L_{22} = 6.046\mathrm{E}{-}8$ H and with $L_{12} = L_{21} = 1.785\mathrm{E}{-}8$ H. However, these values are for one turn in each coil. Each L_{jk} must be multiplied by $N_j N_k$, which is 100 in this problem, to obtain actual inductances. Note that the mutual inductances are much less than self inductances, and are only slightly different from the DC case of Example 6.3. Note also that due to the symmetry of this problem, both self inductances are the same, and both mutual inductances are the same. To obtain reactance components of impedance, all inductances must be multiplied by $\omega = 2\pi f = 2\pi 60 = 377$.

The matrix also contains resistances $R_{11} = R_{22} = 8.916\mathrm{E}{-}8$ Ω and $R_{12} = R_{21} = 6.246\mathrm{E}{-}8$ Ω. Again, these values are for one turn in each coil. Each R_{jk} must be multiplied by $N_j N_k$, which is 100 in this problem, to obtain the actual resistances. These resistances account for eddy current power loss in the conducting steel.

PROBLEMS

6.1 Redo the flux computation of Example 6.1 but find the flux in the steel along the horizontal symmetry plane.

6.2 Recalculate the force of Example 6.2 at the same current and voltage but with a new relation:

$$\lambda = 0.3I^{0.7}/x \qquad (P6.2.1)$$

6.3 Prove that an alternative formula for magnetic force is:

$$F_{mag} = \frac{\partial W_{mag}}{\partial x}\,|\lambda = \text{const.} \qquad (P6.3.1)$$

Then use it to obtain the same force value found in Example 6.2.

6.4 Redo Example 6.3 with the steel changed to the nonlinear "steel_1010."

6.5 Redo Example 6.3 with the number of turns of the lower coil changed to 20.

6.6 Redo Example 6.3 with the upper coil moved up by 4 mm.

6.7 From the results of Example 3.1 using the reluctance method, find the inductance of the coil assuming its number of turns is 100.

6.8 From the results of Example 3.2 using the reluctance method, find the inductance of the coil for both positions assuming the number of turns is 100.

6.9 Given the complex flux linkage $\lambda = 2 + j3$ weber-turns and current $I = 0.5$ A, find the impedance, resistance, reactance, and inductance (for eight turns and 60 Hz).

6.10 Redo Example 6.5 with the steel changed to the nonlinear "steel_1010."

6.11 Redo Example 6.5 with the number of turns of the lower coil changed to 20.

6.12 Redo Example 6.5 with the upper coil moved up by 4 mm.

REFERENCES

1. Sen PC. *Principles of Electric Machines and Power Electronics*, 2nd ed. New York: John Wiley & Sons; 1997.
2. Fitzgerald AE, Kingsley C Jr, Umans SD. *Electric Machinery*, 6th ed. New York: McGraw-Hill; 2003.
3. Slemon GR. *Magnetoelectric Devices*, New York: John Wiley & Sons; 1966.
4. Strangeway RA, Petersen OD, Gassert JD, Lokken RJ. *Contemporary Electric Circuits: Insights and Analysis*, 2nd ed. Upper Saddle River, New Jersey: Pearson Prentice Hall; 2006.

ACTUATORS

CHAPTER 7

Magnetic Actuators Operated by DC

Magnetic actuators use magnetic fields to transform electrical energy into mechanical energy. Thus they are a type of *transducer*, which is any device that transforms one type of energy into another [1]. However, the mechanical motion of the magnetic actuators in this book is over a limited range of motion. This limited range is commonly assumed for actuators, unlike electric motors and other rotary electromagnetic machines, which have large unlimited motion over multiple revolutions. This part of the book, Part II, is devoted to magnetic actuators.

This chapter discusses in detail the most common types of magnetic actuators operated by direct current, in the following order: solenoid actuators, voice coil actuators, other linear actuators, proportional actuators, rotary actuators, magnetic bearings and couplings, and magnetic separators.

7.1 SOLENOID ACTUATORS

Solenoid actuators have a *solenoidal coil*, which is a coil of wire wound to a shape that is sometimes cylindrical. All solenoid actuators also have a steel *armature*, a term for the moving part. In solenoid actuators, the armature moves along a straight line and thus produces *linear motion*. A common use of solenoid actuators is to move a mechanical switch for large voltages and currents. Since the current supplied to the solenoid coil is much smaller than the current being switched, the solenoid and switch form a *relay* or *contactor.*

7.1.1 Clapper Armature

A solenoid actuator with a *clapper* armature functions similar to a clapping hand. The armature acts like one hand that is drawn by a magnetic force to the second hand. Because the second "hand" is stationary, it is called the *stator.*

Like a clapping hand, the clapper armature usually contacts the stator at two areas. In one area, magnetic flux enters the armature, while it leaves the armature through the other area, producing useful magnetic force over both areas. Since magnetic flux

Magnetic Actuators and Sensors, Second Edition. John R. Brauer.
© 2014 The Institute of Electrical and Electronics Engineers, Inc. Published 2014 by John Wiley & Sons, Inc.

must always both enter and leave the armature (or any other volume as discussed in Chapter 1), the clapper armature is often more compact than other designs. The clapper may be either flat, like a flat hand, or may have ridges, like a cupped hand.

The coil may be excited by either a DC voltage (and current) or by an AC voltage (and current). This chapter deals only with actuators operated by DC, while AC actuators are discussed in the next chapter. However, any such voltage (and current) must be turned on and off at some time in the past and future, and such transient effects are discussed in Chapter 9.

An example of a DC clapper solenoid is shown in Figure 7.1a. It is somewhat similar to the Eaton Cutler-Hammer AC solenoid [2] shown in Chapter 4. Its magnetic flux pattern computed by finite-element analysis is shown in Figure 7.1b.

As shown in the solenoid of Figure 7.1, the clapper and the stator can be made of steel laminations. Laminating the steel minimizes eddy current effects that reduce force and increase power loss. If laminations are used, they are usually stacked to form a geometry that must be two dimensional (2D), or planar.

Instead of planar geometry, axisymmetric geometry is very common. Solenoidal coils are often cylindrical and thus axisymmetric. If these coils are surrounded by axisymmetric steel, an axisymmetric solenoid is created. However, since axisymmetric geometries cannot be laminated at low cost, axisymmetric solenoids are commonly made of solid steel.

The magnetic force produced by a solenoid actuator usually varies considerably with airgap and with coil current. Assuming for illustrative purposes that the steel permeability is infinite (and thus the MMF drop in the steel is zero) and that the two airgaps each have length g and poles of equal area, Ampere's law gives the following formula for H in both airgaps:

$$H = NI/(2g) \qquad (7.1)$$

and thus the flux density in both airgaps is:

$$B = \mu_0 NI/(2g) \qquad (7.2)$$

The normal (perpendicular) magnetic pressure from Maxwell's stress tensor of Chapter 5 is equal to the flux density squared divided by twice the permeability of air. Assuming that the two pole surface "seen" areas are each the same S, then the total magnetic force is:

$$F = (\mu_0 NI)^2 (2S)/(4g^2 2\mu_0) = \mu_0 N^2 I^2 S/4g^2 \qquad (7.3)$$

Note that this approximate inherent magnetic force is inversely proportional to the square of the airgap, and thus varies tremendously with airgap. It is also proportional to the square of the coil current, and thus increases greatly with increasing current. When this force is plotted versus x (armature position), a *pull curve* is produced. A set of pull curves for various currents on one plot is often used by actuator designers and users, as will be shown in Example 7.1 and in Chapter 15.

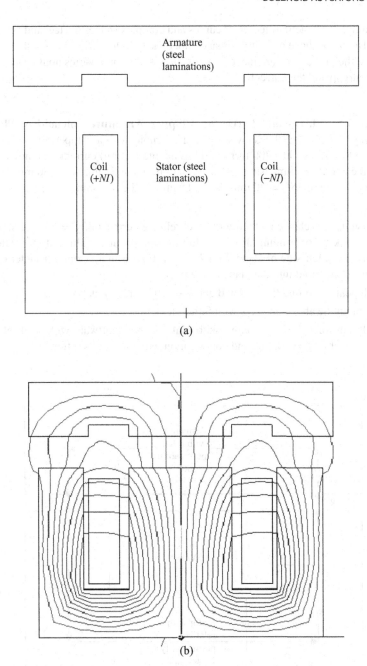

FIGURE 7.1 Typical clapper-type solenoid actuator. (a) geometry, where steel is made up of thin laminations lying in the plane of the page and stacked in the direction out of the page, (b) computer display of flux lines obtained by finite-element analysis.

In reality, due to actual steel B–H curves and complex shapes of steel and air, most solenoid actuators have pull curves that do not closely follow (7.3). For most solenoid actuators, the pull curves are indeed curves, that is, the force varies nonlinearly with both the airgap and the current.

Example 7.1   Fluxes and Forces on Clapper Armature Solenoid of Planar Geometry   Figure E7.1.1 shows a planar solenoid with a clapper armature. The stator winding shown has 200 turns and has an end (return) current path not shown outside the core. The dimensions are $w = 10$ mm, $Al_1 = 5$ mm, $Al_2 = 30$ mm, $Al_3 = 5$ mm, $Sl_1 = 15$ mm, $Sl_2 = 30$ mm, $Sl_3 = 15$ mm, and $g = 2$ mm.

(a) Use the reluctance method with steel relative permeability = 2000, assuming no leakage or fringing fluxes, to find the approximate values of the vector **B** and **F** on left side of clapper for $I = 2$ A. Find **F** in newtons per meter depth of the solenoid into the plane of the page.

(b) Repeat (a) to find the vector **B** and **F** on right side of clapper.

(c) Obtain the above answers with finite-element software.

(d) Obtain force F for the above with finite-element software with steel relative permeability = 10,000, and comment on why the results differ from (c).

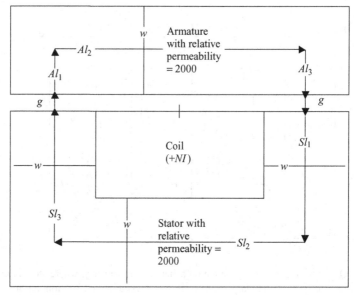

FIGURE E7.1.1   Example 7.1 clapper solenoid actuator of planar geometry.

(e) Repeat (d) for the gap varying from 4 mm to 0.5 mm in 0.5 mm increments and the current varying from 2 to 4 A in 1 A increments.

Solution

(a) The total reluctance is the sum of the stator reluctance $\mathcal{R}_s$, the left airgap reluctance $\mathcal{R}_{gL}$, the armature reluctance $\mathcal{R}_a$, and the right airgap reluctance $\mathcal{R}_{gR}$. Finding each from length over (permeability times area) gives:

$$\mathcal{R}_s = (Sl_1 + Sl_2 + Sl_3)/[(2000 \times 12.57E-7)(w \times 1)]$$
$$= (0.06)/(25.14E-6) = 2387$$
$$\mathcal{R}_{gL} = g/[(12.57E-7)(w \times 1)] = (0.002)/(12.57E-9) = 159,109$$
$$\mathcal{R}_a = (Al_1 + Al_2 + Al_3)/[(2000 \times 12.57E-7)(w \times 1)]$$
$$= (0.040)/(25.14E-6) = 1591$$
$$\mathcal{R}_{gR} = g/[(12.57E-7)(w \times 1)] = (0.002)/(12.57E-9) = 159,109$$

The flux is then NI/(total reluctance) $= (200 \times 2)/(322,196) = 1.2415E-3$ Wb. The flux density in each airgap (and in most of the steel) is the flux divided by $(w \times 1) = 0.12415$ T. From the right-hand rule, the direction of **B** in the left gap is in the $-y$ direction. The pressure on each airgap is thus $B^2/(2 \times 12.57E-7) = 6131$ Pa, which when multiplied by the airgap area of 0.01 m^2 gives the force on the left gap $= 61.31$ N. The force is always attractive, and thus the armature force is in the $-y$ direction.

(b) By symmetry, **B** is reversed in direction but force **F** is the same on the right side of the clapper. Thus the total force by the reluctance method is 122.62 N. For a solenoid depth of d meters into the page, the force must be multiplied by d.

(c) For the airgap shown of 2 mm, a model was quickly made using Maxwell from ANSYS, with modeling steps for magnetostatics described in previous chapters. The computed force on the armature for $I = 2$ A is 135.97 N/m depth. This is larger than the force estimated by the reluctance method, partly because the reluctance method does not account for fringing near the poles.

(d) The higher permeability raises the Maxwell computed force to 143.33 N. The small increase is due to the fact that the steel reluctance is small compared with the airgap reluctance for the 2 mm gap.

(e) The force curves obtained using Maxwell's parametric capability are shown in Figure E7.1.2. These pull curves show that the force is highly nonlinear (even for the assumed linear steel B–H curve of constant permeability), increasing greatly with small gap and high current.

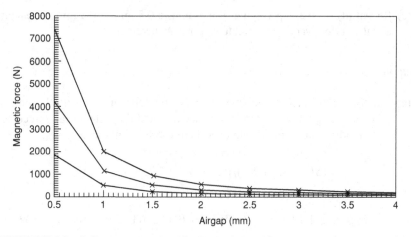

FIGURE E7.1.2   Pull curves computed for Example 7.1. The curves are for currents of 2, 3, and 4 A in order from the lowest curve to the highest curve. The force (N) on the vertical axis is plotted against the airgap (mm) on the horizontal axis.

Example 7.2   Fluxes and Forces on Clapper Armature Solenoid of Axisymmetric Geometry   Figure E7.2.1 shows an axisymmetric solenoid with a clapper armature. The number of turns $N = 2000$ and the current $I = 1$ A. The dimensions are $g = 2$ mm, $w_a = 8$ mm, $R_1 = 15$ mm, $R_2 = 25$ mm, $R_3 = 30$ mm, $Z_1 = 8$ mm, and $Z_2 = 23$ mm.

Use the reluctance method with steel relative permeability $= 2000$, assuming no leakage or fringing fluxes, to find the approximate values of the following.

(a) **B** and **F** on inside of clapper.

(b) **B** and **F** on outside of clapper.

(c) Obtain the above answers with finite-element software.

(d) Obtain the answers for the above with finite-element software with steel relative permeability $= 10,000$, and comment on why the results differ.

Solution

(a) Path lengths for the reluctance method are seen to be: $Al_1 = 4$ mm, $Al_2 = 20$ mm, $Al_3 = 4$ mm, $Sl_1 = 19$ mm, $Sl_2 = 20$ mm, $Sl_3 = 19$ mm, $w_{si} = 15$ mm, $w_{sa} = 8$ mm, $w_{so} = 5$ mm. The total reluctance is the sum of the stator reluctance $\mathcal{R}_s$, the left airgap reluctance $\mathcal{R}_{gL}$, the armature reluctance $\mathcal{R}_a$, and the right airgap reluctance $\mathcal{R}_{gR}$. Finding each from length divided by the quantity (permeability times area) gives:

$$\mathcal{R}_s = (Sl_1)/\left[(2000 \times 12.57\text{E--}7)\pi \left(R_3^2 - R_2^2\right)\right]$$
$$+ (Sl_2)/[(2000 \times 12.57\text{E--}7)\pi (R_1 + R_2)w_{sa}]$$
$$+ (Sl_3)/\left[(2000 \times 12.57\text{E--}7) \left(\pi R_1^2\right)\right]$$

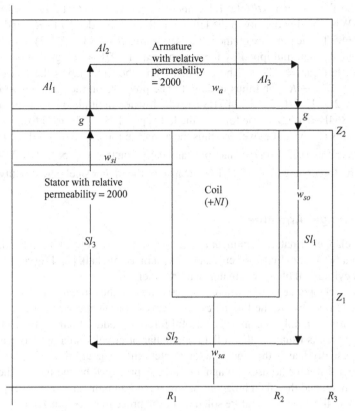

FIGURE E7.2.1 Example 7.2 clapper solenoid actuator of axisymmetric geometry. As customary, the axis of symmetry is the vertical line on the left border.

$$= (0.019)/[(2.514\text{E}{-}3)\pi(0.03^2 - 0.025^2)]$$
$$+ (0.02)/[(2.514\text{E}{-}3)\pi(0.04)(0.008)]$$
$$+ (0.019)/[(2.514\text{E}{-}3)\pi(0.015)^2] = 8748 + 7913 + 10{,}692$$
$$= 27{,}353$$

$$\mathscr{R}_{gL} = g/\left[(12.57\text{E}{-}7)\left(\pi R_1^2\right)\right] = (0.002)/[(12.57\text{E}{-}7)(\pi 0.015^2)]$$
$$= 2.251\text{E}6$$

$$\mathscr{R}_a = (Al_1 + Al_2 + Al_3)/[(2000 \times 12.57\text{E}{-}7)(w_a)\pi(R_1 + R_2)]$$
$$= (0.028)/(2.514\text{E}{-}3)(1.\text{E}{-}3) = 11{,}138$$

$$\mathscr{R}_{gR} = g/\left[(12.57\text{E}{-}7)\pi\left(R_3^2 - R_2^2\right)\right] = (0.002)/[(12.57\text{E}{-}7)(864\text{E}{-}6)]$$
$$= 1.842\text{E}6$$

The flux is then NI/(total reluctance) = $(2000 \times 1)/(4.132\text{E}6) = 48.41\text{E}$–
5 Wb. The flux density in the left gap is the flux divided by (πR_1^2) and thus is
0.685 T. The pressure on the left airgap is thus $B^2/(2 \times 12.57\text{E}{-}7) = 187\text{E}3$ Pa,
which when multiplied by the airgap area of 707E–6 m^2 gives the force on
the left gap = 132 N. The flux density in the right gap is the flux divided
by $\pi(R_3^2 - R_2^2)$ and thus is 0.56 T. The pressure on the right airgap is thus
$B^2/(2 \times 12.57\text{E}{-}7) = 124.7\text{E}3$ Pa, which when multiplied by the airgap area
of 864E–6 m^2 gives the force on the left gap = 108 N. The total force is 240 N
and is always attractive, and thus the armature force is in the $-z$ direction.

(b) B is about 0.82 T on the inner pole and 0.45 T on the outer pole. $F = 279.41$ N.

(c) The force increases to 285.1 N because of the higher steel permeability.

7.1.2 Plunger Armature

Unlike a clapper armature, a plunger armature is not flat or cupped like a hand, but is
basically a brick or cylinder, often shaped like a piece of chalk [3]. Figure 7.2 shows
a typical cylindrical plunger actuator in a DC solenoid.

Unlike a clapper armature, where magnetic force in the direction of linear motion
is produced both where the flux enters and leaves, the plunger armature has useful
magnetic force on only one area. The useful force is produced only at the end of the
plunger, which is usually either a flat rectangular or circular area or a conical area.
The flux usually leaves this end area, and enters the armature through the sides of
the plunger. Little or no useful magnetic force is produced by the flux on the sides
of the plunger, and thus the plunger armature solenoid might have less force per unit
volume than the clapper armature solenoid of the preceding Section 7.1.1.

FIGURE 7.2 Photo of typical plunger-type axisymmetric solenoid actuator. A sector of the
stator has been sawed away to show the coil cross section.

While the plunger is often a pure cylindrical shape as shown in Figure 7.2, it can also be shaped for "constant force" as the armature moves. Such "constant force" or "proportional" solenoids will be discussed in later sections of this chapter.

If the plunger is cylindrical as in Figure 7.2, then the solenoid is usually axisymmetric, and is thus made of solid steel. However, if the plunger is a rectangular brick shape, then it can be inexpensively laminated and placed in a laminated stator to form a planar solenoid. As mentioned in Chapter 2, eddy currents are reduced when steel is laminated. Reducing eddy currents gives reduced losses and increased magnetic forces, as will be further discussed in Chapters 8 and 9.

Finite-element analysis of axisymmetric plunger armature solenoids has been combined with *design optimization methods* to mathematically obtain optimum designs [4]. Design optimization software has been developed over recent years [5] and is becoming increasingly powerful and available. A simpler way, however, to optimize design of magnetic devices is to run multiple finite element analyses with varying design parameters such as pole width and armature length, and choose the design that produces the highest force. Such multiple runs are called *parametric finite-element analysis*. Design optimization will be further discussed in Chapter 15.

Example 7.3   Fluxes and Forces on Planar Plunger Armature Solenoid   Figure E7.3.1 shows a planar solenoid with a plunger armature. The number of turns $N = 1000$ and the current $I = 2$ A. The dimensions are $w = 8$ mm, $g = 4$ mm, $g_s = 1$ mm, $Al_1 = 4$ mm, $Al_2 = 22$ mm, $Sl_1 = 8$ mm, $Sl_2 = 20$ mm, $Sl_3 = 34$ mm, and $Sl_4 = 15$ mm.

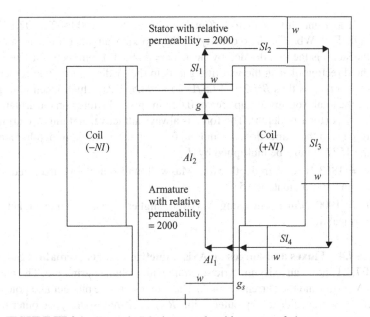

FIGURE E7.3.1   Example 7.3 plunger solenoid actuator of planar geometry.

(a) Use the reluctance method with relative permeability = 2000 in the steel, assuming no leakage flux nor fringing flux, to find the approximate values of **B** and **F** on the end of the plunger.

(b) Compute the above **B** and **F** using finite-element software.

(c) Obtain the answers for the above with finite-element software with steel relative permeability = 10,000, and comment on why the results differ.

Solution

(a) There are two flux paths, left and right, of equal lengths, areas, and reluctances. The total reluctance of the right path is the sum of the stator reluctance $\mathcal{R}_s$, the working airgap reluctance $\mathcal{R}_g$, the armature reluctance $\mathcal{R}_a$, and the side airgap reluctance $\mathcal{R}_{gS}$. Finding each from length over (permeability times area) gives:

$$\mathcal{R}_s = (Sl_1 + Sl_2 + Sl_3 + Sl_4)/[(2000 \times 12.57\text{E}{-}7)(w \times 1)]$$

$$= (0.077)/[(2.514\text{E}{-}3)(0.008)] = 3829$$

$$\mathcal{R}_g = g/[(12.57\text{E}{-}7)(w \times 1)] = (0.004)/(10.06\text{E}{-}9) = 397,614$$

$$\mathcal{R}_a = (Al_1 + Al_2)/[(2000 \times 12.57\text{E}{-}7)(w \times 1)]$$

$$= (0.026)/[(2.514\text{E}{-}3)(0.008)] = 1293$$

$$\mathcal{R}_{gS} = g_s/[(12.57\text{E}{-}7)(w \times 1)] = (0.001)/(10.06\text{E}{-}9) = 99,443$$

The right flux is then $NI/$(total reluctance) $= (1000 \times 2)/(502,179) = 3.983\text{E}{-}3$ Wb. The flux density in the working airgap g (and in most of the steel) is the flux divided by $(w \times 1) = 0.498$ T. From the right-hand rule, the direction of **B** in the working gap is in the $-y$ direction. The pressure on each airgap is thus $B^2/(2 \times 12.57\text{E}{-}7) = 98,582$ Pa, which when multiplied by the total working airgap area of 0.016 m² gives the force on the armature $= 1577$ N (for 1 m depth). The force is always attractive, and thus the armature force is in the y direction. To find the force on the solenoid of depth d meters, the 1577 N must be multiplied by d.

(b) $F = 1863$ N for 1 m depth using Maxwell, and B at the plunger end varies around approximately 0.51 T.

(c) $F = 1900$ N/m depth using Maxwell, higher because of the higher steel permeability.

Example 7.4 Fluxes and Forces on Axisymmetric Plunger Armature Solenoid
Figure E7.4.1 shows an axisymmetric solenoid with a plunger armature. The number of turns $N = 400$ and the current $I = 4$ A. The dimensions are plunger and pole outer radius $R_P = 20$ mm, side airgap outer radius $R_{PG} = 22$ mm, stator yoke outer radius

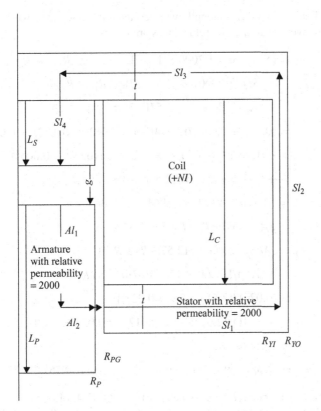

FIGURE E7.4.1 Example 7.4 plunger solenoid actuator of axisymmetric geometry.

$R_{YO} = 70$ mm, stator yoke inner radius $R_{YI} = 66$ mm, working airgap $g = 10$ mm, plunger length $L_P = 42$ mm, stator stopper length $L_S = 16$ mm, coil axial length $L_C = 46$ mm, and stator yoke axial thickness $t = 12$ mm (on top; same on bottom).

(a) Use the reluctance method, assuming no leakage flux, to find the approximate values of **B** and **F** in the working airgap on the end of the plunger with steel relative permeability = 2000.

(b) Compute the above **B** and **F** using finite-element software.

(c) Obtain the answers for the above with finite-element software with steel relative permeability = 10,000, and comment on why the results differ.

Solution

(a) The total reluctance is the sum of the stator reluctance $\mathcal{R}_s$, the working airgap reluctance $\mathcal{R}_{gW}$, the armature reluctance $\mathcal{R}_a$, and the side airgap reluctance

$\mathscr{R}_{gS}$. Each is found from length over permeability times area, where the areas are the average areas. The reluctances are then:

$$\mathscr{R}_s = (Sl_1 + Sl_3)/[(2000 \times 12.57\text{E--}7)[t \times \pi(R_{PG} + R_{YI})]$$
$$+ (Sl_2)/[(2000) \times 12.57\text{E--}7)(\pi(R_P^2 - R_{YI}^2)]$$
$$+ (Sl_4)/[(2000) \times 12.57\text{E--}7)(\pi R_P^2)]$$
$$= (0.046 + 0.058)/[2.514\text{E--}3 \times 0.012 \times \pi(0.022 + 0.066)]$$
$$+ (0.046 + 0.012)/[2.514\text{E--}3 \times \pi(0.07^2 - 0.066^2)]$$
$$+ (0.016 + 0.006)/[2.514\text{E--}3(\pi 0.02^2)]$$
$$= 12{,}470 + 13{,}500 + 6964 = 32{,}934$$
$$\mathscr{R}_{gW} = g/[(12.57\text{E--}7)(\pi R_P^2)] = 6.3307\text{E}6$$
$$\mathscr{R}_a = (Al_1)/[(2000) \times 12.57\text{E--}7)(\pi R_P^2)]$$
$$+ (Al_2)/[(2000 \times 12.57\text{E--}7)[t \times \pi(R_P)]$$
$$= (0.026)/[2.514\text{E--}3 \times \pi 0.022)]$$
$$+ (0.01)/[2.514\text{E--}3 \times 0.012 \times \pi 0.02] = 8230 + 5276$$
$$= 13{,}506$$
$$\mathscr{R}_{gS} = (R_{PG} - R_P)/[(12.57\text{E--}7)(2\pi R_P t)] = 1.055\text{E}6$$

The flux is then $NI/$(total reluctance) $= 1600/(7.426\text{E}6) = 215.5\text{E--}6$ Wb. The flux density in the working airgap (and in most of the steel) is the flux divided by $(\pi R_P^2) = 0.1715$ T. From the right-hand rule, the direction of **B** in the left gap is in the $-z$ direction. The pressure on the working airgap is thus $B^2/(2 \times 12.57\text{E--}7) = 11{,}694$ Pa, which when multiplied by the airgap area of (πR_P^2) square meters gives the force $= 14.7$ N. The force is always attractive, and thus the armature force is in the $+z$ direction.

(b) The computed $F = 19.34$ N using Maxwell. The airgap B on the end of the plunger is approximately 0.170 T.

(c) The computed F using Maxwell is now 19.66 N, with airgap $B = 0.174$ T. These are both raised because of the higher steel permeability.

7.2 VOICE COIL ACTUATORS

Instead of forces on steel, Lorentz force on current-carrying coils is used in many actuators. They are called *voice coil actuators* because of their common use in loudspeakers.

From the Lorentz force equation of Chapter 5, the force on an N-turn coil of average turn length l is:

$$F = NBIl \tag{7.4}$$

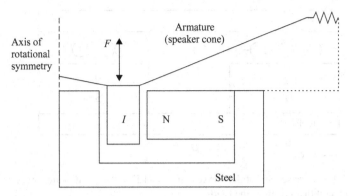

Axis of rotational symmetry

F

Armature (speaker cone)

I N   S

Steel

FIGURE 7.3 Typical voice coil actuator, shown driving a loudspeaker. The movable voice coil carries the current I and is subjected to the magnetic field from a permanent magnet with north (N) and south (S) poles.

where B is the magnetic flux density perpendicular to the coil direction and F is perpendicular to both B and the coil direction. The directions follow the right-hand rule as shown in Figure 7.3 for a typical voice coil in a loudspeaker. Note that when the current direction is reversed, then (assuming that the direction of B is unchanged) the direction of the force is reversed. The reversible force of a voice coil actuator is an advantage over the force of moving iron actuators of the preceding sections, which is directed to attract the iron armature toward the iron stator regardless of the direction of the current. Another advantage over solenoid actuators is that voice coil force is much more independent of armature position and is proportional to current. A minor disadvantage is that the voice coil current must be supplied via a flexible lead, and thus the proper stranded wire must be selected for reliable long-term operation.

Because coils are most easily wound on cylindrical bobbins, voice coil actuators often have axisymmetric geometry. No laminations are needed because the magnetic flux density B is constant and is almost always produced by a permanent magnet. There is some B produced by the voice coil current, but it is usually negligibly small compared to the B produced by the permanent magnet. An advantage of permanent magnets is that their B is produced without any current or associated power loss and temperature rise.

Besides loudspeakers, another very popular use of voice coil actuators for DC (not AC) operation is for computer disk drive head actuation. The current applied is a DC step or pulse to position the voice coil armature at the desired radius (track) on the disk. Often the coil is on a pivot, like old-fashioned turntable arms, as shown previously in Figure 1.3 of Chapter 1. Each of the two radial sectors of the coil is subjected to a permanent magnet field. The field direction is upward in one sector and downward in the other, and since the current direction is opposite in the two coil sectors, their Lorentz forces add to produce a torque on the coil, as shown in Figure 7.4. Note that the force F on each coil remains in the same x direction as long as the coil moves for a "short stroke" of less than about half of the *pole pitch p*. If the motion is not limited to much less than the pole pitch, then for the force (and/or torque) to remain

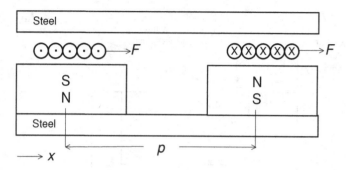

FIGURE 7.4   Cross section of disk drive actuator of Figure 1.3, showing forces F on both halves of the coil that positions the heads.

in the same x direction, the current in the coils must be *commutated*, that is, switched in direction. Thus permanent magnet motors with continuous unlimited motion are commutated either electronically or by mechanical switching via brushes. This book is devoted to actuators and sensors of limited motion with no commutation.

The time to move the coil arm to the desired position is called *seek time* and should be as small as possible to enable high speed data access for computation [6]. Because voice coils are made of copper or aluminum, both of which are lighter than steel or iron, voice coil actuators are preferred over moving iron solenoid actuators for disk head operation. Also, because voice coil force is proportional to current and is relatively independent of position, feedback control based on linear systems can be readily applied.

Example 7.5   Force of Voice Coil Actuator   A voice coil has a resistance of 20 Ω and 400 turns. A permanent magnet provides a DC magnetic flux density of 0.8 T in the radial r direction. The average radius r of the voice coil is 15 mm. Given 12 V DC applied to the voice coil, find the steady-state force produced.

Solution   From Ohm's law, the DC current $I = 12/20 = 0.6$ A. The average length of a coil turn is 2π times the average radius $= 0.015$ m, giving $l = 0.0942$ m. Thus the desired $F = 400(0.8)(0.6)(0.0942) = 18.1$ N.

7.3   OTHER ACTUATORS USING COILS AND PERMANENT MAGNETS

Besides solenoid and voice coil actuators, other actuators are available that use both permanent magnets and coils. The advantage of using permanent magnets is that the **B** they produce does not require current or power loss that coils do. The **B** of the permanent magnets interacts with the **B** of coils to produce the force. In some cases the permanent magnets produce *latching force* to hold the armature in a certain latched condition without requiring any input current.

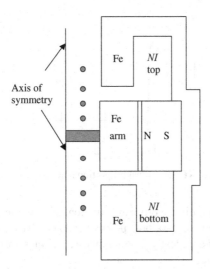

FIGURE 7.5 Actuator with both permanent magnet and coils in stator. The armature labeled "arm" moves either up or down.

Many different designs are possible. Figure 7.5 shows a design from a paper by Lequesne [7]. It is a long-stroke actuator with one radially magnetized permanent magnet, a steel or iron armature, and two coils. The coils are wound and connected so that they both carry current in the same direction. For example, if they both carry current out of the page, then the lower pole of the moving iron armature has higher flux than the upper pole, and the armature experiences a downward force. Reversal of the current gives an upward force, and no current gives zero (balanced) force on the armature. Thus the armature experiences bidirectional force. The force varies with position due to the changing of both airgaps.

Other possible designs include the tubular linear actuator of Figure 1.5, where the stator contains permanent magnets. In some actuators the armatures are permanent magnets [8,9], but the design must carefully avoid breakage of the fairly brittle permanent magnet material. Besides permanent magnets, materials that are magnetostrictive are occasionally used in actuators, but are more often used in sensors as will be described in Chapter 11.

7.4 PROPORTIONAL ACTUATORS

In many applications, proportional actuation is desired. Such actuation ideally produces a position x that is directly proportional to current or voltage, for example:

$$x = kI \tag{7.5}$$

where k is the proportionality constant. Note that x is independent of inertia, load forces, and other external forces. Such ideal proportional positioning is highly desired for valve operation and other applications.

FIGURE 7.6 Ideal proportional actuator force.

A graphical depiction of proportional actuation is shown in Figure 7.6. It shows force F as a function of x and I for both positive and negative currents I. Note that the slopes for all currents I are ideally infinite, that is, the lines are vertical and indicate no effect of external force.

In the real world, infinite slopes are not possible. Instead, a fairly high value of negative slope can be obtained by use of a mechanical spring. Figure 7.7 shows that the combination of a spring that produces a spring force:

$$F_s = k_x x \qquad (7.6)$$

along with a magnetic actuator that produces a force F_{mag} that is independent of x and proportional to I, obtains a total force F that approximates the desired set of proportional force lines. Note that the bigger the spring constant, the steeper are the total force lines, which is desirable. However, the bigger the spring, the larger is the magnetic force required for a specified total force, since the spring and magnetic forces are in opposition.

The required magnetic force that is independent of x and proportional to I is produced by what is commonly called a *constant force magnetic actuator*. From the

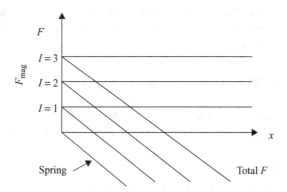

FIGURE 7.7 Nonideal proportional actuator produced by constant magnetic force and opposing spring force.

various types of magnetic actuators discussed in preceding sections, the type that inherently provides constant force is the voice coil actuator.

However, solenoid actuators can also be specially designed for proportional actuation. Without special design, solenoid actuators tend to produce force that varies greatly with position according to (7.3). To obtain force that is independent of position and proportional to current, much design work is required, usually involving electromagnetic finite-element analysis of various possible shapes of steel armatures and pole pieces. With such design work, the magnetic force can be made approximately "constant", that is, independent of position x over a fairly large variation of x. Some nonconstant regions, called *deadbands*, are present and must be avoided during operation. Figure 7.8 shows a typical design of a "constant force" plunger-type solenoid actuator. As pointed out by Lequesne [10], the plunger is usually conical (not cylindrical), and the stator poles usually are angled as well. The exact angles and shapes must be carefully computed using finite-element software or (with less certainty) reluctance methods. Also, steel saturation must occur at the proper level to produce a force that is proportional not to the current square of equation (7.3) but to the first power of current.

Since steady-state current (from Ohm's law) is V/R, a proportional actuator also has steady-state position proportional to applied voltage. Thus proportional actuators can be readily controlled by linear feedback control systems.

The lowest cost way to vary applied voltage is to use a chopper circuit to obtain *pulse–width modulation (PWM)*. A given DC voltage, such as battery voltage in an automobile, is switched on and off. Figure 7.9 shows a typical PWM voltage waveform. In this case the battery voltage is 24 V and the switching frequency is 1 kHz. Also in the figure, the "duty cycle" is approximately 33%, that is, that the

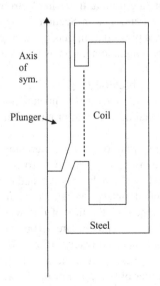

FIGURE 7.8   Plunger solenoid shaped for approximately constant force.

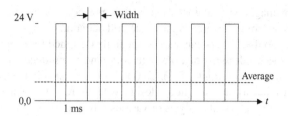

FIGURE 7.9   Typical PWM voltage waveform.

voltage is "on" about one-third of the time. The average, or DC, voltage, equals duty cycle times battery voltage, and thus is 8 V in this case.

The choice of switching frequency f involves trade-offs. Its period, $T = (1/f)$ should be much faster than the desired actuation time, yet not so high as to produce significant eddy currents. The switching frequency produces corresponding pulses in magnetic force, which is not necessarily bad. The force pulsations, called *dither*, can often help overcome static friction, often called *stiction*.

7.5 ROTARY ACTUATORS

Many applications need rotary motion, that is, motion along a circular arc instead of the straight line of linear actuators. Rotary actuators are also sometimes called *torquers*, because they produce torque. This book is restricted to limited motion, and thus rotary actuators or torquers are studied, but not electric motors in general, because they produce motion that is many times 360°/min.

Rotary magnetic actuators are used at times in electrohydraulic servo valves and in computer disk drives. As mentioned in Section 7.3, voice coil actuators in disk drives sometimes operate on a pivoted arm, thus producing rotary motion.

The most common rotary magnetic actuator is the *step motor*. It is also called a *stepper motor* or *stepping motor*.

Step motors get their name because they provide rotary motion in incremental steps. They are very popular because their incremental motion is easily controlled by microprocessors or other digital devices that produce digital pulses, that is, steps of DC current.

Figure 7.10a shows one quadrant of a typical step motor. Like most step motors, the rotor (armature) is on the inside and is made up solely of steel laminations. Over all four quadrants, the rotor shown here has 8 teeth, and the outer stator has 12 teeth. When a step of current is applied to the proper stator winding, the rotor wants to align its teeth for minimum reluctance for the flux of that winding. Hence they are also called *variable reluctance motors*. Thus Figure 7.10b shows that for flux produced by current in the stator slots at approximately 20° and 70°, the rotor tends to move to the position shown. For highest torque, the airgap between stator and rotor is usually quite small, on the order of 0.1 mm. For the 12 stator teeth and 8 rotor teeth shown, each step (increment) in motion is (360/8)–(360/12) = 15°. Other numbers

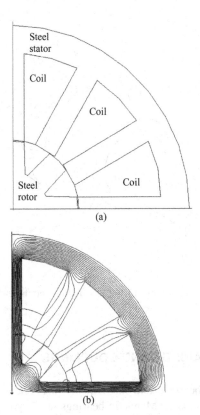

(a)

(b)

FIGURE 7.10   Typical step motor. (a) geometry of one quadrant, (b) Computer display of calculated flux lines.

of teeth can be chosen, but the rotor and stator must never have the same number of teeth.

Step motors can also be operated continuously, that is, over multiple revolutions. Under continuous operation they are commonly called *switched reluctance motors.* Again their advantage is the simple rotor construction, requiring only steel laminations and no coil or permanent magnet.

Another rotary actuator with a rotor made only of steel is shown in Figure 7.11. Its stator has a solenoidal coil embedded in steel that provides three-dimensional (3D) flux paths. The design shown in Figure 7.11 has been used by Sandia National Laboratories [11] and has four teeth on both the stator and the rotor. Its maximum rotation with four poles is less than 90°. Its flux paths and torques have been computed by 3D finite-element analysis [11]. Many design variations are possible.

Rotary actuators and motors are often used to drive gears that convert their motion to straight-line motion. Gears commonly used include ball screws and rack-and-pinions. However, as mentioned in Chapter 1, a linear (straight line) actuator has the advantages of reduced maintenance, reduced friction, and longer life.

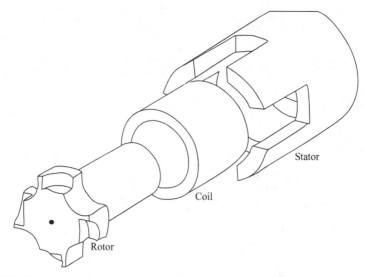

FIGURE 7.11   Rotary actuator with solenoidal winding and three-dimensional flux paths.

7.6 MAGNETIC BEARINGS AND COUPLINGS

Moving objects require bearings, and *magnetic bearings* feature long life due to their avoidance of mechanical wear. Magnetic bearings are a type of magnetic actuator designed to suspend a rotating axle with no mechanical contact. Benefits include higher reliability with little or no maintenance, reduced frictional losses, no contaminating or flammable lubricants, reduced machine vibration, and improved monitoring and diagnostics. Recent advances in magnetic bearing technology, including miniaturization, simplicity, and integration have overcome many of these limitations [12].

Figure 7.12 shows the basic layout of a typical active magnetic bearing system. Stationary electromagnets are positioned around the rotating machine. Typically, two radial magnetic bearings are used to support and position the shaft in the radial directions and one thrust bearing is used to support and position the shaft along the longitudinal (axial) direction. A shaft that is completely supported by magnetic bearings is said to provide support along five axes because the bearings react to motion along the three translational axes and two angular axes. Ideally, the magnetic bearing offers no rotational resistance.

An active magnetic bearing consists of a stator containing the electromagnets and position sensors, and the rotor, which rotates with the shaft. During operation, each magnetic bearing rotor is ideally centered in the corresponding stator so that contact does not occur. The position of the shaft is controlled using a closed-loop feedback system. The position sensors detect the local displacements from the shaft, and these signals are sent to a digital controller. The controller processes these signals and computes how to redistribute the currents in the electromagnets to restore the shaft

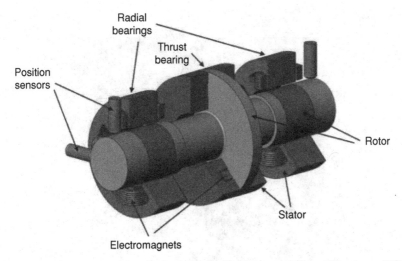

Radial
bearings

Thrust
bearing

Position
sensors

Rotor

Stator

Electromagnets

FIGURE 7.12   Magnetic bearing basic layout. Used by permission of Synchrony [12].

to its centered position. Amplifiers in the controller then readjust the currents in the electromagnets accordingly, typically 15,000 times per second.

Like other kinds of bearings, magnetic bearings provide stiffness and damping. However, unlike other bearings, magnetic stiffness and damping vary as a function of disturbance frequency. It is often convenient to model the bearing as a transfer function with an amplitude and phase that vary with frequency. The optimization of this transfer function is a critical step in ensuring that the magnetic bearing performance has adequate stability and force rejection capability over a range of frequencies. The stiffness and damping can be optimized by simply changing the control algorithm.

The load capacity of a radial bearing is the product of the rotor circumference, the active length, and the equivalent bearing pressure. Because the bearing pressure of a magnetic bearing is limited to below 20 bar from Figure 5.7, many times less than the bearing pressure of oil-lubricated fluid film bearings, the size will in general be greater for the same load capacity. Also, the end windings and position sensors increase the length of the magnetic bearing beyond its active length.

Through recent design innovations, the size of radial magnetic bearings has been reduced by more than 30% [12]. The bearing pressure for radial bearings has been improved by increasing the amount of electrical steel at the bore of the stator where the force is created. At the same time, the outer circumference of the stator has been reduced by splitting the flux paths and isolating the electromagnets. Finally, the length of the radial magnetic bearing has been reduced by developing position sensors that can be integrated into the electromagnets.

The controller for magnetic bearings has recently been miniaturized. The position sensing uses frequency modulation to produce a digital counter. A processor handles network communications, performs digital processing, and generates timing

FIGURE 7.13   Fusion® magnetic radial and thrust bearings with integrated control electronics. The radial bearing has a load capacity of 1330 N, and the thrust bearing has a load capacity of 4448 N. Used by permission of Synchrony [12].

signals for the amplifiers. The controller is now often small enough to be completely integrated into the magnetic bearing. The short distance from the controller to the magnetic coils means that EMI (to be discussed in Chapter 13) is greatly reduced. Typically the controller is supplied with a DC voltage between 48 and 300 V, and each coil magnet has a dedicated power amplifier designed to supply its power factor that is lagging due to its inductance.

The "health" of bearings is important to prevent failure, and can easily be monitored with magnetic bearings. Their built-in position sensors monitor shaft position and activate alarms if vibrations grow too large. Adding up all these advantages, magnetic bearings often outshine mechanical bearings in performance and price.

Figure 7.13 shows magnetic bearings with built-in controllers. These bearings are now smaller than past magnetic bearings that required a large external controller [13]. Each bearing is powered with 48 V DC and includes a dedicated Ethernet port for high speed communications and health monitoring. The small size and simplified mechanical and electrical interface makes it very easy to integrate the bearing into rotating machines such as motors, pumps, fans, and turbines [12].

Instead of the electromagnets of Figures 7.12 and 7.13, some magnetic bearings use permanent magnets [14]. Permanent magnet bearings developed at the Lawrence Livermore National Laboratory [15] are called *passive* because their permanent magnets levitate the rotating member without electromagnets. Such passive bearings do, however, require special passive *stabilizers* to overcome the inherent instability of interacting permanent magnets that was predicted in 1842 by Samuel Earnshaw. One type of stabilizer [15] uses Halbach magnet arrays as discussed in Chapter 5.

Some magnetic bearings also serve as *magnetic couplings*, which transfer torque between two rotating members through seals and walls without mechanical contact [16]. Magnetic couplings can greatly reduce undesirable transfer of vibration and noise. The couplings can be either synchronous (maintaining the identical speed for both rotating members) or asynchronous with the driven member having a different speed. Asynchronous eddy current couplings produce reduced speed and have the disadvantage of high power loss compared to synchronous couplings.

Also required for many rotating systems are magnetic brakes and clutches, which are often operated by clapper armature solenoids [17, 18]. So-called *magnetic particle* brakes and clutches contain ferromagnetic particles, which experience forces as explained in Section 5.8.

7.7 MAGNETIC SEPARATORS

Another type of actuator using permanent magnets and/or coils is the *magnetic separator*. Its stator produces magnetic force on multiple "armatures" consisting of ferromagnetic particles of various sizes and shapes. A practical way to calculate the force acting on small ferromagnetic particles is to use the equations for force per unit volume derived previously in Section 5.8.

In applications in the mining and recycling industries, a fluid (air, other gas, or liquid) containing both nonmagnetic and magnetic objects is fed to a magnetic separator. The fluid can sometimes have enough ferrous particles to be considered a *ferrofluid* [19]. The separator has permanent magnets and/or coils carrying either conventional or superconducting current, and is used to separate out the magnetic objects or particles, often by their attraction to steel wires [20].

Example 7.6 Separator Using a Steel Wire   The classic paper by Oberteuffer [20] examines magnetic separation forces due to cylindrical ferromagnetic wires placed in a uniform DC magnetic field. A typical wire and large particle are shown in Figure E7.6.1. Use finite-element software to compute the magnetic flux lines of the figure, and then (using the equations of Section 5.8 in the software) compute the magnetic force for the particle diameters 0.1, 1.0, 2.0, and 3.0 μm. Compare the computed forces for both cylindrical and spherical particles with results computed using simplified equations derived by Oberteuffer [20].

Solution   Rather than sketch by hand the magnetic flux lines as in Figure 4 of Oberteuffer [20], in Figure E7.6.1 the flux lines have been computed using the finite-element software Maxwell. The relative permeability of both the wire and the particle is assumed to be 1000, that is $\mu_{rp} = 1000$. Since Figure E7.6.1 is a 2D planar analysis, the particle (like the wire) is assumed to be a cylinder normal to the paper. Note that, as

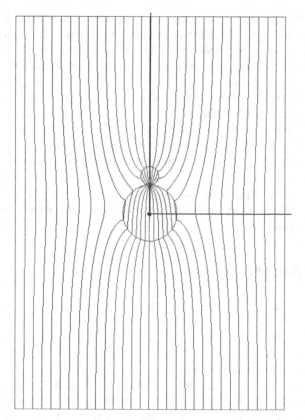

FIGURE E7.6.1   Ferromagnetic wire of radius 10 μm below a ferromagnetic particle of radius 3.3 μm, with flux lines computed by finite-element software. Similar to Oberteuffer [20], both objects are immersed in a uniform magnetic field. Here the field is assumed to be **B** = 1 T in the vertical y direction outside the modeled region of size 100 by 140 μm.

might be expected, the flux pattern is not symmetric above and below the wire because the large particle alters the field. Unfortunately, Oberteuffer's derived formulas do not account for such alteration by the particle; they only include the alteration of the field by the wire itself. Also, his formulas assume that the magnetic flux density at the *center* of the particle acts on the entire particle.

Based on the above assumptions, Oberteuffer derives a formula for the vertical magnetic force on the particle of Figure E7.6.1 which can be written as [20]:

$$F_{my} = -2v_p \left(1 - \frac{1}{\mu_{rp}}\right) \left(H_0 + H_0 \frac{a^2}{r^2}\right) B_0 \frac{a^2}{r^3} \qquad (\text{E7.6.1})$$

where v_p is the volume of the particle of relative permeability μ_{rp} located at radius r from the center of the wire of radius a in a y-directed magnetic field in air of intensity

H_0 and flux density B_0. Dividing both sides of (E7.6.1) by volume v_p, one obtains the *force density* expression:

$$f_{my} = -2\left(1 - \frac{1}{\mu_{rp}}\right)\frac{B_0}{\mu_0}\left(1 + \frac{a^2}{r^2}\right)B_0\frac{a^2}{r^3} \qquad (E7.6.2)$$

which is in units of N/m³ and shall be called the Oberteuffer force density from now on.

To compare Oberteuffer's approximate force density (E7.6.2) with finite-element computations using (5.32) of Section 5.8, the geometry of Figure E7.6.1 is here examined for the case $B_0 = 1$ T. The relative permeability μ_{rp} of both the particle and the wire is assumed to be 1000. Various particles are considered, but all particles are assumed centered at radius 13.3 μm from the center of the wire as shown in Figure E7.6.1. Substituting $r = 13.3E{-}6$ and $a = 10.E{-}6$ in (E7.6.2) along with $B_0 = 1$ and $\mu_{rp} = 1000$ and $\mu_0 = 12.57E{-}7$ obtains $f_{my} = -105.75E9$ N/m³.

Finite-element force computations using (5.34) and the model of Figure E7.6.1 were carried out as follows. The flux density B needed for (5.34) was computed using the particle permeability changed to that of air, and then a postprocessor calculator macro was developed for Maxwell to carry out the operation of (5.34) to compute and display the force density. A typical display of force density is shown in Figure E7.6.2, which is available in color on the internet but shown in black and white in print [21].

To find the total force acting on a ferromagnetic particle, the volume integrals of Section 5.8 can be carried out. Both the integrals of (5.38) and Problem 5.15 are

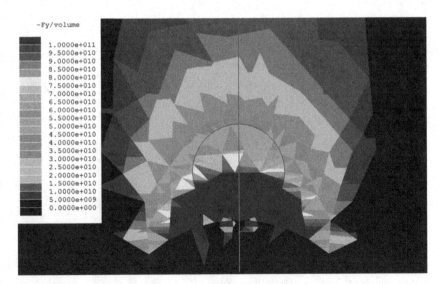

FIGURE E7.6.2 Zoomed plot of geometry of Figure E7.6.1 at the region on top of the wire with particle of 3.3 μm radius, showing the magnetic force density in the downward direction displayed in contours of N/m³ using the scale shown. For color display see website http://ieeexplore.ieee.org/ for [21].

TABLE E7.6.1   Magnetic Forces Computed by Force Density Integration on Particles Centered at Radius 3.3 μm From Wire of Radius 10 μm, Both With Relative Permeability 1000 in 1 T Field

Particle Radius (μm)	Cylindrical Particle (N/m)	Spherical Particle (N)
0.1	3.072E–3	401.9E–12
1.0	0.3128	427.0E–9
2.0	1.2314	3.503E–6
3.3	3.591	14.57E–6

here carried out on the magnetic force density displayed in Figure E7.6.2 for various particles of radius in the micrometer and nanometer range. The resulting forces are listed in Table E7.6.1. Because of known accuracy problems with Maxwell stress computations [22], only the 0.1 μm radius cylindrical case was checked by Maxwell stress, which obtained force in the range of 10E–3 N/m to 3.6E–3 N/m depending on the finite-element mesh used.

Dividing the forces of Table E7.6.1 by the volumes of the cylindrical and spherical particles, the computed average force densities are obtained and plotted versus particle radius in Figure E7.6.3. Note that they are all slightly smaller than the theoretical density obtained by the approximate Oberteuffer formula. Note, however, that in all cases the finite-element computations agree reasonably well with Oberteuffer, within −8.46 to −0.12%.

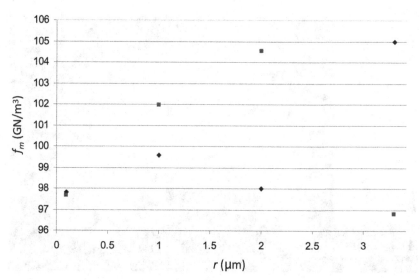

FIGURE E7.6.3   Plot of volume-average magnetic force densities computed in Table E7.6.1 versus radius for cylinders (diamonds) and spheres (squares) as data points compared to approximate Oberteuffer density constant value of 105.75E9 N/m³. Note that the vertical scale is zoomed; the difference between computed data points and Oberteuffer varies from −8.46% to −0.1%.

Besides mineral separation using magnetic field gradients [23–25], a rapidly grow-ing area of magnetic separation is for medical devices and biomolecular screening including DNA testing. High field gradient magnetic separation is used to separate red blood cells from blood and cancer cells from bone marrow. Other biomedical mag-netic actuators are being investigated to target drug delivery and to remove toxins [26]. As mentioned previously, permanent magnets are often used as a field source for magnetic separators. They are being investigated for biomolecular screening using standard biomolecular testing apparatus.

Example 7.7   Force Density of a Permanent Magnet Separator on a Biomolec-ular Microplate   For drug discovery, DNA/RNA testing, detection of proteins/antigens/genes, and other types of biomolecular screening, a standard tool is the microplate [27], which typically contains 96 wells as shown in Figure E7.7.1. One

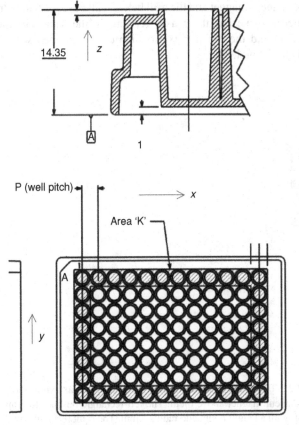

FIGURE E7.7.1   View of 96-well microplate for biomolecular screening [27]. The lower view shows the entire microplate in the horizontal *xy* plane. The upper view is a detailed section showing one well in the *xz* plane. Dimensions are in mm, and standard well pitch is 9 mm. Microplate materials have the magnetic permeability of air.

screening method is to place in each well, a solution with suspended permeable microparticles, often spherical beads. By coating the permeable microparticles with different proteins or other biological molecules [26, 28] in each well, magnetic separation enables biomolecular screening. Thus such microparticles can serve as magnetic sensors [29, 30]. Spherical microparticle diameters are often in the range of 1–4 μm, but may be reduced to nanometers. Here the magnetic force density acting on such particles is to be computed with finite-element software in one of the wells of Figure E7.7.1 with a permanent magnet separator immediately above it. The permanent magnet is first made of ferrite and next made of neodymium iron boron.

Solution   The top region of one well of Figure E7.7.1 is shown in Figure E7.7.2 along with an adjacent well. Also shown is the magnetic flux plot computed using Maxwell for a cylindrical permanent magnet placed just above the well of primary interest. The flux lines are computed using axisymmetric finite-element analysis because the permanent magnet and the well below it all have axisymmetric geometry. The axisymmetric plane has the z axis vertical (in the same direction as z axis of Figure E7.7.1) and the radial r axis (which can be aligned with the x axis of Figure E7.7.1) horizontal.

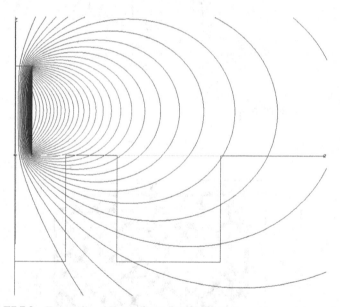

FIGURE E7.7.2   Two wells of Figure E7.7.1 modeled in the axisymmetric rz plane with a cylindrical permanent magnet above them. The left rectangle outlines the top of the well of primary interest, with a height 6 mm and radius 3 mm. The right rectangle models an adjacent well top of the same size (height 6 mm and diameter 6 mm). The centerlines of both wells are separated by 9 mm, agreeing with the standard microplate of Figure E7.7.1. Also shown are the flux lines computed by Maxwell axisymmetric finite-element analysis.

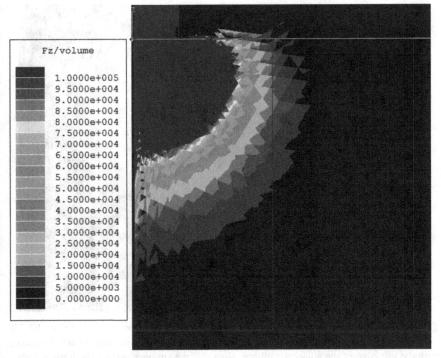

FIGURE E7.7.3 The vertical magnetic force density in N/m³ computed for Figure E7.7.2 below the magnet made of ferrite. This zoomed plot shows the entire primary well top along with the left edge of the adjacent well top.

The computed vertical z-directed magnetic force density distribution is displayed in Figures E7.7.3 and E7.7.4 for two different permanent magnet separators. The vertical magnetic force density is upward against gravity, and thus must exceed gravitational and buoyancy forces for successful separation. The gravitational force density is:

$$f_{gz} = -\rho g \tag{E7.7.1}$$

which is in the $-z$ direction, where ρ is the mass density and g is the acceleration of gravity. Since in SI units $g = 9.8$ kg m/s² and $\rho = 1000\rho_s$ kg/m³, where ρ_s is specific gravity, the gravitational force density is:

$$f_{gz} = -9.8E3\rho_s \tag{E7.7.2}$$

For iron, specific gravity $\rho_s = 7.8$, giving $f_{gz} = -7.644E4$ N/m³. For microparticles placed in solution in the microplate wells of Figures E7.7.1 and E7.7.2, ρ_s is less than that of iron because such particles are made of iron oxide mixed with polymers. Because of this mixture, the permeability of the microparticles is also reduced from that of iron.

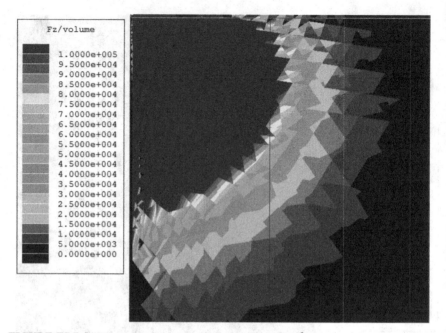

FIGURE E7.7.4 The vertical magnetic force density in N/m^3 computed for Figure E7.7.2 below the magnet made of NdFeB. This zoomed plot shows the entire primary well top along with the left edge of the adjacent well top. Note that the stronger magnet produces a higher force density.

Microparticle specific gravity and permeability are either unknown or proprietary and thus cannot be reported here. However, since both are substantially lower than that of iron, both the gravitational force density of (E7.7.2) and the magnetic force density of (5.32) are reduced from those of iron. Thus for comparison both forces are computed here using the properties of iron. The relative permeability of the iron is assumed to be 1000 or higher, since such high μ_{rp} is shown by (5.32) to produce force variations of 0.1% or less.

For the permanent magnet separator of Figure E7.7.2 to function properly, the magnetic force density in the primary well must exceed downward force density. The downward force density is the vector sum of gravitational, buoyancy, viscous, and surface tension force densities. Here the gravitational force is assumed to dominate. Since the gravitational force density is of magnitude 7.644E4 N/m^3, the magnetic force density will be displayed in the range from 0 to 1.E5 N/m^3.

Figures E7.7.3 and E7.7.4 display the vertical magnetic force densities computed using (5.32) for two different permanent magnet separators in Figure E7.7.2. In both cases the permanent magnet is assumed to have radius 1 mm and length 5 mm, as shown in Figure E7.7.2.

In Figure E7.7.3 the permanent magnet is assumed to be made of ceramic 5 ferrite with coercive field intensity $H_c = 1.91$E5 A/m and relative permeability $= 1.08$.

The vertical magnetic force density display shows that the 1.E5 N/m^3 zone is much smaller than the primary well. Thus over most of the well the magnetic force density is insufficient to overcome the gravitational force density.

In Figure E7.7.4 the permanent magnet is instead assumed made of neodymium iron boron (NdFeB) with $H_c = 8.9E5$ A/m and relative permeability $= 1.1$. The vertical magnetic force density now has a 1.E5 N/m^3 zone that is large enough to encompass most of the primary well top, without intruding into the adjacent well. Thus the magnetic force density overcomes the gravitational force density in most of the primary well while not interfering with the adjacent well, resulting in proper magnetic separation in the microplate.

In actual microplate magnetic separation, it is known that NdFeB permanent magnets with $H_c \sim 8.9E5$ A/m are required. Thus the computed results of Figures E7.7.3 and E7.7.4 agree qualitatively with observed behavior. The finite-element method computations can readily analyze many other designs of magnetic separators made of permanent magnets, steel, and other materials.

PROBLEMS

7.1 Assume the planar solenoid with a clapper armature shown in Figure E7.1.1 has the following dimensions: $w = 12$ mm, $Al_1 = 6$ mm, $Al_2 = 40$ mm, $Al_3 = 6$ mm, $Sl_1 = 20$ mm, $Sl_2 = 40$ mm, $Sl_3 = 20$ mm, $g = 1$ mm. It also has $N = 300$ and $I = 2$ A.

Use the reluctance method, assuming no leakage or fringing fluxes, to find the approximate values of the following.

(a) Vector **B** and **F** on left side of clapper.

(b) Vector **B** and **F** on right side of clapper.

(c) Obtain the above answers with Maxwell finite-element software by using high permeability material (relative permeability $= 2000$) in the steel.

(d) Obtain the answers for the above with finite-element software with steel relative permeability $= 10,000$, and comment on why the results differ.

(e) Repeat (c) for $I = 2$, 3, and 4 and the airgap $g = 0.5$ mm, 1 mm, 1.5 mm, 2 mm, 2.5 mm, 3 mm, 3.5 mm, and 4 mm and plot the resulting pull curves.

7.2 The axisymmetric solenoid with a clapper armature shown in Figure E7.2.1 has the following dimensions: $g = 1$ mm, $w_a = 7$ mm, $R_1 = 15$ mm, $R_2 = 25$ mm, $R_3 = 30$ mm, $Z_1 = 7$ mm, and $Z_2 = 23$ mm. It also has $N = 1500$ and $I = 1$ A.

Use the reluctance method, assuming no leakage or fringing fluxes, to find the approximate values of the following.

(a) **B** and **F** on inside of clapper.

(b) **B** and **F** on outside of clapper.

(c) Obtain the above answers with finite-element software by using high permeability material (relative permeability = 2000) in the steel.

(d) Obtain the answers for the above with finite-element software with steel relative permeability = 10,000, and comment on why the results differ.

7.3   The planar solenoid with a plunger armature shown in Figure E7.3.1 has the following dimensions: $w = 8$ mm, $g = 3$ mm, $g_s = 1.5$ mm, $Al_1 = 4$ mm, $Al_2 = 22$ mm, $Sl_1 = 8$ mm, $Sl_2 = 20$ mm, $Sl_3 = 34$ mm, and $Sl_4 = 15$ mm. It also has $N = 800$ and $I = 2$ A.

(a) Use the reluctance method, assuming no leakage flux, to find the approximate values of **B** and **F** on the end of the plunger.

(b) Compute the above **B** and **F** using finite-element software with the relative permeability = 2000 in the steel.

(c) Obtain the answers for the above with finite-element software with steel relative permeability = 10,000, and comment on why the results differ.

7.4   The axisymmetric solenoid with a plunger armature shown in Figure E7.4.1 has the following dimensions: plunger and pole outer radius $R_P = 18$ mm, side airgap outer radius $R_{PG} = 20$ mm, stator yoke outer radius $R_{YO} = 70$ mm, stator yoke inner radius $R_{YI} = 66$ mm, working airgap $g = 8$ mm, plunger length $L_P = 42$ mm, stator stopper length $L_S = 16$ mm, coil axial length $L_C = 46$ mm, and stator yoke axial thickness $t = 12$ mm (on top, same on bottom). It also has $N = 400$ and $I = 3$ A.

(a) Use the reluctance method, assuming no leakage flux, to find the approximate values of **B** and **F** on the end of the plunger.

(b) Compute the above **B** and **F** using finite-element software, with the relative permeability = 2000 in the steel.

(c) Obtain the answers for the above with finite-element software with steel relative permeability = 10,000, and comment on why the results differ.

7.5   A voice coil has a resistance of 15 Ω and 300 turns. A permanent magnet provides a DC magnetic flux density of 0.6 T in the radial r direction. The average radius r of the voice coil is 20 mm. If 10 volts DC is applied to the voice coil, find the steady-state vector force produced.

7.6   A voice coil lies from radius 10 mm to 14 mm. It has 200 turns and a resistance of 20 Ω. It is subjected to a DC magnetic flux density of 0.5 T in the radial r direction. If 6 volts DC is applied to the coil, find its steady-state vector force.

7.7   A proportional actuator is made of a voice coil actuator with $F = 100I$ and a spring with $F_s = 1.E5x$ in opposition. On a graph of total force versus x, for $x = 0$–0.010 m, draw the force lines for I varying from -1 A to $+10$ A in 1 A increments.

7.8   A step motor has 10 stator teeth and 8 rotor teeth. Find its incremental (step) angle of motion.

REFERENCES

1. Jay F (ed.). *IEEE Standard Dictionary of Electrical and Electronics Terms*, 2nd ed. New York, NY: Wiley-Interscience; 1977.

2. Juds MA, Brauer JR. AC contactor motion computed with coupled electromagnetic and structural finite elements. *IEEE Trans Magn* 1995;31:3575–3577.

3. Bessho K, Yamada S, Kanamura Y. Analysis of transient characteristics of plunger type electromagnets. *Electr Eng Jpn* 1978;98:56–62.

4. Yoon SB, Hur J, Chun YD, Hyun DS. Shape optimization of solenoid actuator using the finite element method and numerical optimization technique. *IEEE Trans Magn* 1997;33:4140–4142.

5. Brauer JR (ed.). *What Every Engineer Should Know About Finite Element Analysis*, 2nd ed. New York: Marcel Dekker, Inc.; 1993. Chapter 7, Design Optimization, by R. S. Lahey.

6. Ratliffe RT, Pagilla PR. Design, modeling, and seek control of a voice-coil motor actuator with nonlinear magnetic bias. *IEEE Trans Magn* 2005;41:2180–2188.

7. Lequesne B. Fast acting, long-stroke solenoids with two springs. *IEEE Trans Industry Appl* 1990;26:845–856.

8. Lequesne B. Permanent magnet linear motors for short strokes. *IEEE Trans Industry Appl* 1996;32:161–168.

9. Lequesne B. Fast acting, long-stroke bistable solenoids with moving permanent magnets. *IEEE Trans Industry Appl* 1990;26:401–407.

10. Lequesne B. Finite element analysis of a constant force solenoid for fluid flow control. *IEEE Trans Ind Appl* 1988;24:574–581.

11. Brauer JR, Aronson EA, McCaughey KG, Sullivan WN. Three dimensional finite element calculation of saturable magnetic fluxes and torques of an actuator. *IEEE Trans Magn* 1988;24:455–458.

12. Iannello V. *Advances in Magnetic Bearings*, Brochure from Synchrony Division of Dresser-Rand, Salem, VA, 2009 (excerpted by permission).

13. Chen SL, Chen SH, Yan ST. Experimental validation of a current-controlled three-pole magnetic rotor-bearing system. *IEEE Trans Magn* 2005;41:99–112.

14. Tian L-L, Ai X-P, Tian Y-Q. Analytical model of magnetic force for axial stack of permanent magnet bearings. *IEEE Trans Magn* 2012;48:2592–2599.

15. Bachovchin KD, Hoburg JF, Post RF. Stable levitation of a passive magnetic bearing. *IEEE Trans Magn* 2013;49:609–617.

16. Wu W, Lovatt HC, Dunlop JB. Analysis and design optimization of magnetic couplings using 3D finite element modeling. *IEEE Trans Magn* 1997;33:4083–4085.

17. Kamm LJ. *Understanding Electro-Mechanical Engineering*. New York: IEEE Press; 1996.

18. Kallenbach E, Eick R, Quendt P, Ströhla T, Feindt K, Kallenbach M. In: Teubner BG (ed.), *Elektromagnete*, 2nd ed. Wiesbaden, Germany: Verlag/Springer; 2003 (in German).

19. Rosensweig RE. *Ferrohydrodynamics*. Mineola, NY: Dover Publications; 1985.

20. Oberteuffer JA. Magnetic separation: a review of principles, devices, and applications. *IEEE Trans Magn* 1974;10:223–238.

21. Brauer JR, Cook DL, Bray TE. Finite element computation of magnetic force densities on permeable particles in magnetic separators. *IEEE Trans Magn* 2007;43:3483–3487.

22. Burow DW, Salon SJ. Dependence of torque calculation on mesh in induction machines. *IEEE Trans Magn* 1995;31:3593–3595.

23. Gerber R, Birss R. *High Gradient Magnetic Separators*. New York: John Wiley & Sons; 1984.

24. Smolkin MR, Smolkin RD. Calculation and analysis of the magnetic force acting on a particle in the magnetic field of a separator. Analysis of the equations used in the magnetic methods of separation. *IEEE Trans Magn* 2006;42:3682–3693.

25. Hoffmann C, Franzreb M. A novel repulsive-mode high gradient magnetic separator—I. Design and experimental results. *IEEE Trans Magn* 2004;40:456–461.

26. Hoffman A. Magnetic viruses for biological and medical applications. *Magn Bus & Technol* 2005;24 ff.

27. Standard 2-2004 for Microplates, American National Standards Institute/Society for Biomolecular Sciences (March 28, 2005).

28. Brauer JR. Modeling force density distributions on biomolecular nanoparticles undergoing magnetic separation. *Magnetics Bus & Technol* 2008;26–27.

29. Fodil K, Denoual M, Dolabdjian C, Harnois M, Senez V. Dynamic sensing of magnetic nanoparticles in microchannel using GMI technology. *IEEE Trans Magn* 2013;49:93–96.

30. Eberbeck D, Dennis CL, Huls NF, Krycka KL, Gruttner C, Westphal F. Multicore magnetic nanoparticles for magnetic particle imaging. *IEEE Trans Magn* 2013;49:269–274.

CHAPTER 8

Magnetic Actuators Operated by AC

Because alternating current (AC) voltage is commonly available, many magnetic actuators are designed for AC operation. In most cases, the direct current (DC) actuators of the preceding chapter must be substantially modified for AC operation. Armatures of clapper type and plunger type are still common, but changes to both the steel core and the coils are usually required.

8.1 SKIN DEPTH

Chapter 2 has shown that time-varying magnetic fields induce eddy currents in conducting materials. Since AC currents produce time-varying magnetic fields, eddy currents are a key issue in AC magnetic actuators. An easy way to examine AC eddy current effects is to calculate *skin depth*.

Faraday's law of Chapter 2 can be used to find the skin depth [1]:

$$\delta = \frac{1}{\sqrt{\pi f \mu \sigma}} \tag{8.1}$$

where f is the AC frequency in Hz, μ is the permeability, and σ is the electrical conductivity. Both the steel core and the coils of magnetic actuators are made of conducting materials, and thus the skin depth is often smaller than typical core or wire dimensions. Skin depth in wires will be discussed in Chapter 12, while here skin depth in steel cores is examined. For example, the 60 Hz magnetic flux in Example 6.5 was seen in Figure E6.5.1 to be concentrated near the outer skin of its cylindrical steel core. Such flux concentration on the skin, and lack of flux in the interior, is called *skin effect*. The skin depth of (8.1) is the depth at which flux density and eddy currents have decayed to $1/e$ of their surface value, or 36.8%. The decay from surface into the interior is exponential [1].

Note that skin depth of (8.1) is infinite for DC (0 frequency), but decreases inversely proportional to the square root of frequency. It is also inversely proportional to the

Magnetic Actuators and Sensors, Second Edition. John R. Brauer.
© 2014 The Institute of Electrical and Electronics Engineers, Inc. Published 2014 by John Wiley & Sons, Inc.

product of permeability times conductivity. This product is especially high for steel. Steel in magnetic actuators and sensors typically has permeabilities thousands of times that of air, and has conductivities on the order of 1.E6 S/m. In Example 6.5, the relative permeability of 2000, conductivity of 2.E6 S/m, and frequency 60 Hz in (8.1) gives a skin depth of 1 mm. Note that the flux lines in Figure E6.5.1 are indeed concentrated in the outer 1 mm or so of the 10-mm radius steel core. The inner 9-mm steel region is essentially unused for AC, and it could be removed to save weight and cost.

To carry more flux than just the skin will allow, most AC steel cores are laminated as was shown in Figure 2.6 and described in Chapter 2. The lamination thickness should be less than approximately 0.5–1 mm skin depth. Thus most steel laminations for 60 Hz AC operation are less than 1 mm thick. For example, the Eaton AC magnetic actuator of Figure 4.3 has steel of depth 28.5 mm into the page and is made of approximately 46 laminations, each approximately 0.62 mm thick. Both its stator and its armature are made of laminations that have been punched from sheet steel and then stacked and held together by rivets.

8.2 POWER LOSSES IN STEEL

8.2.1 Laminated Steel

Laminating steel does not completely eliminate steel eddy currents and their losses, but greatly reduces them. As long as the lamination thickness t is considerably less than skin depth, the eddy power loss density can be shown to obey [2] the following formula:

$$\frac{P_e}{v} = \frac{t^2 \omega^2 B^2 \sigma}{24} \tag{8.2}$$

where B is the peak magnetic flux density, $\omega = 2\pi f$, P_e is power in watts, and v is the volume of the conducting material in cubic meters. The above loss equation holds true when the eddy currents cause negligible changes in the magnetic field. If the skin depth is less than the lamination thickness, more complicated relations often apply [2], and the laminations do not carry nearly as much flux.

In addition to eddy current loss, steel carrying AC flux has *hysteresis loss*. As explained in Chapter 5, areas on B–H planes correspond to energy, and hysteresis causes the B–H relation to have more than one single-valued curve. Thus in Figure 8.1, which shows typical B–H loops for AC and H, the energy density lost per cycle is the area enclosed by the B–H loop. Hysteresis loss is significant whenever eddy loss of (8.2) is small due to small lamination thickness. The sum of hysteresis loss and eddy current loss is called *core loss*.

Because the energy density lost per cycle is the area of Figure 8.1, hysteresis loss is proportional to frequency. Also, for typical steel the area of Figure 8.1 is

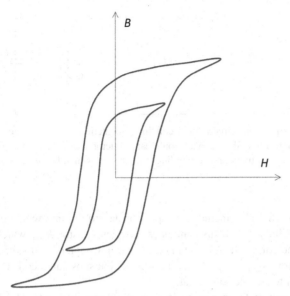

FIGURE 8.1 *B–H* hysteresis loops for AC *H* due to AC. The area enclosed is the energy lost per cycle. Two loops are shown for two different peak AC values.

approximately proportional to the peak *B* to the power *n*, where *n* varies from about 1.5 to about 2.5 [3]. Hence the hysteresis power loss density obeys:

$$\frac{P_h}{V} = K_h f B^n \tag{8.3}$$

where K_h is approximately constant.

Manufacturers of steel laminations for magnetic devices, often called electrical steels, customarily measure and publish *core loss curves*. Such curves are commonly available for 50 and 60 Hz over a range of peak *B* values.

Steel manufacturers also measure and publish *B–H* curves for AC devices called *normal AC B–H curves*. Such a normal curve is the locus of the tips of *B–H* loops such as shown in Figure 8.1. The normal AC curve may differ somewhat from the DC curve.

8.2.2 Equivalent Circuit

As discussed in Example 6.5, eddy current loss (and thus core loss) introduces a resistance in the impedance seen by the coil. Since the losses of (8.2) and (8.3) are proportional to B^2, and V_L across an inductor is proportional to *B* (from Faraday's law), core loss is equivalent to the parallel resistor R_{core} shown in Figure 8.2. Its power loss, like core loss, is proportional to the square of the voltage across the inductor:

$$P_{core} = V_L^2 / R_{core} \tag{8.4}$$

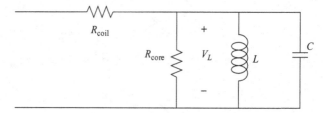

FIGURE 8.2   Equivalent circuit for a coil in a magnetic device. R_{core} accounts for core loss; R_{coil} accounts for coil loss. The nonlinear inductor accounts for flux and Faraday's law. The capacitor accounts for interwinding capacitance, which becomes important at high frequencies.

Note that Figure 8.2 also includes a capacitor due to the interwinding capacitance mentioned in Chapter 6. It also includes a series resistor R_{coil} which equals the resistance of the coil itself. As explained further in Chapter 12, if skin effects in the coil are negligible R_{coil} is the DC coil resistance, but substantial coil AC skin effects may cause R_{coil} to change somewhat.

The impedance of the capacitor is typically negligible at 60 Hz, but becomes more important the higher the frequency. Since the capacitor C and inductor L are in parallel, the coil has a natural resonant frequency:

$$f_r = \frac{1}{2\pi\sqrt{LC}} \tag{8.5}$$

As discussed in Chapter 6, finite-element software usually outputs the impedance seen by each coil. This impedance is a sum of real and imaginary parts in series:

$$Z_s = R_s + jX_s \tag{8.6}$$

and is here assumed to account only for the core loss and inductance of the coil; that is, the coil's DC resistance and capacitance are not included in (8.6). However, Figure 8.2 needs the *parallel* impedance:

$$Z_p = R_{core} + j\omega L = R_p + jX_p \tag{8.7}$$

An advantage of the parallel impedance is that when the coil is excited by a voltage V_s, the core current loss draws an extra current, the current through the parallel resistor.

To find R_p and X_p of Figure 8.2, apply the parallel circuit formula:

$$R_s + jX_s = \frac{(R_p)(jX_p)}{R_p + jX_p} \tag{8.8}$$

Using complex AC circuit methods, (8.8) can be expressed in terms of magnitudes and phase angles as:

$$R_s + jX_s = \frac{R_p X_p \angle 90°}{\left(R_p^2 + X_p^2\right)^{1/2} \angle \arctan \dfrac{X_p}{R_p}} \tag{8.9}$$

$$\left(R_s^2 + X_s^2\right)^{1/2} \angle \arctan \frac{X_s}{R_s} = \frac{R_p X_p}{\left(R_p^2 + X_p^2\right)^{1/2}} \angle 90° - \arctan \frac{X_p}{R_p} \tag{8.10}$$

The magnitudes on both sides must be the same and the angles on both sides must also be the same. Also, in most cases the finite-element results will find $R_s \ll X_s$. Thus from the first term of a series expansion:

$$\arctan \frac{X_s}{R_s} \cong 90° - \frac{R_s}{X_s} \tag{8.11}$$

Assuming the above equality is exact, then equating the angles of (8.10) gives:

$$\frac{R_s}{X_s} = \frac{X_p}{R_p} \tag{8.12}$$

Also, if $R_s \ll X_s$, then from the magnitudes of (8.10), it is approximately true that:

$$X_p = X_s \tag{8.13}$$

Finally, (8.12) and (8.13) obtain:

$$R_p = \frac{X_p^2}{R_s} \tag{8.14}$$

A major problem with AC equivalent circuits for devices made of nonlinear *B–H* materials is that the nonlinearity often causes the voltage and/or the current to become nonsinusoidal. The greater the current and voltage, the greater the saturation, and the more the waveforms can deviate from sinusoidal. Since the inductive reactances X_p and X_s in the above equations are constants, they cannot fully represent nonlinear behavior. However, these reactances can approximately simulate saturation effects when their magnitudes decrease as current and flux rise. These effective reactances are often obtained by AC finite-element software by use of "effective permeability" which decreases with current magnitude [4].

For more exact modeling, and to obtain nonsinusoidal waveforms, nonlinear transient techniques of Chapters 9 and 14 are required. Chapter 14 will also discuss special AC magnetic actuators designed to produce reciprocating (back and forth) force and motion, while this chapter concerns AC actuators that produce force and motion in only one direction.

8.2.3 Solid Steel

The equivalent circuit of Figure 8.2 still applies if some or all of the steel is solid. However, the simple equations for the eddy current loss and hysteresis loss derived above for laminated steel no longer apply. In most cases, the eddy current loss in solid steel greatly exceeds the hysteresis loss, which can thus be neglected. The best way to predict the steel power loss and equivalent circuit parameters is to use the finite-element method.

Solid steel is sometimes used in AC actuators, depending on the type (from Chapter 7):

- *For planar actuators with clapper armatures,* the stator is easily laminated, but often the armature is made of solid steel, especially if the armature is steel plate (or steel scrap) to be lifted by a *lifting magnet.*
- *For axisymmetric actuators with clapper armatures,* lamination is difficult and expensive, and thus both stator and armature steels are usually solid.
- *For planar actuators with plunger armatures,* both stator and armature are usually laminated, but a solid armature is sometimes required due to its greater mechanical rigidity.
- *For axisymmetric actuators with clapper armatures,* lamination is difficult and expensive, and thus both stator and armature steels are usually solid.

The easiest way to design and manufacture an AC actuator is to simply apply AC to an existing DC solenoid. However, such a simple design method is usually unsatisfactory. Besides the need to laminate to reduce eddy power loss, the number of turns usually must be changed to obtain the proper current for the particular AC voltage applied. Further changes are discussed in the remainder of this chapter.

Example 8.1   AC Flux Linkage and Equivalent Circuits of Solenoid of Example 7.1 with a Solid Steel Clapper   Figure E8.1.1 shows the solenoid of Example 7.1 turned upside down to act as an AC lifting magnet. The stator is laminated

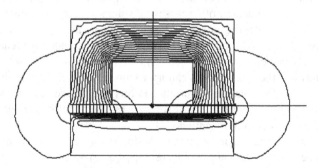

FIGURE E8.1.1   Computer display of AC magnetic actuator with a laminated steel stator and solid steel rotor. Its dimensions are the same as those in Figure E7.1.1. The flux lines are computed for a frequency of 60 Hz.

but the clapper being lifted is solid steel. The stator winding shown has 200 turns carrying 2-A rms AC 60-Hz current. The dimensions are $w = 10$ mm, $Al_1 = 5$ mm, $Al_2 = 30$ mm, $Al_3 = 5$ mm, $Sl_1 = 15$ mm, $Sl_2 = 30$ mm, $Sl_3 = 15$ mm, $g = 2$ mm. Assuming all steel has relative permeability of 2000, and the clapper has conductivity 2.E6 S/m, find the AC fields and series and parallel equivalent circuits using Maxwell 2D planar finite-element software. Assume that the stator has a depth into the page of 1 m in the z direction and that the clapper extends beyond the stator both in the $+z$ and $-z$ directions to allow end region eddy currents to flow with zero end region resistance and zero end region inductance.

Solution   The flux linkage output by Maxwell is $\lambda = 1.814\text{E}{-}5{-}j7.51\text{E}{-}7$ per turn. The series impedance for one turn is then found using (6.31):

$$Z = j\omega\lambda/I \tag{E8.1.1}$$

where $I = 2$ A and $\omega = 377$. The resulting series impedance (for 200 turns) is:

$$Z_s = R_s + jX_s = (200)(141.6\text{E}{-}6 + j3.42\text{E}{-}3) = 28.32\text{E}{-}3 + j684\text{E}{-}3$$

$$\tag{E8.1.2}$$

Note that as expected, the resistive part is much smaller than the reactive part.

To obtain the parallel equivalent circuit of Figure 8.2, recall that $X_p = X_s$, so $X_p = 684\text{E}{-}3$ and $L_p = 684\text{E}{-}3/377 = 1.814\text{E}{-}3$ H. Also, use (8.14) to obtain the parallel core loss resistor:

$$R_p = \frac{X_p^2}{R_s} = 16.52\ \Omega \tag{E8.1.3}$$

Note that the parallel core loss resistor has much higher impedance than the parallel inductive reactor, as expected.

The computed flux lines are shown in Figure E8.1.1. Note the substantial skin effects in the solid steel armature. As mentioned in the problem definition, end region effects are ignored in this 2D model; adding end region effects to a 2D model will be discussed in Chapter 14. Also in that chapter, an example transformer will demonstrate the use of parallel impedances.

8.3   FORCE PULSATIONS

8.3.1   Force with Single AC Coil

Because AC actuators have AC that varies with time, Ampere's law shows that their magnetic fields B also vary with time. Since magnetic reluctance force of Chapter 5

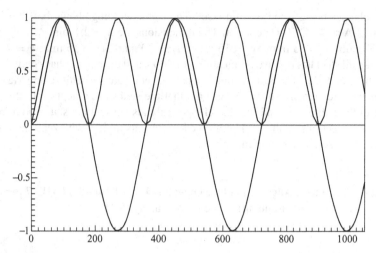

FIGURE 8.3 Current and force waveforms versus time for a single AC in an actuator. The current swings positive and negative, while the force is always positive and has double frequency content.

is approximately proportional to the square of B, the time-varying B causes force variations called *pulsations*.

If eddy currents are small enough to not affect the field, then the magnetic field throughout the actuator is due solely to the current through the energized coil or coils. Often there is only one coil, and thus there is only one AC and all fields are in phase with the current. Since the current goes through zero twice per cycle, the magnetic fields throughout the actuator must go through zero twice per cycle.

For example, the current is often a sine wave:

$$I_1 = \sin(2\pi f t) \qquad (8.15)$$

Figure 8.3 plots this AC sine wave over a few cycles, when the argument of the sine is varied from 0° to about 1000°. From (5.11), the force is then proportional to:

$$F_1 = k I_1^2 = k \sin^2(2\pi f t) = \frac{k}{2}(1 - \cos 4\pi f t) \qquad (8.16)$$

The force versus time waveform is also plotted in Figure 8.3. Note that the force pulsates from its maximum to zero, going through zero twice per cycle of current. The alternating component of the force has twice the electrical frequency. The average force value is one-half the maximum force, and the total force never reverses direction.

The above current and force relations assume that the current is sinusoidal, as often assumed for the reactances of Figure 8.2. In actuality, most AC solenoids are voltage driven, typically by 120 V 60 Hz in the United States. If saturation occurs, then the current may be nonsinusoidal, as mentioned in Subsection 8.2.2. However, in all cases the current goes through zero twice per cycle, and so will the force.

FIGURE 8.4   Photo of stator core with two shading rings in Eaton AC actuator of Chapter 4.

8.3.2   Force with Added Shading Coil

If an additional significant current at a different phase angle exists in the AC solenoid, then it is possible that the force pulsations may be reduced. For example, suppose that one winding has the current of (8.15) and produces the force of (8.16) on its steel pole, but that there is also an additional current:

$$I_2 = \cos(2\pi ft) \qquad (8.17)$$

Suppose also that this current produces an additional force on its own steel pole, giving a total force:

$$F_1 + F_2 = k\left(I_1^2 + I_2^2\right) = k[\sin^2(2\pi ft) + \cos^2(2\pi ft)] = k \qquad (8.18)$$

Note that a trigonometric identity has been used to obtain a magnetic force that is independent of time. Thus the force has no pulsations.

In practice, it is extremely difficult to obtain the constant force of (8.18). However, it is possible to raise the minimum force considerably above zero by a relatively inexpensive modification.

The inexpensive way of reducing the pulsation is to add a *shading coil* or a one-turn version called a *shading ring*. Figure 8.4 shows the aluminum shading rings added to the two poles in the Eaton AC actuator of Chapter 4. According to the time derivative in Faraday's law, the induced voltage in each ring will be 90° out of phase from the magnetic flux passing through it. If the induced current were in phase with that induced voltage, then the phase relation of (8.18) would be satisfied. However, due to self-inductance, the current phase lags further behind. Also, the magnitude of the induced shading ampere-turns will always be less than the magnitude of the ampere-turns in the main coil. Thus a shading coil will not eliminate force pulsations but can reduce them.

While reluctance techniques have been used to calculate shading coil effects [5], the easiest way to calculate them is using finite-element analysis. The effects of the shading coil in the Eaton AC actuator of Figure 8.4 have been calculated using nonlinear transient finite-element analysis, including the effects of armature motion, and will be discussed in Chapter 14. Chapter 14 will also discuss AC actuators where force pulsations (actually force reversals) are desirable to create reciprocating motion.

Since induced currents in solid metals affect the force, armatures made of nonlaminated, high conductivity metal experience large eddy current forces which in some cases can be shown to be *repulsive* instead of attractive. Pure aluminum armatures placed above a steel stator with an AC stator coil can in some cases be *levitated* above the stator. That is, the magnetic repulsive force vertically upward balances the downward force of gravity [6, 7], and can be used to levitate trains.

Example 8.2   Fluxes and Forces on Axisymmetric Plunger Armature Solenoid of Example 7.4 with Added Shading Ring   Figure E8.2.1 shows the axisymmetric solenoid with a plunger armature of Example 7.4 with an added shading ring on its stator pole. The number of turns of the main driven coil remain $N = 400$. The current I is now 4-A rms at 60 Hz. The dimensions remain the same as in Example 7.4, except

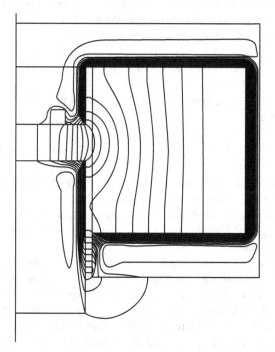

FIGURE E8.2.1   Computer display of geometry and computed flux lines of axisymmetric actuator with shading coil placed on stator pole face.

that now there is an added shading ring. The steel has relative permeability 2000 and conductivity 5.E5 S/m. The ring has an inner radius of 10 mm and a cross section of 4 mm by 4 mm. It is made of copper with conductivity 5.8E7 S/m.

Compute the magnetic fields and the force using the finite-element software Maxwell.

Solution   If Maxwell SV is used, the easiest way to make this model is to modify the model of Example 7.4. The solution type must be changed to "Eddy Current." The shading ring must be added using a rectangular box command. Be sure to change the steel conductivity to the specified nonzero value, and choose copper for the shading ring material. Also be sure to specify "Force" under "Setup Executive Parameters." To allow convergence to small energy error, allow 16 passes under "Setup Solution."

If Maxwell version 16 is used, it is easiest to modify the model of Example 7.4. The solution type must be changed to "Eddy Current." The shading ring must be added using a create box command. Be sure the change the steel conductivity to the specified nonzero value, and choose copper for the shading ring material. Also be sure to assign a "Force" under "Parameters." To allow convergence to a small energy error, allow 16 passes under "Solve Setup" dialog box.

The solution has the magnetic flux line plot shown in Figure E8.2.1 at time equal to 0° phase angle. Note that skin effects are clearly visible. The force output by Maxwell has a time-average value of 8.46 N and an "AC fluctuation" of 8.40 N. Thus the fluctuation or pulsation is large, but somewhat less than the average value. Hence the shading ring prevents the force from going through zero twice per cycle.

8.4   CUTS IN STEEL

A way to improve the performance of AC magnetic devices made of solid steel is to put cuts in the steel. The cuts or slots partly block the eddy current circulation pattern, somewhat like laminations, and thereby reduce power loss and increase flux.

8.4.1   Special Finite-Element Formulation

To compute the magnetic fields and flux in slotted steel with eddy currents, an unusual finite-element formulation called "Eddy Axial" is available in Maxwell 2D (versions 12 or earlier including SV) or by using a full 3D model. Displacement currents are neglected because the frequency applied to actuators is usually 1 kHz or lower.

The formulation begins with Ampere's law with $\mathbf{J} = \sigma\mathbf{E}$ from Chapter 2:

$$\nabla \times \mathbf{H} = \sigma\mathbf{E} \tag{8.19}$$

Premultiplying both sides by the inverse of the conductivity and then taking the curl of both sides:

$$\nabla \times \sigma^{-1}\nabla \times \mathbf{H} = \nabla \times \mathbf{E} \tag{8.20}$$

Applying Faraday's law gives:

$$\nabla \times \sigma^{-1}\nabla \times \mathbf{H} = -\frac{\partial \mathbf{B}}{\partial t} \qquad (8.21)$$

which for sinusoidal fields of angular frequency $\omega = 2\pi f$ is:

$$\nabla \times \sigma^{-1}\nabla \times \mathbf{H} = -j\omega\mu\mathbf{H} \qquad (8.22)$$

where permeability μ is assumed constant, that is, the B–H curve is assumed to be operating in its linear region.

In comparison, the planar eddy current solution of Example 8.1 solves (8.19) by using $H = B/\mu$, obtaining:

$$\nabla \times \mu^{-1}\mathbf{B} = \sigma\mathbf{E} \qquad (8.23)$$

Substituting (2.45) and (2.47), and again assuming sinusoidal fields gives:

$$\nabla \times \mu^{-1}\nabla \times \mathbf{A} = -j\omega\sigma\mathbf{A} \qquad (8.24)$$

which is the differential equation for "Eddy Current" problems such as Example 8.1. In the 2D case, the only component of $\mathbf{A}$ is commonly the z component normal to the plane of the problem as described in Chapter 4.

The Eddy Axial solver uses planar 2D finite elements (of default depth 1 m) in which only one component of $\mathbf{H}$ exists, the component H_z perpendicular to the elements (out of the page). Computing only one unknown H_z at each finite-element node is much more computationally efficient than computing two components of $\mathbf{A}$ (A_x and A_y) for planar eddy currents, as much as $3^2 = 9$ times less expensive. After H_z is computed, the software also computes the eddy current density $\mathbf{J}$ using Ampere's law to give:

$$\mathbf{J} = \nabla \times \mathbf{H} \qquad (8.25)$$

In the 2D case with $\mathbf{H} = H_z\,\mathbf{u}_z$, $\mathbf{J}$ has only x and y components in the plane of the triangular finite elements. Contours of constant H_z correspond to eddy current flow lines.

8.4.2 Loss and Reluctance Computations

The finite-element method of the preceding subsection can be used to analyze a 2D circular model of a cylindrical solid steel pole. While other steel shapes can also be analyzed, the cylinder is selected here because of its common use in axisymmetric magnetic actuators and sensors. Also, because equations have been derived for conductors that are perfect circles without cuts, some finite-element computations can be verified by comparing them with the equations.

Equations for the AC fields in circular conductors have been derived by Ramo et al. [8] as follows. Applying vector identities to (8.22) gives:

$$\nabla^2 \mathbf{H} = j\omega\mu\sigma\mathbf{H} \tag{8.26}$$

For $\mathbf{H} = H_z\,\mathbf{u}_z$ as discussed in the preceding subsection, (8.26) can be written in cylindrical coordinates as [8]:

$$\frac{d^2 H_z}{dr^2} + \frac{1}{r}\frac{d H_z}{dr} - j\omega\mu\sigma H_z = 0 \tag{8.27}$$

where r is the radius, the only coordinate over which H varies.

The solution of (8.27) can be shown [8] to involve Bessel functions. Since (8.27) is a complex equation, Bessel functions of complex numbers are required. It is common to define [8] Bessel real and Bessel imaginary functions:

$$\text{Ber}(v) = \text{Re}[J_o j^{-1/2}v], \quad \text{Bei}(v) = \text{Im}[J_o j^{-1/2}v] \tag{8.28}$$

Using the Ber and Bei definitions, the solution of (8.27) can be shown to be [8]:

$$H_z = H_a[(\text{Ber}(2^{1/2}r/\delta) + j\text{Bei}(2^{1/2}r/\delta))/(\text{Ber}(2^{1/2}r_o/\delta) + j\text{Bei}(2^{1/2}r_o/\delta))] \tag{8.29}$$

where r_o is the outer radius is at which H_a is applied by a coil, and δ is the skin depth.

To compare the results of (8.29) with those of the finite elements of (8.22), a steel cylinder of radius 2.5 mm is selected as shown in Figure 8.5. The steel is assumed to have relative permeability 2000 and electrical conductivity 2.E6 S/m. The H_z applied to the outside of the model of Figure 8.5 is 398.1 A/m at 60 Hz, which corresponds to 1 T flux density just inside the outer radius of the steel cylinder. The computed flux lines are also shown in Figure 8.5 at the instant that the applied 398.1 A/m is peaking.

The computed flux lines in Figure 8.5 are nonuniform, due to the eddy currents in the steel. To compare the computed field distribution with the theory of (8.29), the ratio of the magnitude of H over the applied $H = 398.1$ A/m is plotted in Figure 8.6. Also plotted is the theoretical magnitude of H from (8.29), using Ber and Bei function values interpolated from tables [9]. In all cases the computed value is within 0–1% higher than the theoretical value obtained using (8.29).

As an additional verification of the finite-element formulation, the conductivity of the steel cylinder of Figure 8.5 is changed to zero. The computed flux is 19.65 μWb,

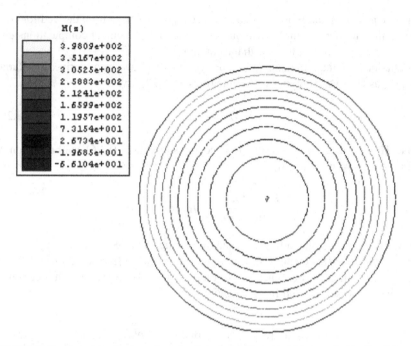

FIGURE 8.5   Computer display of conducting steel cylinder (without cuts) of radius 2.5 mm. The eddy current flow lines are displayed at the instant that the applied outer H of frequency 60 Hz is 398.1 A/m, corresponding to 1 T just inside the outer steel. The contour values show that the real part of H varies from 398.1 down to –66 A/m. The magnetic flux is directed normal to the plane of this figure.

which substituted in (3.8) along with $H = 398.1$ A/m and depth 1 m gives computed DC reluctance $\mathfrak{R}_{DC} = 20.26E6$ A/Wb. This value agrees exactly with the value obtained using the conventional magnetostatic reluctance formula (3.9) with area equal to that of the circle.

As mentioned in Chapter 3, AC eddy currents cause reluctance to become complex. The above magnetostatic reluctance $\mathfrak{R}_{DC}$ can be compared with the 60 Hz reluctance of Figure 8.5. Using the definition of $\mathfrak{R}_{RE}$, the real part of the complex reluctance of (3.16), the software finds $\mathfrak{R}_{RE} = 47.57E6$ A/Wb. This is 234.8% of $\mathfrak{R}_{DC} = 20.26E6$ A/Wb, and is listed as that percentage in the first row of Table 8.1. As expected from Lenz' law, eddy currents cause the AC reluctance to be higher than the magnetostatic reluctance. Thus the finite-element computations have been verified.

To reduce the AC reluctance and associated eddy power loss, cuts have often been made in solid steel poles and armatures [10]. A similar technique is to use *tape wound* steel, as shown in the spiral steel pole of Figure 8.7. The eddy current flow lines computed by the "Eddy Axial" solver of Maxwell are shown. Note that the circular eddy flow patterns are well blocked by the spiraling. Table 8.1 lists the computed power loss and AC reluctance, both of which are greatly reduced by the spiraling. Thus placing cuts in steel poles can greatly improve their AC performance.

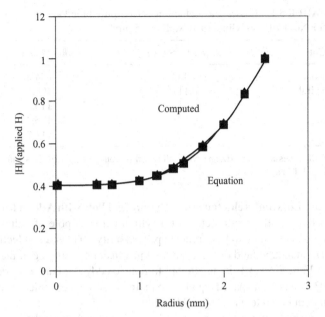

FIGURE 8.6 Ratio of magnitude of H inside steel cylinder over applied H of frequency 60 Hz. The cylinder has a radius of 2.5 mm, a conductivity of 2.E6 S/m, and a relative permeability of 2000. The lower curve is from Equation 8.29, while the upper curve was computed using finite elements.

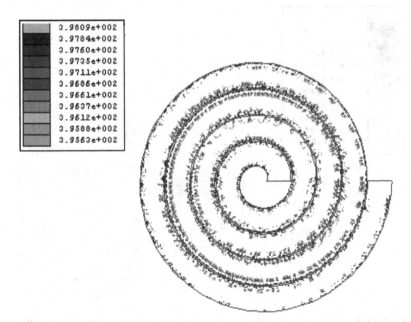

FIGURE 8.7 Computer display of spiral steel pole of radius 2.5 mm and outer $B = 1$ T at 60 Hz, showing computed eddy current flow patterns and contours of constant real H.

TABLE 8.1 Losses and Reluctances Computed for Various (or No) Cuts in Cylinders of Radius 2.5 mm^a

Cuts	Power Loss (W)	$\mathcal{R}_{RE}/\mathcal{R}_{DC}$ (%)
None	0.481	234.8
Spiral	0.111	100.5
4	0.411	115.7
8	0.198	105.2
36	0.021	111.7

^{a}All values are at 60 Hz and with solid steel of conductivity 2.E6 S/m and $B = 1$ T in steel skin.

Example 8.3 Loss and Reluctance of a Cylindrical Pole with Added Radial Slots
Figure E8.3.1 shows the cross section of a cylindrical steel pole of radius 2.5 mm. The steel is again assumed to have relative permeability 2000 and conductivity 2.E6 S/m. A 60-Hz magnetic field of 1 T is applied just inside the surface of the cylinder. As shown in Figure E.8.3.1, in an attempt to reduce eddy currents, four radial slots of width 0.03 mm are cut spaced equally. To maintain some mechanical integrity, the slots are only cut down to a radius of 0.55 mm.

Compute the eddy power loss and real AC reluctance $\mathcal{R}_{RE}$ using the finite-element software Maxwell.

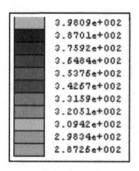
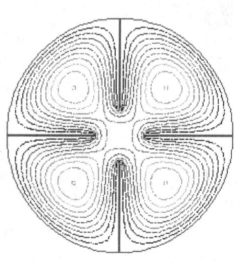

FIGURE E8.3.1 Computer display of steel cylinder with four radial slots and outer $B = 1$ T at 60 Hz, showing computed eddy current flow patterns and contours of constant real H.

Solution Maxwell's "Eddy Axial" is selected as the Solver, along with the default *xy* plane. The applied outer boundary condition is $H_z = 398.1$ A/m to obtain 1 T just inside the steel surface. The resulting plot of contours of real H_z is shown in Figure E8.3.1. The power loss is 0.411 W for 1 m depth and is entered in Table 8.1, where it is seen to be somewhat less than the power loss for no slots. Table 8.1 also lists the computed real AC reluctance as a percentage of the DC reluctance. Note that slotting has significantly reduced the AC reluctance. Table 8.1 also lists computed results with increased numbers of equally spaced slots.

PROBLEMS

8.1 Redo Example 6.5 but at a frequency of 10 Hz.

8.2 Redo Example 6.5 but with a conductivity of 2.E5 S/m.

8.3 Calculate the skin depth at 60 Hz of material with:

(a) A semiconductor with conductivity = 1 S/m and relative permeability = 1.

(b) Copper with conductivity = 5.8E7 S/m and relative permeability = 1.

(c) A high conductivity (5.E6 S/m) steel with relative permeability = 8000.

8.4 Find (in a book containing mathematical series) the complete trigonometric series for (8.11).

8.5 Redo Example 8.1 but with Maxwell's nonlinear "steel_1010."

8.6 Repeat Problem 8.5 but with the current increased to 5 A. Note that Maxwell handles nonlinear *B–H* characteristics in AC eddy current problems, and saturation can affect the equivalent circuit parameters.

8.7 Redo Example 8.2 but with the shading ring moved to an inner radius of 9 mm.

8.8 Redo Example 8.2 but with the shading ring moved to an inner radius of 11 mm.

8.9 Redo Example 8.3 but with eight equally spaced radial slots. Besides obtaining the numerical values in Table 8.1, also obtain a plot of the eddy current flow pattern.

8.10 Redo Example 8.3 but with 36 equally spaced radial slots. Besides obtaining the numerical values in Table 8.1, also obtain a plot of the eddy current flow pattern. Explain why the AC reluctance has not decreased further.

REFERENCES

1. Sadiku M. *Elements of Electromagnetics*, 3rd ed. New York: Oxford University Press; 2001.

2. Brauer JR, Cendes ZJ, Beihoff BC, Phillips KP. Laminated steel eddy current losses versus frequency computed with finite elements. *IEEE Trans Indus Appl* 2000;36:1132–1137.

3. Fitzgerald AE, Kingsley C. *Electric Machinery*, 2nd ed. New York: McGraw-Hill; 1961. pp 346–347.

4. Demerdash NA, Nehl TW. Use of numerical analysis of nonlinear eddy current problems by finite elements in the determination of parameters of electrical machines with solid iron rotors. *IEEE Trans Magn* 1979;15:1482–1484.

5. Juds MA. A nonlinear magnetodynamic model for AC magnets with shading coils, *Proceedings of International Relay Conference*, 1994.

6. Laithwaite ER. *Propulsion without Wheels*, 2nd ed. London: English Universities Press Ltd.; 1970.

7. Laithwaite ER. *Induction Machines for Special Purposes*. New York: Chemical Publishing Co., Inc.; 1966, Chapter 9.

8. Ramo S, Whinnery JR, Van Duzer T. *Fields and Waves in Communication Electronics*. New York: John Wiley & Sons; 1965, pp 292–293.

9. Dwight HB. *Tables of Integrals and other Mathematical Data*. New York: MacMillan Co.; 1961, pp 324–325.

10. Roters HC. *Electromagnetic Devices*. New York: John Wiley & Sons; 1941.

Magnetic Actuator Transient Operation

While the preceding two chapters analyzed magnetic actuators under DC and AC operation, they did not consider the transient effects of turning on (or off) the DC and AC currents or voltages. This chapter is the first to investigate transient operation of actuators. Its techniques will help determine actuator performance for any applied voltage or current versus time.

9.1 BASIC TIMELINE

The basic timeline for transient operation of magnetic actuators is the following.

(1) An energizing circuit turns on, supplying voltage and current.

(2) The coil current rises, partly determined by an electrical time constant.

(3) The magnetic flux density rises, partly determined by a time constant called the magnetic diffusion time or nonlinear infusion time.

(4) The force rises as the magnetic flux density rises.

(5) The force produces acceleration of the armature and/or attached mass.

(6) After a certain time, during which the armature may or may not reach the end of its stroke, the energizing circuit may be turned off. If the current suddenly becomes zero, then the magnetic flux density (and related force) falls, partly determined by the magnetic diffusion time or nonlinear effusion time.

Each of the above steps may introduce a time delay. For many applications, a time delay is undesirable because it slows the speed of response.

This chapter studies the basics of steps 3, 4, 5, and 6 above. Steps 1 and 2 will not be investigated until Chapters 12, 14, and 15, which will examine further aspects of transient operation.

Magnetic Actuators and Sensors, Second Edition. John R. Brauer.
© 2014 The Institute of Electrical and Electronics Engineers, Inc. Published 2014 by John Wiley & Sons, Inc.

9.2   SIZE, FORCE, AND ACCELERATION

Magnetic force is the most important parameter of magnetic actuators, and it determines the mechanical acceleration and speed of response of the actuator. While many aspects of the mechanical performance will not be studied until Chapters 14–16, here some basic relations between force and actuator size are presented.

The force for axisymmetric plunger actuators of the type shown in Figure E7.4.1 is related to the actuator of a more general design shown in Figure 9.1. Both designs have the plunger and its stopper extending to the same radius, but in Figure 9.1 their inner radius may be greater than zero. Thus the cylindrical plunger of radius R_p may be hollow, with its end having area $k_a \pi R_p^2$ instead of πR_p^2, where k_a is an area constant less than or equal to one.

For a cylindrical plunger of length L_p, the optimum design would have the same area for magnetic flux entering through the side and leaving the circular end of the plunger:

$$2\pi R_p L_p = k_a \pi R_p^2 \tag{9.1}$$

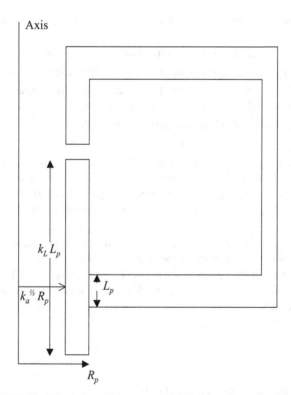

FIGURE 9.1   Dimensional parameters of axisymmetric plunger actuator.

Hence:

$$L_p = k_a R_p / 2 \tag{9.2}$$

would be the minimum plunger length. However, in many cases, such as in the actuator of Figure 9.1, the actual plunger length is longer. Denoting the actual length as $k_L L_p$, where k_L is a constant greater than or equal to one, the plunger volume is:

$$v = k_L L_p \left(k_a \pi \, R_p^2 \right) = k_L \left(\pi k_a^2 \, R_p^3 \right) / 2 \tag{9.3}$$

The acceleration is:

$$a = F / m = P \left(\pi k_a R_p^2 \right) / (\rho \, v) \tag{9.4}$$

where P is the magnetic pressure and ρ is the mass density of the plunger material. It has been shown in Chapter 5 that for typical steel the maximum magnetic pressure $P = 20.\mathrm{E}5 \mathrm{N/m}^2$. Therefore the maximum acceleration of the cylindrical plunger is:

$$a = 20.\mathrm{E}5 \left(\pi \, k_a \, R_p^2 \right) / \left(\rho \left(\pi \, k_L k_a^2 \, R_p^3 \right) / 2 \right) \tag{9.5}$$

$$a = 40.\mathrm{E}5 / (\rho \, k_L k_a \, R_p) \tag{9.6}$$

For steel $\rho = 7800 \ \mathrm{kg/m}^3$, and thus:

$$a = 513 / (R_p k_L k_a) \tag{9.7}$$

Thus to maximize acceleration, $k_L k_a R_p$ should be minimized; that is, the smaller the plunger, the faster it will move, assuming there are no other forces than the magnetic force. This speedier acceleration of small objects is commonly observed in nature; for example, a 2-g hummingbird accelerates much faster than a 10-kg turkey.

If the actuator of Figure 9.1 is the Bessho actuator of Figure 5.2, then the above dimensions are known. With $R_p = 20\mathrm{E}{-}3$ m, the area constant $k_a = 1$, and the length constant $k_L = 2\pi(0.02)(0.142)/\pi(0.02)^2 = 14.2$. Then (9.7) gives $a = 1806 \ \mathrm{m/s}^2$. If this acceleration exists over the entire stroke s, then:

$$s = \tfrac{1}{2} \, a t^2 \tag{9.8}$$

or

$$t = (2s / a)^{0.5} \tag{9.9}$$

For example, if $s = 10\mathrm{E}{-}3$ m, (9.9) gives $t = 3.33$ ms. Note that this is the *minimum possible stroke time* for the Bessho actuator with no opposing forces or masses on the armature. Actual stroke times either experimentally observed [1] or computed using time-stepping finite-element analysis [2] are 51 ms or greater.

For other types of actuators, such as the clapper types of Chapter 7, similar simple equations can be developed. Thus the maximum force and minimum operation times can be estimated.

The above minimum stroke time assumes that the magnetic pressure is always the maximum. Thus it requires high current, which usually requires a significant rise time, and zero time for the magnetic field to diffuse through the plunger. The actual time for the magnetic field to diffuse is the subject of the next sections. Throughout this chapter, unless otherwise indicated, the hysteresis loop of the steel is assumed to be negligibly small.

9.3   LINEAR MAGNETIC DIFFUSION TIMES

Magnetic diffusion time τ_m is a useful parameter for predicting eddy current delay effects on transient magnetic fields and related parameters such as inductance and force. This section examines diffusion times in conducting materials with linear B–H curves, that is, with constant magnetic permeability μ.

9.3.1   Steel Slab Turnon and Turnoff

The *diffusion equation* is derived starting with Maxwell's equations which obtained (8.21) in Chapter 8:

$$\nabla \times \sigma^{-1} \nabla \times \mathbf{H} = -\frac{\partial \mathbf{B}}{\partial t} \tag{9.10}$$

which for constant uniform conductivity σ and permeability μ becomes:

$$\frac{1}{\mu\sigma} \nabla \times \nabla \times \mathbf{B} = -\frac{\partial \mathbf{B}}{\partial t} \tag{9.11}$$

Using vector identities we obtain the *vector diffusion equation* [3]:

$$\frac{1}{\mu\sigma} \nabla^2 \mathbf{B} = \frac{\partial \mathbf{B}}{\partial t} \tag{9.12}$$

Figure 9.2 shows a C-core similar to that of Figure E3.1.1 except that its gap is now filled with a conducting slab. As in Example 8.1, the conducting material is assumed to extend in the $+z$ and $-z$ directions out of the page to obtain zero end region resistance for its eddy currents which elsewhere flow only in the $+z$ and $-z$ directions. As in Figure E3.1.1, throughout the entire 2D problem region the fields are assumed to be independent of z. The slab is assumed to have $\mathbf{B}$ that only varies in the

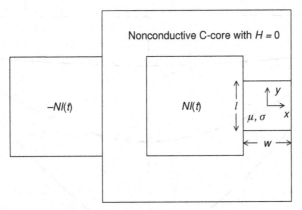

FIGURE 9.2 Conducting slab filling the gap of a planar C-core made of nonconducting high permeability ($H = 0$) material. The fields and eddy currents in the slab are assumed to vary only in the x direction, and the eddy currents are assumed z-directed except at the slab ends which have zero voltage drops. The current $I(t)$ is first assumed to be turned on at time zero.

x direction and only has a y component. Thus in the conducting region of Figure 9.2, (9.12) becomes the *1D slab diffusion differential equation* [3]:

$$\frac{1}{\mu\sigma}\frac{\partial^2 B_y}{\partial x^2} = \frac{\partial B_y}{\partial t} \tag{9.13}$$

The current $I(t)$ in Figure 9.2 is initially assumed to be zero before time zero and to be turned on as a step (with zero rise time) at time zero. In this "turnon" case, B throughout the slab must be zero at time zero, and then must gradually increase. After a sufficiently long time all transient eddy currents in the slab will die out and B_y will be uniform in the slab with its final (DC) value B_F obeying Ampere's law. With these boundary conditions for the diffusion equation (9.13), its solution is:

$$B_y = B_F\left[1 - \sum_{n=\text{odd}}\left(\left(\frac{4}{n\pi}\right)\sin\left(n\pi\left(\frac{1}{2} - \frac{x}{w}\right)\right)e^{-n^2 t/\tau_m}\right)\right] \tag{9.14}$$

where the *diffusion time* is:

$$\tau_m = \mu\,\sigma(w/\pi)^2 \tag{9.15}$$

where w is the width of the conducting slab as shown in Figure 9.2. When the step rise in current and $H(t)$ is applied, the magnetic flux density B diffuses inward (infuses) from both slab sides (coil regions), for example, moving in the x direction from its left side to the right and from its right side to the left, as shown in Figure 9.3. The fundamental Fourier component of the flux reaches 63% of its final value in the middle of the conducting slab at $t = \tau_m$. Figure 9.3 plots the magnetic flux density

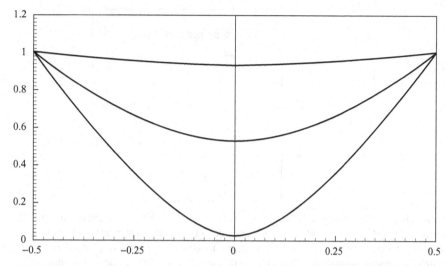

FIGURE 9.3 Computer display of analytical results using (9.14) for diffusion into a steel slab with constant permeability. The top curve is for $t = 3\,\tau_m$, the middle curve is for $t = \tau_m$, and the lower curve is for $t = 0.3\,\tau_m$. The vertical axis is $B(t)/B_F$, and the horizontal axis is x/w.

versus the position of (9.14) at various times, showing the inward diffusion. Note that the *total* flux density in the middle of the conducting slab at $t = \tau_m$ reaches 53.2% of the final value, a change from the exponential 63% due to only the fundamental component.

If the DC coil current has existed over many diffusion times and is suddenly switched off, the current and $H(t)$ then undergo a step decrement. The magnetic flux density B diffuses outward (effuses) to both slab sides, decaying most slowly in the center of the slab. Thus the magnetic force will also eventually decay to zero. However, if the steel is a hard magnetic material as discussed in Chapter 5, then it will have significant coercive field intensity H_c and residual flux density B_r, and there may be a significant permanent magnet force. For soft steels, however, the permanent magnet force is usually less than 1% of the steady-state force when energized. For a current step decrement at time zero, the flux density of (9.14) becomes:

$$B_y = B_o \sum_{n=\text{odd}} \left[\left(\frac{4}{n\pi} \right) \sin \left(n\pi \left(\frac{1}{2} - \frac{x}{w} \right) \right) e^{-\frac{n^2 t}{\tau_m}} \right] \qquad (9.16)$$

where B_o is the initial flux density (in the y direction). At time $t = \tau_m$ the maximum flux density B_y in (9.16) is 47% of B_o and is located at $x = 0$, in the middle of the slab.

Magnetic infusion and effusion in a steel slab can be likened to a dry rectangular sponge that is immersed in a pail of water. The water will infuse inward from the outer sponge walls until the entire sponge is wet. Then if the sponge is set out in air, its water will effuse outward with the outer walls drying first and the inside drying last.

Example 9.1   Magnetic Diffusion into Linear Steel Slab During Turnon and Turnoff   Figure E9.1.1 shows a large planar magnetic device made up of one coil, solid steel, and laminated steel. All steel is of depth 1 m. There is no airgap between the solid armature and the laminated stator. The laminations have relative permeability of 100,000 and zero conductivity. Figure E9.1.2 shows a 1D slice through the right half of Figure E9.1.1 with width $w = 0.447$ m. The slice is used because diffusion time of (9.15) is a 1D concept. The conductivity of the armature is assumed to be 1.7E6 S/m, and its B–H curve is assumed to obey an equation for cast steel of Chapter 2:

$$H = (49.4e^{1.46B^2} + 520.6)B \qquad (E9.1.1)$$

which for low B–H has a relative permeability of 1344.

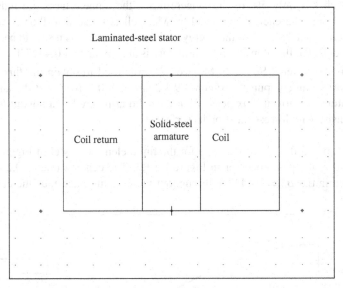

FIGURE E9.1.1   Computer display of large magnetic device. Its depth is 1 m. The solid armature is of (x, y) dimensions 0.447 by 1.0 m. Both coil regions are of dimension 0.6055 by 1.0 m. The laminated stator has overall dimensions 2.49 by 2.0 m.

Armature core $T1$ winding   ---- Stator yoke ----

FIGURE E9.1.2   Computer display of slice of Figure E9.1.1 for use in determining diffusion. The rotor core has half width equal to the dimension $T1 = 0.2235$ m in Figure E9.1.1. The armature has conductivity 1.7E6 S/m and relative permeability of 1344. The stator has zero conductivity and relative permeability of 100,000. The computed magnetic flux lines are shown at $t = 100$ s after the winding current decrement.

Assuming $NI = 264$ ampere-turns has existed for a long time in Figure E9.1.1, and is switched off at time zero, the following are to be found.

(a) Use (9.15) to calculate the diffusion time.

(b) Use Maxwell's magnetostatic solver to obtain the flux line plot and energy at time zero for the entire device and for the slice.

(c) Use the 2D planar transient finite-element solver of Maxwell (not included in the student version SV) to find the flux line plot at $t = 100$ s, and also graph B versus x at $t = \tau_m$, and compare it with (9.16).

Solution

(a) The diffusion time τ_m for Figure E9.1.2 is found using (9.15) with $w = 0.447$ m to be 58.14 s.

(b) For the entire device, Maxwell obtains an energy of 25.74 J. For the slice, which has only 5% of the ampere-turns, the ampere-turns are reduced to 13.2, and the energy computed by Maxwell is reduced to 0.65 J, which is approximately 2.5% of the energy of the entire device, as it should be. In both models, the flux density in the armature is approximately 0.4427 T.

(c) The "Transient" Solver in Maxwell 2D was used to obtain the flux density distribution computed in Figure E9.1.2 at $t = 100$ s. Note that the laminated stator has uniform flux density but the solid armature has a nonuniform flux density with lowest value on its surface.

The computed flux density at 58 s for the finite-element model of Figure E9.1.1 is graphed by the postprocessor in Figure E9.1.3. The center value of 0.2073 T is 47% of the initial $B_o = 0.4427$ T throughout the armature, and thus the computed

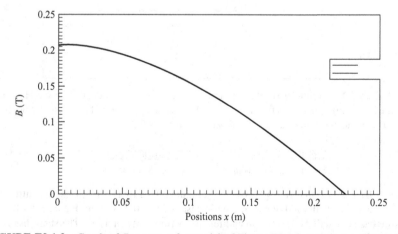

FIGURE E9.1.3   Graph of B versus x for model of Figure E9.1.2 at time $t = 58$ s, which is close to the computed diffusion time using (9.16). Note that the dark theoretical curve of (9.12) lies on top of the lighter curve computed by transient finite-element analysis.

diffusion time agrees closely with the 58.14 s predicted by (9.15). In Figure E9.1.3 the curve of finite-element results lies on top of the analytical curve from (9.16).

9.3.2 Steel Cylinder

Because many actuators, including the Bessho actuator of Figure 5.2, have cylindrical plunger armatures and cylindrical stoppers made of solid steel, magnetic diffusion in cylinders is important. The eddy current induced in axisymmetric objects such as cylinders circles around the z axis of symmetry and thus there are no end effects (unlike in the previous planar slab). The vector magnetic diffusion equation (9.12) for cylindrical coordinates of radius r and axial position z becomes the *1D cylindrical diffusion differential equation*:

$$\frac{1}{\mu\sigma}\frac{1}{r}\frac{\partial}{\partial r}\left(r\frac{\partial B_z}{\partial r}\right) = \frac{\partial B_z}{\partial t} \tag{9.17}$$

where the factors $(1/r)$ and r come from the expression for curl in cylindrical coordinates. The solution for this differential equation for boundary conditions (as for the previous slab case) of zero flux density until time zero, when a step increment in current is applied, is the equation [3]:

$$B_z = B_F\left[1 - \sum_{i=1,2,3} C_i J_o\left(v_i\frac{r}{R}\right)e^{-t/\tau_{mi}}\right] \tag{9.18}$$

where B_F is the final flux density long after the step current is turned on, R is the outer radius of the cylinder, and J_o is the Bessel function of the first order with the first three roots $v_1 = 2.4048$, $v_2 = 5.5201$, and $v_3 = 8.6537$. The constant coefficients in (9.18) can be shown to be $C_1 = 1.602$, $C_2 = -1.066$, $C_3 = 0.851$, followed by less important ones. Using (9.18), at time $t = \tau_m$ the flux density along the cylinder axis has the fundamental component of 63% of the final value B_F, but the *total* flux density at that time along the axis is only 41.62% of B_F. In (9.18) τ_{mi} is the same as τ_m for $i = 1$, and in general obeys:

$$\tau_{mi} = \mu\sigma(R/v_i)^2 \tag{9.19}$$

Thus the diffusion time τ_m for a cylinder with constant permeability is defined using the first root to give [3]:

$$\tau_m = \mu\sigma R^2/(2.4048)^2 \tag{9.20}$$

For the Bessho actuator of Figure 5.2, the conductivity of all steels is specified as 1.7E6 S/m [1]. There are two different *B–H* curves specified. One is for the armature and stopper of Figure 5.2 and the other is for the yoke. The cylindrical armature *B–H* curve is given in Table B2.2 and Figure B2 of Appendix B; it begins with a constant permeability of 630 times that of air. The resulting magnetic diffusion time τ_m using

(9.20) is 93 ms. However, nonlinear finite-element computations [2] have obtained closing times for the Bessho actuator with the above steel as low as 51 ms, so the diffusion time τ_m of (9.20) appears to be longer than actual nonlinear diffusion time.

If the DC coil current has existed over many diffusion times and is suddenly switched off, the current and $H(t)$ then undergo a step decrement. The magnetic flux density B diffuses outward toward the surface of the cylinder, so instead of (9.18) the flux density for $H(t)$ turned off at $t = 0$ is:

$$B_z = B_o \left[\sum_{i=1,2,3} C_i J_o \left(v_i \frac{r}{R} \right) e^{-t/\tau_{mi}} \right] \tag{9.21}$$

where B_o is the initial flux density (in the z direction). At time $t = \tau_m$ the maximum flux density B_z in (9.21) is 58.38% of B_o and is located at zero radius. Magnetic infusion and effusion in a steel cylinder can be likened to a dry roll of paper towels (with zero interior radius) that is immersed in water. The water will infuse inward from the outer radius until the entire roll is wet. Then if the roll is set out in air, its water will effuse outward with the outer radius drying first and the inside drying last.

9.4   NONLINEAR MAGNETIC INFUSION TIMES

The formulas for magnetic diffusion time τ_m in all the preceding section contain permeability μ. Many magnetic devices contain steel operated in the nonlinear region of its B–H curve, and thus the value of μ cannot be determined. The effect of a nonlinear B–H curve on diffusion time must be investigated. It will be found that the nonlinear diffusion time not only depends on the magnitude of the applied H, but also depends on whether the applied H is being turned on or being turned off. When current and H are turned on, the magnetic field B diffuses inward with a diffusion time called *infusion* time. This section examines infusion times, while the next section will examine *effusion* times which apply when current and H are turned off and the magnetic field B leaves or effuses from the conducting material.

9.4.1   Simple Equation for Steel Slab with "Step" B–H

To obtain an approximate formula for τ_m in nonlinear planar slabs, first assume a "step" B–H curve [4] with infinite slope from $B = 0$ to B_m and zero slope for $B > B_m$. This simple B–H curve has been shown to respond to an applied field intensity $H_o(t)$ by producing a traveling flux density wavefront of position $x_o(t)$ given by [4]:

$$x_o(t) = \left[\int H_o(t) dt / (1/2 \, \sigma \, B_m) \right]^{1/2} \tag{9.22}$$

where $1/2$ is placed in the denominator because the applied $H_o(t)$ here starts at zero, not $-B_m$ as assumed in [4]. Assuming a planar slab of width w with $H_o(t)$ stepped to H_o on both slab sides ($x = 0$ and $x = w$), then $x_o(t) = w/2$ at time $t = \tau_m$. Then the integral is replaced by $H_o \tau_m$, resulting in the formula for nonlinear planar magnetic diffusion time with step $B-H$:

$$\tau_m = \sigma w^2 B_m / (8H_o) \qquad (9.23)$$

9.4.2 Transient Finite-Element Computations for Steel Slabs

The infusion time τ_m for devices with nonlinear $B-H$ curves may be computed using transient finite-element analysis of the type used in Example 9.1. Various $B-H$ curves may be input, both the preceding "step" curve and curves of actual steels.

Example 9.2   Magnetic Infusion in Nonlinear Rectangular Inductor   The first example of applying (9.23) is to the rectangular inductor shown in Figure E9.2.1. The entire rectangular core of this inductor is 45 mm wide and 90 mm high and assumed made of solid steel. The magnetic field and eddy currents are assumed to be invariant

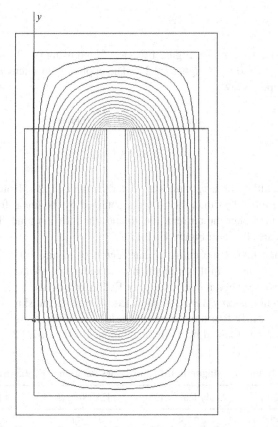

FIGURE E9.2.1   Rectangular planar inductor, showing typical magnetostatic flux lines computed by 2D planar finite-element analysis. The solid steel is a hollow rectangle consisting of four legs of width 20 mm, with copper windings on its left and right legs carrying currents that produce magnetic flux as shown. The window in the center of the legs is of size 5 by 50 mm.

with dimension z (out of the page), and thus the fields vary only in the xy plane. To achieve such independence of z, the depth of the inductor into the page must be much greater than its dimensions in the xy plane. The eddy currents in the steel core of conductivity 1.7E6 S/m flow in the $\pm z$ direction and must sum to zero according to Kirchhoff's current law. Each of the two coils in Figure E9.2.1 has 660 turns.

Then

   (a) The nonlinear infusion time is to be found using (9.23) assuming a step curve at $B_m = 2$ T for coil currents of 0.5, 1, and 2 A. Also, find its magnetostatic flux lines using Maxwell and constant steel relative permeability of 2000.

   (b) Using Maxwell 2D or other available nonlinear transient finite-element software, confirm the results of (9.23) and then use the actual armature steel B–H curve given in [1]. To save computer time, a slice model can be used as in Example 9.1.

Solution

   (a) The infusion time τ_m for Figure E9.2.1 is found using (9.23) as follows. The width $w = 0.020$ m in Figure E9.2.1. The applied field intensity H_o is found using Ampere's law:

$$(H_o)(2 \times 0.050) = 2(660)I \qquad \text{(E9.2.1)}$$

$$H_o = 13,200\,I \qquad \text{(E9.2.2)}$$

The resulting three τ_m values are listed in Table E9.2.1. Note that all three values are much less than the 54.6 ms obtained by the linear formula (9.15), and that the higher the current, the smaller the infusion time. The computed flux lines are shown in Figure E9.2.1.

   (b) Rather than model the entire planar region of Figure E9.2.1, because the inductor dimension in the y direction is considerably greater than that in the x direction, the slice model of Figure E9.2.2 can be used. Its dimension in the y direction is only 2 mm. Because eddy currents and skin effects must be modeled, a fairly fine finite-element mesh is required, and thus the reduced computer time obtained by the reduced model of Figure E9.2.2 is helpful.

TABLE E9.2.1   Nonlinear Magnetic Infusion Times for Figure E9.2.2 (ms)

I (A)	H_o (A/m)	Analytical (9.17)	FEA (Step B–H)	FEA (Real B–H)
0.5	6600	25.8	25.0	18.9
1.0	13,200	12.9	12.5	9.0
2.0	26,400	6.4	6.2	5.0

$-NI|$ Core leg of width w $|+2NI|$ Core leg of width w $|-NI$

$\rightarrow x$

FIGURE E9.2.2 Two millimeter high section of Figure E9.2.1. Computed flux lines are for $t = 4$ ms and real $B\text{–}H$ curve. Note that the flux lines are diffusing inward from the four outer steel core surfaces with windings of conductors N containing current $I = 1$ A.

The windings in Figure E9.2.2 contain a step current $I(t)$ and number of conductors N. The N in Figure E9.2.2 must be (2 mm)/(50 mm) times the 660 turns of Figure E9.2.1, and thus $N = 26.4$.

Transient finite-element analysis was first used to obtain the infusion time for low H_o values, for which the relative steel permeability is 630. Ansoft's Maxwell

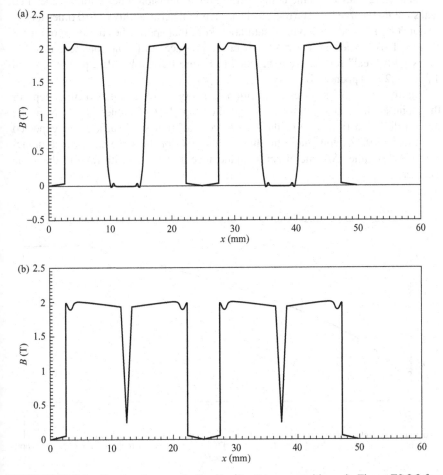

FIGURE E9.2.3 Computed magnetic flux density B versus position x in Figure E9.2.2 for $I = 2$ A and step $B\text{–}H$ curve at times: (a) $t = 3$ ms, (b) $t = 6$ ms.

software computed the time for the flux density in the center of both steel legs to reach approximately half of its final value as 56 ms. This agrees well with 54.6 ms obtained using the linear formula (9.15).

Next, nonlinear transient analyses were performed. Since the slope of the $B–H$ curve is used at each time step in Newton's method of Chapter 4, the infinite slope assumed for the step $B–H$ curve of (9.23) must be replaced by a large finite slope. Here B_m is assumed to be 2 T and the slope from $B = 0$ to 1.93 T is assumed to be 100,000 times that of air. Above 1.93 T, the slope is gradually decreased to that of air at 2.07 T and higher. This step $B–H$ curve is graphed in Figure B1 with data points listed in Table B2.1 of Appendix B.

Table E9.2.1 lists the values of τ_m for a range of H_o. The theoretical values of (9.23) are compared with finite-element computed values with the approximate step $B–H$ curve. Note that the computed times agree closely with those of (9.23).

Table E9.2.1 also lists the computed nonlinear infusion times with the specified real steel $B–H$ curve [1]. The real steel has an initial permeability of 630 times that of air for $B \ll 1$ T, and has a typical saturation knee that approaches the slope of air for $B \gg 2$ T. As expected, the real $B–H$ curve produces different nonlinear infusion times. This "real" $B–H$ curve is graphed in Figure B2 and has data points listed in Table B2.2 of Appendix B.

Figure E9.2.3 plots B versus position x in Figure E9.2.2 at two time steps of the finite-element analysis for 2 A and the step $B–H$ of Table E9.2.1. Note the inward diffusion (infusion) of the step B from all four steel surfaces, as expected [4]. Figure E9.2.4 plots magnetic flux versus time for the same case. Since inductance is flux times N/I, the effective inductance also varies with time in the same manner.

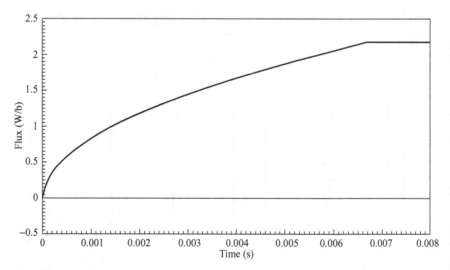

FIGURE E9.2.4 Computed magnetic flux versus time for $I = 2$ A and step $B–H$ curve in Figure E9.2.2 with depth into page assumed to be 1 m.

9.4.3   Simple Equation for Steel Cylinder with "Step" B–H

To obtain an approximate formula for infusion time τ_m in nonlinear steel cylinders, first assume a "step" B–H curve as in the preceding section. Assuming the step curve for a steel cylinder of radius R, it has been found [4] that the magnetic field diffuses into the cylinder following a step wavefront at moving position $r_0(t)$. A formula [4] for the wavefront $r_0(t)$ derived for a field switched between $-B_m$ and $+B_m$ is here altered to account for a field switched from 0 to $+B_m$ by an applied $H_o(t)$:

$$2 \int H_o dt / [\tfrac{1}{2} \sigma R^2 B_m] = \lambda(\ln \lambda - 1) + 1 \qquad (9.24)$$

where:

$$\lambda = r_0^2(t)/R^2 \qquad (9.25)$$

The above two formulas are here applied to determine magnetic infusion time. Because the current is a step applied at time zero, H_o is also a step, and thus the left hand side of (9.24) has its integral over time replaced by $H_o \tau_m$. The value of λ is zero, since $r_0(t) = 0$ when $t = \tau_m$. Thus (9.24) becomes:

$$4H_o \tau_m / [\sigma R^2 B_m] = 1 \qquad (9.26)$$

giving [5]:

$$\tau_m = [\sigma R^2 B_m]/[4H_o] \qquad (9.27)$$

9.4.4   Transient Finite-Element Computations for Steel Cylinders

Transient finite-element analysis of the type used in Examples 9.1 and 9.2 can be used to compute diffusion into nonlinear steel cylinders. Various B–H curves may be input, both the preceding "step" curve of (9.27) and curves of actual steels.

Further studies of transient behavior of magnetic actuators will be presented in Chapters 14–16. They will show that mechanical motion can greatly affect the current waveform, magnetic flux, and magnetic force.

Example 9.3   Magnetic Infusion in Nonlinear Bessho Plunger Actuator   An example of applying (9.27) is to the Bessho axisymmetric plunger actuator of Figure 5.2. Rather than modeling the entire actuator, a thin (10 mm) slice as shown in Figure E9.3.1 is to be modeled with 2D finite elements. The slice represents the armature (here assumed stationary with zero airgap), the coil, and the yoke. The purpose of the slice is to model only the variation of B with radial position r.

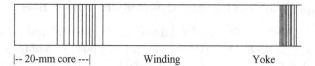

|-- 20-mm core ---| Winding   Yoke

FIGURE E9.3.1   Slice (height 10 mm) of Bessho magnetic actuator, showing computed magnetic flux lines at $t = 20$ ms with actual $B–H$ and $I = 0.5$ A.

Then

(a) The nonlinear infusion time is to be found using (9.27) assuming a step curve at $B_m = 2$ T for coil currents of 0.5 and 2 A. Also, find its magnetostatic flux lines using Maxwell and constant steel relative permeability of 630.

(b) Using Maxwell 2D or other available nonlinear transient finite-element software, confirm the results of (9.27) and then use the actual armature steel $B–H$ curve given in [1] for 0.5 and 2 A coil currents.

Solution

(a) The infusion time τ_m for Figure E9.3.1 is found using (9.27) as follows. The radius $R = 0.020$ m from Figure 5.2. The applied field intensity H_o is found using Ampere's law for Figure 5.2:

$$H_o = NI/l \qquad (E9.3.1)$$

where $N = 3300$. The path length l is assumed to be the 250-mm axial length of the coil window, meaning that the cylindrical core of radius 20-mm drops all ampere-turns (the outer yoke is assumed to have negligible MMF drop). The assumed B_m in steel is 2 T. Thus $I = 0.5$ A gives $H_o = 6600$ A/m. Substituting in (9.27) with the steel $\sigma = 1.7E6$ S/m gives $\tau_m = 51.5$ ms, which is listed in Table E9.3.1. The other analytical value listed in Table E9.3.1 is for the current quadrupled to 2 A, for which (9.27) obtains one-fourth the diffusion time, $\tau_m = 12.9$ ms. Note that both values are much less than the 93 ms obtained by the linear formula (9.20), and that the higher the current, the smaller the diffusion time.

TABLE E9.3.1   Bessho Actuator Nonlinear Magnetic Infusion Times (ms)

Current (A)	Analytical (9.21)	Finite Element (Step $B–H$)	Finite Element (Real $B–H$)
0.5	51.5	53.0	42.0
2.0	12.9	13.4	11.5

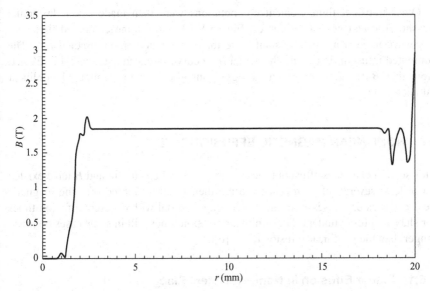

FIGURE E9.3.2 Flux density B (T) versus radius (mm) in core computed by nonlinear transient finite elements with step $B–H$ curve at $t = 51$ ms.

(b) For the step $B–H$ curve, finite-element analyses of Figure E9.3.1 were made with input currents of 0.5 and 2 A. Figure E9.3.2 shows the computed magnetic flux density waveshape in the core at a typical instant; note that it is a steep step to 2 T as expected. The time when the wavefront reaches zero radius is the magnetic diffusion time entered in Table E9.3.1. The computed diffusion (infusion) times agree well with the times predicted using (9.27).

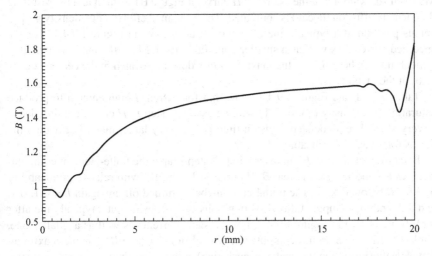

FIGURE E9.3.3 Flux density B (T) versus radius (mm) in core computed by nonlinear transient finite elements with actual $B–H$ curve at $t = 40$ ms.

Other transient finite-element computations were then made using the actual Bessho core and yoke $B-H$ curves. Figure E9.3.3 shows the computed flux density waveshape at a typical instant; note that it now has a more gradual rise. The computed infusion times with the actual $B-H$ curves are also listed in Table E9.3.1. Note that $B-H$ curve shape has a significant effect on the computed nonlinear infusion time.

9.5   NONLINEAR MAGNETIC EFFUSION TIME

This section examines effusion times, which apply when current and H have existed for at least several infusion times and are then suddenly turned off. The magnetic field B leaves or effuses from the conducting material with different effusion times for slabs and for cylinders. The nonlinear effusion times will in some cases be much longer than the nonlinear infusion times [6].

9.5.1   Planar Effusion in Nonlinear Steel Slab

Example 9.4   Magnetic Effusion in Nonlinear Rectangular Inductor   The first example of nonlinear effusion is the rectangular inductor made of a conducting core that was shown in Figure E9.2.1. Its current is assumed to have flowed for a long time before it is turned off at time zero. The slice model of Figure E9.2.2 is to be used for initial currents of 0.1, 0.5, and 2 A with several different $B-H$ curves to compute the flux decay.

Solution   The finite-element model of Figure E9.1.2 is analyzed as follows [6] using Maxwell software. First, the "step" $B-H$ curve of Figure B1 is input and the resulting flux density effusion (decay) is computed. The computed effusion flux density graphs versus position at a typical time after turnoff are shown in Figure E9.4.1. For the specified currents as well as a smaller one, Table E9.4.2 lists B values computed at typical times, where B_{max} is the maximum flux density (located in the center of each slab) at the given time.

Table E9.4.1 and Figure E9.4.1 show that for current I high enough to produce saturating flux density (above 2 T) with an assumed step $B-H$ curve, the flux decay is very slow. The associated effusion time is thus very large, much larger than the 55 ms for constant permeability.

Because real steel $B-H$ curves are not of step shape, the finite-element computations were repeated for the real $B-H$ curve of Figure B2 with relative permeability 630 for B below 0.55 T. The initial currents being turned off are again $I = 0.1, 0.5,$ and 2 A and the computed flux density graphs versus position at a typical time after turnoff are shown in Figure E9.4.2. For the same currents as well as a smaller one, Table E9.4.2 lists B values computed at typical times, where B_{max} is the maximum flux density (located in the center of each slab) at the given time.

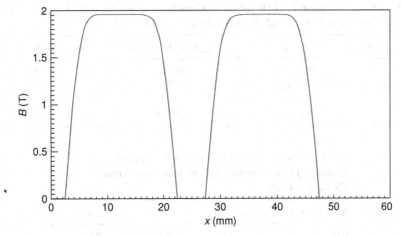

FIGURE E9.4.1 Computed effusion flux density distributions for "step" *B–H* curve in Figure E9.2.2 at time 50 ms after currents of 0.1, 0.5, and 2 A are turned off. The three distributions lie on top of one another.

TABLE E9.4.1 **Effusion Flux Densities (T) Computed Versus Time for Planar Turnoff of Figures E9.2.2 and E9.4.1 with Nonlinear Step *B–H* Curve of Table B2.1**

$I(t = 0)$ in A	$B(t = 0)$ in T	$B_{max}(t = 0.05)$ in T
0.001	0.166597	0.16596
0.1	2.0349	1.9598
0.5	2.0655	1.9599
2.0	2.0926	1.960

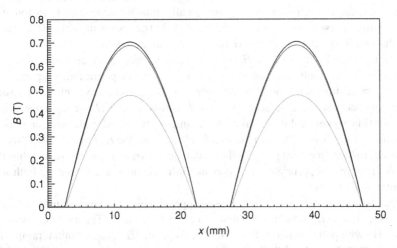

FIGURE E9.4.2 Computed effusion flux density distributions for real *B–H* curve in Figure E9.2.2 at time 50 ms after currents of 0.1, 0.5, and 2 A are turned off. The lowest curve is for the lowest current and the highest curve is for the highest current.

TABLE E9.4.2 Effusion Flux Densities (T) Computed Versus Time for Planar Turnoff of Figures E9.2.2 and E9.4.2 with Real Nonlinear B–H Curve of Table B2.2

$I(t = 0)$ in A	$B(t = 0)$ in T	$B_{max}(t = 0.05)$ in T	$B_{max}(t = 0.05)/B(0)$
0.01	0.10437	0.0532	0.510
0.1	0.97481	0.4775	0.490
0.5	1.6496	0.6909	0.419
2.0	1.8563	0.70568	0.380

TABLE E9.4.3 Magnetic Effusion Times (ms) Computed Using Finite-Element Analysis of Figures E9.2.2 and E9.4.2

$I(t = 0)$ in A	H_o in A/m	$\mu_r = 630$	Real B–H
0.01	132	55	55
0.1	1320	55	55
0.5	6600	55	47
2.0	26,400	55	42

Table E9.4.2 and Figure E9.4.2 show that the flux density is now decaying more rapidly with the real B–H curve. The associated effusion times are listed in Table E9.4.3.

Table E9.4.3 shows that the unsaturated results (for low current) agree with constant permeability results, as expected. Constant permeability results agree with the theory, showing equal infusion and effusion times. Table E9.4.3 also shows that with the real B–H curve, saturation produces only a gradual and small decrease in effusion time. The change in effusion time is much smaller than the huge reductions found in infusion time caused by saturation in Example 9.2. The results in Table E9.4.1 show that the step B–H curve is not appropriate for effusion modeling.

The reason why the step B–H curve is not appropriate for effusion models is evidently its extremely high permeability at low H. As permeability approaches infinity, linear diffusion time also approaches infinity, leading to nonlinear effusion times approaching infinity. Since real B–H curves at low H usually approach a constant (linear) permeability, real effusion times vary only slightly with saturation. In contrast, for infusion (turnon) the step B–H curve has the B/H ratio (called "secant permeability" or "apparent permeability") that decreases as H increases, and thus the step B–H curve is appropriate for turnon, and infusion times are similar for both step and real B–H curves.

Because a step B–H curve is too gross an approximation to obtain accurate effusion times, the next B–H curve investigated is a ramp B–H curve. The curve is shown in Figure B3 with points listed in Table B2.3 of Appendix B. It has an initial ramp with slope of a relative permeability of 630 that agrees with the initial permeability of the real B–H curve of Table B2.2. Near 2 T, the ramp is transitioned to the same slope μ_o of the step curve of Figure B1 with $B_m = 2$ T. Using this ramp B–H curve in the slice

TABLE E9.4.4   Effusion Flux Densities (T) Computed Versus Time for Planar Turnoff of Figure E9.2.2 with Ramp _B–H_ Curve of Table B2.3

$I(t = 0)$ in A	$B(t = 0)$ in T	$B_{max}(t = 0.05)$ in T	$B_{max}(t = 0.05)/B(0)$
0.01	0.1045	0.0534	0.512
0.1	1.0452	0.5336	0.511
0.5	2.0528	1.0050	0.490
2.0	2.0871	1.0080	0.483

model of Figure E9.2.2 for several currents, the computed results are summarized in Table E9.4.4.

Comparing Table E9.4.4 with Table E9.4.2, it is seen that turnoff with the ramp _B–H_ curve agrees only somewhat with the real curve. Comparing Tables E9.4.4, E9.4.2, and E9.4.1, the ramp _B–H_ curve does yield somewhat more accurate turnoff results than the step _B–H_ curve. For truly accurate effusion prediction, however, the real _B–H_ curve must be used. Thus the simple formula (9.23) for step _B–H_ curves is only applicable for infusion time, and is not applicable for effusion time. In addition, if any formulas were to be derived for nonlinear saturable effusion time, they would have to contain multiple variables to account for actual _B–H_ curve shape, not just the single variable B_m of step _B–H_ curves.

9.5.2   Axisymmetric Effusion in Nonlinear Steel Cylinder

Example 9.5   Magnetic Effusion in Nonlinear Bessho Plunger   An axisymmetric example of nonlinear effusion is due to turnoff of current flowing in the cylindrical plunger slice model shown in Figure E9.3.1. The current is assumed to have flowed for a long time before it is turned off at time zero. The model of Figure E9.3.1 includes the cylindrical plunger of the Bessho actuator, and is to be analyzed using Maxwell finite-element software first for constant permeability and then for initial currents of 0.1, 0.5, and 2 A. Since the preceding Example 9.4 showed that step and ramp _B–H_ curves do not give accurate effusion times, here only the real _B–H_ curve will be used in Maxwell finite-element software.

Solution   The axisymmetric finite-element model pictured in Figure E9.3.1 is first analyzed with constant relative permeability equal to 630 and with 10 A current turned off at $t = 0$. At $t = 0$, the computed flux density throughout the cylindrical plunger is 0.7557 T. At $t = 93$ ms, the computed flux lines are shown in Figure E9.5.1 and the flux density versus radius is graphed in Figure E9.5.2. It shows that surface flux density is approximately zero, but the flux density at zero radius is 0.43391 T. The ratio of $(0.43391/0.7557) = 57.4\%$, reasonably close to the 58.8% of (9.21).

For the real nonlinear _B–H_ curve of Figure B2 and Table B2.2 of Appendix B, the computed B versus radius at $t = 90$ ms is graphed in Figure E9.5.3 for the three

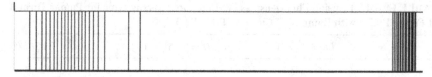

FIGURE E9.5.1   Slice model (height 10 mm) of Bessho actuator with computed flux lines 93 ms after current turnoff with constant steel relative permeability of 630.

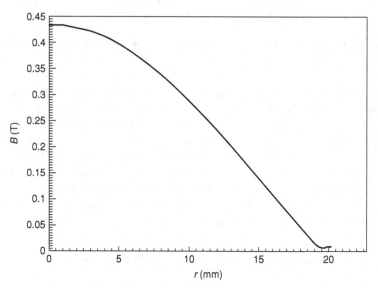

FIGURE E9.5.2   Computed effusion flux density versus radius 93 ms after current turnoff in Bessho cylindrical slice model with constant relative permeability of 630, as in Figure E9.5.1.

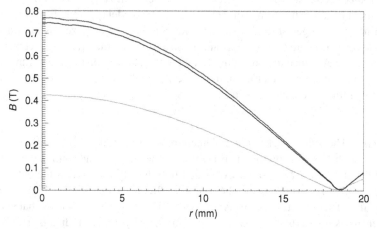

FIGURE E9.5.3   Computed effusion flux density distributions for real *B–H* curve in Bessho actuator at $t = 90$ ms after currents of 0.1, 0.5, and 2 A are turned off. The lowest curve is for the lowest current and the highest curve is for the highest current.

TABLE E9.5.1   Effusion Flux Densities (T) Computed Versus Time for Axisymmetric Cylinder Turnoff of Figure E9.5.1 with Real B–H Curve

$I(t = 0)$ in A	$B(t = 0)$ in T	$B_{max}(t = 0.09)$ in T	$B_{max}(t = 0.09)/B(0)$
0.01	0.0997	0.06742	0.676
0.1	0.7829	0.4250	0.543
0.5	1.6209	0.7476	0.461
2.0	1.8453	0.7698	0.417

TABLE E9.5.2   Magnetic Effusion Times (ms) Computed Using Finite-Element Model of Cylinder of Figure E9.5.1

$I(t = 0)$ in A	H_o in A/m	$\mu_r = 630$	Real B–H
0.01	132	93	93
0.1	1320	93	84
0.5	6600	93	70
2.0	26,400	93	58

input currents. For the same currents as well as a smaller one, Table E9.5.1 lists B values computed at typical times, where B_{max} is the maximum flux density (located at zero radius) at the given time. For currents above 0.01 A, the slice model assumes the actual yoke B–H curve, whereas at the lowest current the yoke is assumed to have relative permeability of 10,000 as in Figures E9.5.1 and F9.5.2. Because the cross-sectional area of the yoke is proportional to the difference between the squares of its outer and inner radii, its area is 36% larger than the cylindrical plunger, and thus its MMF drop is much less than that of the plunger. Also, because the yoke has radial thickness of only 4 mm, much smaller than the 20-mm plunger radius, its flux diffuses much faster than the plunger flux. Thus the yoke has only a minor effect on MMF drops and the results, and it is not included in the calculated applied H values in Tables E9.5.1 and E9.5.2.

Table E9.5.2 lists the computed effusion times for the various input currents. It shows that the unsaturated results (for low current) agree with constant permeability results, as expected. Constant permeability results agree with theory, showing effusion time of 93 ms, the same as the linear infusion time. Table E9.5.2 also shows that saturation produces only a gradual and rather small decrease in effusion time. The reduction in effusion time caused by saturation is much smaller than the huge reductions in saturated infusion time.

9.6   PULSE RESPONSE OF NONLINEAR STEEL

In many cases, currents and surface **H** fields are pulsed [7–9]. The excitation is turned on and then off. Thus the magnetic flux densities and the flux in conducting steel are expected to first rise with an infusion time and then to fall with an effusion time. Since the infusion and effusion times in nonlinear steel may differ considerably, the flux rise and fall times may differ if the steel has flux densities in its nonlinear region.

Example 9.6   Flux Rise and Fall in Conducting Slabs   The coil current in the planar steel slabs of Figure E9.2.2 is zero before $t = 0$, when it is switched to 2 A. At $t = 30$ ms the current is switched back to zero. Find the flux linkage of the coil versus time for $0 < t < 80$ ms using the actual steel B–H curve.

Solution   The Maxwell 2D planar transient finite-element solution produced the flux linkage graph in Figure E9.6.1. The expected infusion time is 5 ms from Table E9.2.1, while the expected effusion time is 42 ms from Table E9.4.3. These two times correspond respectively with the rise and fall times observable in Figure E9.6.1. Due to saturation, the fall time is over eight times the rise time.

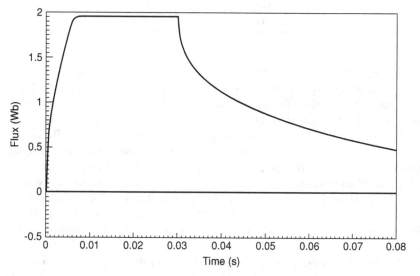

FIGURE E9.6.1   Computed flux linkage versus time for the slabs of Figure E9.2.2 with the real B–H curve and $I = 2$ A turned on at $t = 0$ and turned off at $t = 30$ ms.

Example 9.7   Flux Rise and Fall in a Conducting Cylinder   The coil current in the axisymmetric slice model of the Bessho actuator in Figure E9.3.1 is zero before $t = 0$, when it is switched to 2 A. At $t = 30$ ms the current is switched back to zero. Find the flux linkage of the coil versus time for $0 < t < 90$ ms using the actual (real) steel B–H curve.

Solution   The Maxwell 2D axisymmetric transient finite-element solution produced the flux linkage graph in Figure E9.7.1. The expected infusion time is 11.5 ms from

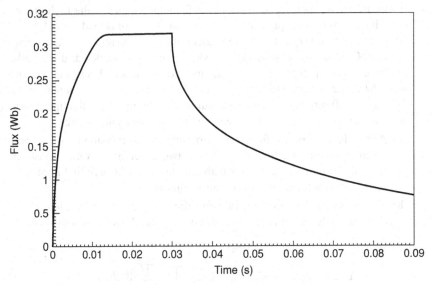

FIGURE E9.7.1   Computed flux linkage versus time for the cylindrical slice model of Figure E9.5.1 with the real *B–H* curve and *I* = 2A turned on at *t* = 0 and turned off at *t* = 30 ms.

Table E9.3.1, while the expected effusion time is 58 ms from Table E9.5.2. These two times correspond respectively with the rise and fall times observable in Figure E9.7.1. The fall time is over five times the rise time due to saturation.

PROBLEMS

9.1   Find the maximum acceleration and minimum operation time for a smaller version of the Bessho actuator, in which all dimensions are reduced by 50%.

9.2   Find the analytic diffusion time for the steel with the properties of Example 9.1a but with its width reduced by a factor of (a) 10, (b) 100, and (c) 1000.

9.3   Redo Example 9.1b and 9.1c using Maxwell but with its width reduced by a factor of 10.

9.4   Redo Example 9.1b and 9.1c using Maxwell but with its width reduced by a factor of 100.

9.5   Redo Example 9.1b and 9.1c using Maxwell but with its width reduced by a factor of 1000.

9.6   Put a 1-mm airgap at the upper armature/stator interface of Problem 9.3 and repeat the computations of Problem 9.3. Show that the airgap produces signifi-cant changes in the flux pattern.

9.7   Figure P9.7.1 shows a sector of a toroidal inductor with a circular (cylindrical) core. It extends into the page to a depth of 1 m. The core is made of solid steel and has inner radius $r_i = 10$ mm and outer radius $r_o = 30$ mm. Only a 4° sector of the 360° of the circular inductor is shown and to be modeled; this smaller model saves computer time. Inside the inner radius is a coil carrying ampere-turns NI in the $+z$ (out) direction, where $NI = 4.608$ or 9.215 ampere-turns. Outside the 30-mm core radius, to a radius of 31.6 mm, the coil ampere-turns are returned as $-NI$. The eddy currents in the core must sum to zero.

(a) Apply the nonlinear infusion time formula to find infusion times for both given ampere-turns. Note that because two different H_o values exist due to its differing inner and outer path lengths, you will only find a range of infusion times for each value of ampere-turns.

(b) Use Maxwell 2D to obtain infusion times by nonlinear transient finite-element analyses with the real B–H curve of the Bessho armature steel.

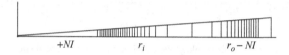

$$+NI \qquad\qquad r_i \qquad\qquad\qquad r_o - NI$$

FIGURE P9.7.1   Four-degree sector of circular inductor with inner core radius $r_i = 10$ mm and outer core radius $r_o = 30$ mm. The flux lines at 10 ms are computed with windings of $NI = 4.608$ A and with the real B–H curve. Note that the flux lines are diffusing inward from both steel core surfaces.

9.8   Figure P9.8.1 shows a laminated core inductor mounted in a solid-steel cabinet. The cabinet is made of solid-steel panels on all of its six sides. Inductors are often placed inside such cabinets to house them mechanically and to confine their magnetic fields and thereby reduce electromagnetic interference. The cabinet (also called a box or a housing) is known to often increase the inductance. However, the inductance increase is delayed by the magnetic diffusion time of the solid-steel panels.

The inductor of Figure P9.8.1 has N_T total turns placed in the center of a steel cabinet. Because the inductor core is laminated or made of composite ferromagnetic material with extremely low electrical conductivity, the electrical conductivity of the core is assumed to be zero. Thus eddy currents exist only in the solid-steel panels. Figure P9.8.2 is a 2-mm high slice of the right half of Figure P9.8.1. It includes only the right half of the center leg, the right coil half and the right panel, which has an eddy current that does not sum to zero. The eddy current in the right panel is returned through the other five cabinet panels.

Because the coil is only on the inside of the steel panel and its eddy currents can be returned through the other five cabinet panels, diffusion only occurs into the inside of the panel. Thus w must be considered to be twice the panel thickness T, giving:

$$\tau_m = \sigma\,(2T)^2\,B_m/(8H_o) = \sigma\,T^2\,B_m/(2H_o) \qquad (P9.8.1)$$

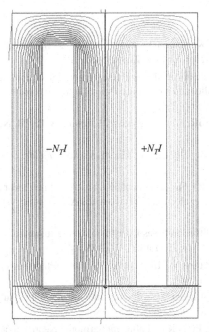

FIGURE P9.8.1 Computer display of laminated inductor core surrounded by solid-steel panels on all four sides, showing typical magnetostatic (no eddy current) flux lines computed by Maxwell. The overall dimensions are 95 mm high and 60 mm wide. The four panels are all 10 mm thick. The laminated center core is 20 mm thick.

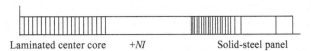

Laminated center core $+NI$ Solid-steel panel

FIGURE P9.8.2 Two millimeter high section of right half of inductor of Figure P9.8.1 with computed flux lines at time $t = 6$ ms and $I = 0.5$ A for real $B–H$ curve. Note that the flux lines are diffusing into the panel from the coil ampere-turns NI, but that the flux lines in the laminated core are uniform due to its lack of eddy currents.

The core material is here assumed to be laminated SAE 1010 steel with the $B–H$ curve given in Maxwell. The outer shell is assumed to be made of solid steel with conductivity 1.7E6 S/m. The number of conductors N in Figure P9.8.2 is assumed to be 26.4.

(a) Find the analytic infusion times for a step $B–H$ curve at 2 T and currents of 0.5, 1, and 2 A.

(b) Verify the results of (a) using Maxwell 2D or other software, and then repeat using the actual Bessho armature $B–H$ curve [1].

REFERENCES

1. Bessho K, Yamada S, Kanamura Y. Analysis of transient characteristics of plunger type electromagnets. *Electr Eng Japan* 1978;98:56–62.
2. Brauer JR. Magnetic actuator models including prediction of nonlinear eddy current effects and coupling to hydraulics and mechanics. *Proc Congresso Brasileiro de Eletromagnetismo*, Gramado, Brazil, November 2002.
3. Woodson HH, Melcher JR. *Electromechanical Dynamics*, Part II. New York: John Wiley & Sons; 1968.
4. Mayergoyz ID. *Nonlinear Diffusion of Electromagnetic Fields*. San Diego, CA: Academic Press; 1998.
5. Brauer JR, Mayergoyz ID. Finite element computation of nonlinear magnetic diffusion and its effects when coupled to electrical, mechanical, and hydraulic systems. *IEEE Trans Magn* 2004;40:537–540.
6. Brauer JR. Magnetic diffusion times for infusion and effusion in nonlinear steel slabs and cylinders. *IEEE Trans Magn* 2007;43:3181–3188.
7. Bendre A, Divan D, Kranz W, Brumsickle W. Are voltage sags destroying equipment? *IEEE Industry Appl Mag* 2006;12:12–21.
8. Bíró O, Preis K. An efficient time domain method for nonlinear periodic eddy current problems. *IEEE Trans Magn* 2006;42:695–698.
9. Ausserhofer S, Bíró O, Preis K. An efficient harmonic balance method for nonlinear eddy current problems. *IEEE Trans Magn* 2007;43:1229–1232.

PART III

SENSORS

CHAPTER 10

Hall Effect and Magnetoresistive Sensors

As mentioned in Chapter 1, magnetic sensors use magnetic fields to sense motion. Input energy in the form of mechanical energy and/or electromagnetic energy is converted to electrical energy, and thus magnetic sensors are a type of transducer. Usually magnetic sensors output a signal voltage along with little or no current. The magnetic fields sensed may vary over an extremely broad range as shown in Figure 10.1.

Magnetic sensors using the Hall effect are very common, small, and inexpensive. This chapter examines the behavior, accuracy, and construction of typical Hall sensors. Along with the Hall effect, this chapter also investigates a related parameter called *magnetoresistance*. Note that Figure 10.1 shows that Hall and magnetoresistive (MR) sensors may sense fields as low as approximately 1.E–5 T, and related giant magnetoresistance (GMR) sensors can sense even smaller magnetic fields. While the previous chapters have shown that magnetic actuators usually operate in the range of 1–2 T, magnetic sensors often operate over a much greater range of flux densities.

10.1 SIMPLE HALL VOLTAGE EQUATION

The Hall effect was discovered in 1879 by E. H. Hall. He found that a voltage is developed across a current-carrying semiconductor in a direction normal to the current flow and proportional to the magnetic flux density that is normal to both the current and the voltage difference. The Hall voltage across terminals spaced in the y direction is:

$$V_y = k_H B_z J_x d_y \tag{10.1}$$

where d_y is the terminal (electrode) spacing shown in Figure 10.2, which also shows the current density J_x flowing in the x direction, along with the flux density B_z. The Hall voltage is developed because the current experiences the Lorentz force of

Magnetic Actuators and Sensors, Second Edition. John R. Brauer.
© 2014 The Institute of Electrical and Electronics Engineers, Inc. Published 2014 by John Wiley & Sons, Inc.

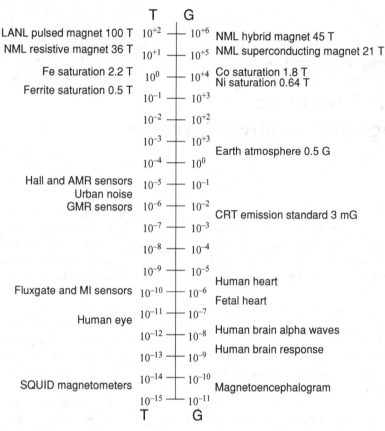

FIGURE 10.1 Range of magnetic flux densities in teslas (T) and gauss (G). The highest values are produced in magnets made by Los Alamos National Laboratory (LANL) or the National Magnet Lab (NML), which have coils that are either superconducting, resistive (using conventional Cu or Al wire), or hybrid (both coil types).

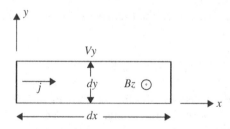

FIGURE 10.2 Hall effect voltage in a narrow semiconductor bar.

Chapter 2. It can be shown that the Hall electric field $\mathbf{E}$ in volts per meter balances the motional electric field of (2.36) [1].

The Hall coefficient k_H in (10.1) can be shown for a narrow bar with $d_x \gg d_y$ to equal approximately [1]:

$$k_H = -\frac{1}{ne} \tag{10.2}$$

where n is the electron concentration in electrons per cubic meter and e is the charge on an electron $= 1.602\mathrm{E}{-}19$ C. The electron concentration varies with the semiconductor material used. Typically, silicon is used, because it is so widely available, thermally stable, inexpensive, and has conductivity in the semiconductor range. In fact, many Hall sensors are built within CMOS or other integrated silicon circuits. The semiconductor range of conductivity is within several orders of magnitude of 1 S/m, and can be varied by doping with impurities. If doped so as to contain holes, the Hall constant of (10.2) becomes:

$$k_H = \frac{1}{pe} \tag{10.3}$$

where p is the hole concentration in holes per cubic meter.

Practical Hall sensors often have many large electrodes and a wide range of geometries, which can make the accuracy of (10.1) poor. To analyze a wide range of geometries and materials, a numerical technique is required that will solve for current flow in Hall sensors [2].

10.2   HALL EFFECT CONDUCTIVITY TENSOR

Current flow is governed by Ohm's law for fields given in Chapter 2. When conductivity σ is anisotropic (dependent on direction), then it must be expressed as a matrix called a tensor, and Ohm's law becomes:

$$\begin{bmatrix} J_x \\ J_y \\ J_z \end{bmatrix} = \begin{pmatrix} \sigma_{xx} & \sigma_{xy} & \sigma_{xz} \\ \sigma_{yx} & \sigma_{yy} & \sigma_{yz} \\ \sigma_{zx} & \sigma_{zy} & \sigma_{zz} \end{pmatrix} \begin{bmatrix} E_x \\ E_y \\ E_z \end{bmatrix} \tag{10.4}$$

Because an applied magnetic field produces a Lorentz force on moving electrons, the above conductivity tensor is indeed anisotropic. Fermi gas theory of free electrons [3] experiencing $\mathbf{B}$ in the z direction produces the following matrix elements of (10.4):

$$\sigma_{xx} = \sigma_{yy} = \frac{\sigma_o}{1 + (\omega_c \tau)^2} \tag{10.5}$$

$$\sigma_{xy} = \frac{-\omega_c \tau \sigma_o}{1 + (\omega_c \tau)^2} \tag{10.6}$$

$$\sigma_{yx} = \frac{\omega_c \tau \sigma_o}{1 + (\omega_c \tau)^2} \tag{10.7}$$

$$\sigma_{xz} = \sigma_{xz} = \sigma_{xz} = \sigma_{xz} = 0 \tag{10.8}$$

$$\sigma_{zz} = \sigma_o \tag{10.9}$$

where ω_c is the cyclotron frequency for a free electron, τ is the collision time, and σ_o is the *bulk conductivity* [3]:

$$\sigma_o = ne^2 \tau / m \tag{10.10}$$

where m is electron mass $= 9.1085E{-}21$ kg. The cyclotron frequency of the free electrons due to the force of the magnetic field is:

$$\omega_c = eB/m \tag{10.11}$$

Because (10.6) and (10.7) differ for $B > 0$, Hall effect and magnetoresistance create an *unsymmetric* conductivity tensor.

The Hall voltage equation (10.1) can be derived by making the following assumptions:

(1) Assuming $E_z = 0$, then (10.4)–(10.9) give:

$$J_y = \frac{\sigma_o}{1 + (\omega_c \tau)^2}(\omega_c \tau E_x + E_y) \tag{10.12}$$

(2) Assuming $J_y = 0$, then (10.12) and (10.11) give:

$$E_y = -\omega_c \tau E_x = -\frac{eB}{m}E_x \tag{10.13}$$

Next, (10.4)–(10.9) give:

$$J_x = \frac{\sigma_o}{1 + (\omega_c \tau)^2}(E_x - \omega_c \tau E_y) \tag{10.14}$$

Substituting (10.13) into (10.14) gives:

$$J_x = \sigma_o E_x \tag{10.15}$$

(3) Assuming E_y is uniform in y, then (10.13) gives:

$$V_y = E_y d_y = -d_y \frac{eB_z \tau}{m}E_x \tag{10.16}$$

Substituting (10.15) and then (10.10) obtains the Hall voltage:

$$V_y = -d_y \frac{eB_z \tau}{m\sigma_o} J_x = -d_y \frac{B_z J_x}{ne} \tag{10.17}$$

Note that (10.17) agrees with (10.1) and (10.2). However, the above three assumptions have to hold true for the simple equation (10.1) to apply. These assumptions are valid only for narrow bars with infinitesimal Hall voltage sensing electrodes.

Example 10.1 Simple Hall Equation Applied to Semiconducting Bar   A wide bar of semiconductor material as shown in Figure 10.2 has dimensions $8 \times 10 \times 0.25$ mm in x, y, and z, respectively. The measured current I in the x direction is 2.5 mA for 1 V applied across the current-carrying electrodes at $x = 0$ and $x = d_x$.

(a) Calculate the bulk resistance and σ_o.

(b) A narrow bar of the same material and current density placed in a magnetic field of 1 T obtains a Hall electric field of -23.8 V/m. Find the Hall coefficient and the conductivity tensor.

Solution

(a) The bulk resistance is $R = V/I = 1/.0025 = 400\ \Omega$. Analogous to reluctance of Chapter 3, resistance of a bar obeys:

$$R = l/(\sigma S) \tag{E10.1.1}$$

where as in Chapter 3, l is length and S is cross-sectional area. Substituting the length and area of the wide bar obtains the bulk conductivity $\sigma_o = 8$ S/m, which is within the wide conductivity range of semiconductors.

(b) From (10.17), (10.1), and (10.2), the Hall electric field for the narrow bar is:

$$E_y = k_H B_z J_x \tag{E10.1.2}$$

In terms of resistivity ρ, the inverse of conductivity, where $1/\sigma_o = 1/8 = 0.125$:

$$\begin{bmatrix} E_x \\ E_y \end{bmatrix} = \begin{pmatrix} 0.125 & -k_H B_z \\ k_H B_z & 0.125 \end{pmatrix} \begin{bmatrix} J_x \\ J_y \end{bmatrix} \tag{E10.1.3}$$

Given the -23.8 V/m measurement for $J_x = I/S = 1000$ A/m^2, the off-diagonal terms can be filled in to give:

$$\begin{bmatrix} E_x \\ E_y \end{bmatrix} = \begin{pmatrix} 0.125 & 0.0238 \\ -0.0238 & 0.125 \end{pmatrix} \begin{bmatrix} J_x \\ J_y \end{bmatrix} \tag{E10.1.4}$$

The Hall coefficient is thus $k_H = -0.0238$. The tensor conductivity is the inverse of the above resistivity tensor matrix:

$$\begin{pmatrix} \sigma_{xx} & \sigma_{xy} \\ \sigma_{yx} & \sigma_{yy} \end{pmatrix} = \begin{pmatrix} \rho_{xx} & \rho_{xy} \\ \rho_{yx} & \rho_{yy} \end{pmatrix}^{-1} \tag{E10.1.5}$$

which means that the product of the resistivity and conductivity tensors is unity:

$$\begin{pmatrix} \rho_{xx} & \rho_{xy} \\ \rho_{yx} & \rho_{yy} \end{pmatrix} \begin{pmatrix} \sigma_{xx} & \sigma_{xy} \\ \sigma_{yx} & \sigma_{yy} \end{pmatrix} = \begin{pmatrix} 1 & 0 \\ 0 & 1 \end{pmatrix} \tag{E10.1.6}$$

The conductivity tensor can be easily found in the case when (10.5)–(10.8) make diagonal terms equal and off-diagonal terms equal but opposite in sign:

$$\begin{pmatrix} \rho_{xx} & \rho_{xy} \\ -\rho_{xy} & \rho_{xx} \end{pmatrix} \begin{pmatrix} \sigma_{xx} & \sigma_{xy} \\ -\sigma_{xy} & \sigma_{xx} \end{pmatrix} = \begin{pmatrix} 1 & 0 \\ 0 & 1 \end{pmatrix} \tag{E10.1.7}$$

Multiplying out the left side and using Cramer's rule gives:

$$\sigma_{xx} = \rho_{xx}/(\rho_{xx}^2 + \rho_{xy}^2) \tag{E10.1.8}$$

$$\sigma_{xy} = -\rho_{xy}/(\rho_{xx}^2 + \rho_{xy}^2) \tag{E10.1.9}$$

which for the resistivity tensor of (E10.1.4) gives the desired conductivity tensor:

$$\begin{pmatrix} \sigma_{xx} & \sigma_{xy} \\ \sigma_{yx} & \sigma_{yy} \end{pmatrix} = \begin{pmatrix} 7.72013 & -1.46991 \\ 1.46991 & 7.72013 \end{pmatrix} \tag{E10.1.10}$$

10.3 FINITE-ELEMENT COMPUTATION OF HALL FIELDS

To predict Hall voltages for arbitrary geometries, including bars that are not narrow, as well as for arbitrary materials, a numerical method is needed. Similar to the finite elements of Chapter 4 for magnetic fields, here finite elements are applied to current flow problems with anisotropic tensor conductivity.

10.3.1 Unsymmetric Matrix Equation

From Ohm's law, voltages and currents are related by the matrix equation:

$$[C]\{V\} = \{I\} \tag{10.18}$$

where $[C]$ is the conductance matrix. Chapter 4 showed that finite elements have shape functions. For a shape function N of any type, it can be shown that the conductance matrix is of the form:

$$[C] = \sum dv [\nabla N]^T [\sigma][\nabla N] \qquad (10.19)$$

where the summation is over volume integrals of individual finite elements. Note that the tensor conductivity appears, postmultiplied by the gradient of N and premultiplied by the matrix transpose of the gradient of N. Because the Hall effect conductivity tensor of (10.4) is unsymmetric, the conductance matrix $[C]$ is also unsymmetric for nonzero magnetic fields. Thus the finite element software used to solve (10.18) must have the capabilities of inputting and solving an unsymmetric matrix.

10.3.2   2D Results

Finite-element software that solves DC current flow problems with unsymmetric conductivity tensors has been used to analyze Hall effect and the related magnetoresistance. Here 2D finite-element analysis is applied to several examples.

Example 10.2   Hall Voltage Computed for Wide 2D Semiconductor Bar of Example 10.1   The wide bar of semiconductor material of Example 10.1 has the conductivity tensor determined in that example. Use the unsymmetric conductivity tensor in 2D finite-element software to compute the actual Hall electric field in the wide bar along with the resistance of the bar.

Solution   The results are shown in Figure E10.2.1 and included in Table 10.1. Note that quadrilateral finite elements are used, all squares 1 mm on a side. The voltage contours in Figure E10.2.1a are "S" shaped due to Hall effect, which produces the

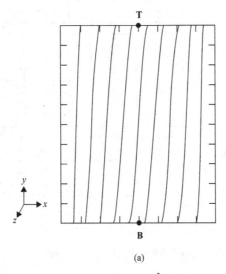

(a)

FIGURE E10.2.1   Finite-element model with 1–mm² elements and computed results for Hall sensor of dimensions 8 mm in x and 10 mm in y. **B** is in the z direction and applied **J** in x direction. The top electrode is indicated by T and the bottom electrode is indicated by B. (a) voltage contours, (b) Hall electric field in y direction, (c) current density vectors.

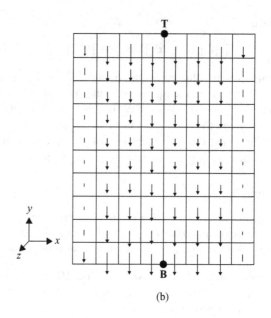

(b)

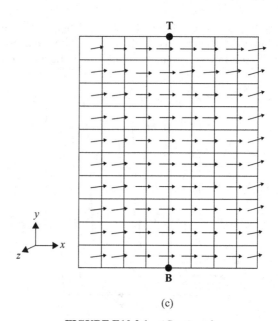

(c)

FIGURE E10.2.1 (*Continued*)

TABLE 10.1   Calculated Hall Fields and Magnetoresistance R_M

Model Dimensions (mm)	A_{el} (mm²)	V_B (V)	V_T (V)	$E_{y,av}$ (V/m)	R_M (Ω)	R_M (%)
2D 8 × 10 × 0.25	0	0.43304	0.56696	−13.39	408.51	102.13
2D 8 × 2 × 0.25	0	0.47635	0.52365	−23.65	2011.5	100.58
3D 8 × 10 × 2	2	0.44575	0.55425	−10.85	48.978	102.26
3D 8 × 2 × 2	2	0.48489	0.51511	−15.11	187.66	100.98

y components of electric field in Figure E10.2.1b. The current density has both x and y components as shown in Figure E10.2.1c. The top row of Table 10.1 shows that the actual Hall electric field is −13.39 V/m, considerably reduced from the −23.8 V/m obtained for the narrow rod as described in Example 10.1. Table 10.1 lists the voltage V_T in the top Hall electrode and the voltage V_B in the bottom Hall electrode, whose difference is the Hall voltage. Note also in the top row of Table 10.1 that the resistance is now 408.51 Ω, an increase from the 400 Ω without a magnetic field. The increased resistance is called magnetoresistance.

Example 10.3   Hall Voltage Computed for More Narrow 2D Semiconductor Bar
The semiconductor material with the conductivity tensor of Examples 10.1 and 10.2 is used in a bar of size 8 × 2 × 0.25 mm. Use the unsymmetric conductivity tensor in 2D finite-element software to compute the actual Hall voltage and magnetoresistance.

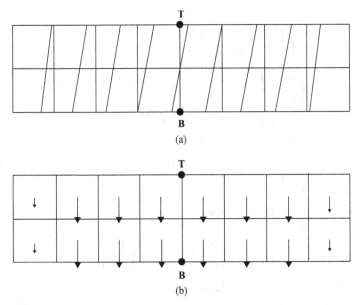

FIGURE E10.3.1   Finite-element model with 1–mm² elements and computed results for Hall sensor of dimensions 8 mm in x and 2 mm in y. **B** is in the z direction and applied **J** in the x direction. The top electrode is indicated by T and the bottom electrode is indicated by B. (a) voltage contours, (b) Hall electric field in y direction.

Solution   The results are shown in Figure E10.3.1 and included in Table 10.1. The voltage contours in Figure E10.3.1a are slanted due to Hall effect, which produces the *y* components of electric field in Figure E10.3.1b. The second row of Table 10.1 shows that the actual Hall electric field is −23.65 V/m, very close to the −23.8 V/m obtained for the narrow rod as described in Example 10.1. Note also in the second row of Table 10.1 that the resistance is now 2011.5 Ω, a small percentage increase from the 2000 Ω without a magnetic field.

Example 10.4   Effect of Finite Hall Electrodes on Current Flow   The wide bar of semiconductor material of Example 10.1 is placed where there is no magnetic field. Maxwell is to be used to find its voltage contours and current flow pattern for two cases. The first case has no Hall electrodes. The second case has two Hall aluminum electrodes of size 1 mm in *x* and 1 mm in *y*. In both cases the applied magnetic field is assumed to be zero, and the power loss is also to be found.

Solution   First, change the Solver in Maxwell to "DC Conduction." Next, enter the 8 × 10-mm rectangular box for the geometry, and also use boxes for both electrodes. Under "Setup Materials," since there is no magnetic field, the bulk conductivity of 8 S/m is entered. The electrodes are first made the same 8 S/m, and later changed to aluminum. Finally, boundary conditions of 1 V on the left edge and 0 V on the right edge are specified.

The computed voltage contours without aluminum electrodes are shown in Figure E10.4.1a, and the computed current density arrows are in Figure E10.4.1b. The

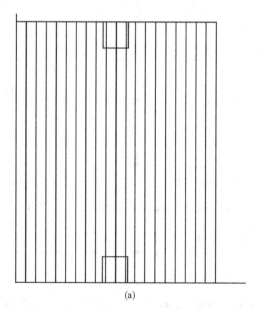

(a)

FIGURE E10.4.1   Wide semiconductor bar of Example 10.4 with no applied magnetic field. (a) computed voltage contours, (b) computed current density arrows.

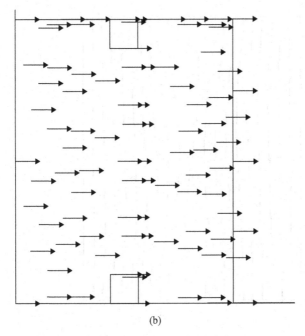

(b)

FIGURE E10.4.1   (*Continued*)

computed power loss is exactly 10 W for 1 m depth. Since for 1 m depth the resistance from (E10.1.1) is 0.1 Ω, the expected power loss is $V^2/0.1 = 10$ W, agreeing with the computed power loss.

The computed voltage contours with aluminum electrodes are shown in Figure E10.4.2a, and the computed current density arrows are in Figure E10.4.2b. Both patterns show that the aluminum electrodes perturb the field. The computed power loss is now 10.7365 W for 1 m depth. The power loss has increased for 1 V because some of the semiconductor is replaced by aluminum, reducing the resistance.

10.3.3   3D Results

To compute the voltage contours, Hall electric field and voltage, and resistance of sensors with 3D geometries, 3D finite elements are used to assemble the matrix equation of (10.18). The shape function of (10.19) for the first-order 3D finite elements used here obtains a voltage within an element obeying:

$$V(x, y, z) = \sum_{k=L,M,N,O} V_k(1 - a_k x)(1 - b_k y)(1 - c_k z) \qquad (10.20)$$

Thus the shape function of (10.19) is:

$$N = (1 - a_k x)(1 - b_k y)(1 - c_k z) \qquad (10.21)$$

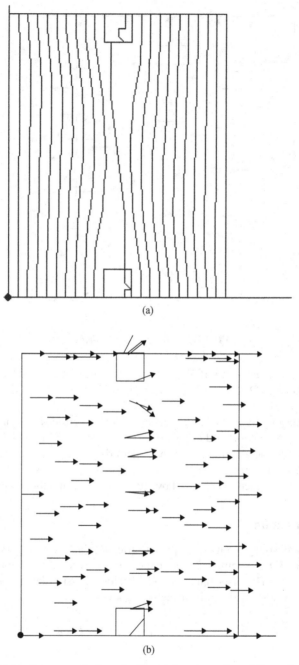

(a)

(b)

FIGURE E10.4.2 Computer display of wide semiconductor bar of Example 10.4 with no applied field but with 1×1-mm aluminum electrodes. (a) computed voltage contours, (b) computed current density arrows.

The 3D finite elements in the examples below are all hexahedrons, which are six-sided solids. In the examples below the hexahedrons are all cubes, but they could be any shape not distorted greatly from a rectangular box. Other 3D finite elements include the tetrahedrons discussed in Chapter 4.

Example 10.5   Hall Voltage Computed for Wide 3D Semiconductor Bar with Finite Hall Electrodes   The wide bar of semiconductor material of Example 10.1 has the conductivity tensor determined in that example. The Hall electrodes are thin highly conducting plates of area 2×1 mm as shown in Figure 10.1, which shows a bar thickness of 2 mm. Use the unsymmetric conductivity tensor in 3D finite-element software to compute the actual Hall voltage contours and electric field in the wide bar along with the resistance of the bar.

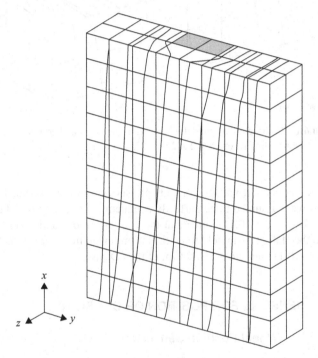

FIGURE 10.1   Finite-element model with 1–mm^3 elements and computed voltage contours for Hall sensor of dimensions $8 \times 10 \times 2$ mm.

Solution   The computed voltage contours are shown in Figure 10.1, in which the electrodes were assumed to be perfectly conducting. Other computed results are listed in row 3 of Table 10.1. Notice that the Hall electrodes and other 3D effects have reduced the Hall electric field to -10.85 V/m, even further from the predicted -23.8 V/m of (10.1).

Example 10.6 Hall Voltage Computed for Narrower 3D Semiconductor Bar with Finite Hall Electrodes   The narrower bar of semiconductor material of Example 10.2 is made 2 mm thick in the third (z) dimension. It has thin highly conducting plates of area 2×1 mm for Hall electrodes, as shown in Figure E10.6.1. Use the unsymmetric conductivity tensor in 3D finite-element software to compute the actual Hall voltage contours and electric field in the narrower bar along with the resistance of the bar.

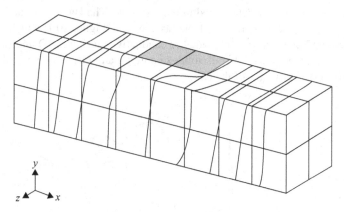

FIGURE E10.6.1   Finite-element model with 1–mm³ elements and computed voltage contours for Hall sensor of dimensions $8 \times 2 \times 2$ mm.

Solution   The computed voltage contours are shown in Figure E10.6.1, in which the electrodes were assumed to be perfectly conducting. Other computed results are listed in row 4 of Table 10.1. Notice that the Hall electrodes and other 3D effects have reduced the Hall electric field to -15.11 V/m, much further from the predicted -23.8 V/m of (10.1) than the -23.65 V/m of the 2D case.

10.4 HALL SENSORS FOR POSITION OR CURRENT

10.4.1 Toothed Wheel Position Sensor

The Hall sensor element can be used in a magnetic circuit to sense position. The most popular magnetic circuit has the toothed wheel, which may be a gear, shown in Figure 10.3. The sensor of Figure 10.3 is commonly used in the automotive industry for wheel position and speed sensing. Four wheel sensing is essential for the following automotive safety systems.

- Anti-lock braking, in which zero or low speed of one or two wheels indicates sliding on ice etc. and therefore triggers rapid pumping actuation of the brakes.
- Traction control, in which high speed of one or two wheels indicates spinning on ice etc. and therefore triggers changes in engine torque applied to the wheels.

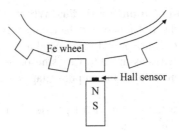

FIGURE 10.3 Toothed wheel position sensor with Hall effect magnetic field detector mounted on a permanent magnet with N and S poles.

- Brake force distribution, in which higher speed of certain wheels carrying a heavier cargo load indicates that they need greater braking force.
- Stability control, in which zero or low speed of one or two wheels during turning indicates a side skid and therefore triggers changes in engine torque applied to the wheels.

In most of the above systems, feedback control as shown in Figure 1.4 is used with both the magnetic sensor and an actuator. The actuator may be purely magnetic or may also involve an electrohydraulic system to be discussed in Chapter 16.

In Figure 10.3 the magnetic field is supplied by a permanent magnet stator. In some cases, a coil winding on a steel stator can be used instead, as analyzed in Example 3.2. As described in Example 3.2, the reluctance changes as the steel wheel teeth rotate past the stator. This variable reluctance causes the magnetic flux and flux density to change. For simplicity, assume the Hall sensor is subjected to a sinusoidal B change with wheel angular position θ of the type:

$$B(\theta) = B_{pk} \sin n_T \theta \qquad (10.22)$$

where n_T is the number of teeth on the wheel.

Assume also that the above flux density is in the z direction of Figure 10.2 and (10.1), which applies along with a current density J_x and Hall sensor dimension d_y. Then (10.22) and (10.1) give the sensor output voltage:

$$V_y = k_H J_x d_y B_{pk} \sin n_T \theta \qquad (10.23)$$

Because the above four coefficients are all constants, the sensor output voltage is directly proportional to $\sin(n_T \theta)$. Thus the position θ is sensed over a tooth "pitch", and speed is also easily sensed by taking the time derivative of the output voltage.

In actuality, the simple sinusoid assumed in (10.22) is only an approximation. Finite-element analysis can determine the exact variation with position for any tooth shape. To sense absolute position, a particular tooth can be removed or the tooth pattern can be made nonuniform.

Example 10.7   Hall Sensor Output for Toothed Wheel of Example 3.2   A small Hall sensor is inserted in the airgap between the teeth of Example 3.2. Assume that the Hall sensor is supplied a current and is arranged as in Figure 10.2 to sense the magnetic flux density, which is assumed to vary sinusoidally between the values of Example 3.2. Find the expression for the output voltage as a function of position.

Solution   From Example 3.2, the flux density varies from 0.06285 T to 0.6285 T. Assuming a sinusoidal variation gives:

$$B(\theta) = 0.06285 + (0.6285 - 0.06285)\sin n_T\theta \qquad \text{(E10.7.1)}$$

Substituting in (10.1) gives the output voltage:

$$V_y = k_H J_x d_y (0.06285 + 0.5656\sin n_T\theta) \qquad \text{(E10.7.2)}$$

where the three coefficients are all constants.

10.4.2   Position Sensors using Multipole Magnets

Instead of a toothed wheel, position sensors can be made using Hall sensor elements placed near moving multipole permanent magnets [4]. Sensors have been made for position range as great as 5 m. The magnets can be made of flexible thin material that is glued to the moving surface. In this way one makes an *encoder*, a sensor that determines position. Figure 10.4 shows typical encoders using multipole permanent magnets.

Besides encoders for linear position sensing, even more popular are encoders for sensing rotary position, that is, shaft angle. Multiple permanent magnets are mounted on the shaft, and then a nearby Hall element or elements are used to make a rotary encoder. This encoder can then be used to control commutation of rotary motors such as brushless DC motors [5]. Since such motors usually have a permanent magnet rotor, sometimes no additional permanent magnets are needed.

10.4.3   Hall Effect Current Sensors

Since Ampere's law states that magnetic intensity is proportional to current, Hall magnetic field sensors can be used to sense current. With the aim of reducing currents

FIGURE 10.4   Multipole permanent magnets for linear position and rotary position encoders, with rectangular Hall sensors vertically separated by typically less than 15 mm [4].

and associated power loss, it is increasingly desirable to use current sensors as meters throughout circuits used in industrial plants.

An advantage of Hall sensors for current metering is that isolation from voltages is inherent thanks to their magnetic coupling [6]. Each Hall sensor should be aligned with the expected direction of the magnetic field, which is circumferential for a current-carrying wire as shown in Figure 2.4. If two nearby wires carrying opposite but equal currents replace the one wire in Figure 2.4, then the Hall sensor can be used to make a ground-fault circuit interrupter; the field sensed becomes large only if the current in one wire flows elsewhere through a human or an animal. Instead of Hall sensors, other magnetic field sensors including magnetoresistive sensors may be used to meter current [7], including Chattock coils as will be presented in the next chapter.

10.5   MAGNETORESISTANCE

Both Hall effect and related magnetoresistance are extremely popular methods of sensing magnetic fields. In the past few years devices using magnetoresistance, commonly abbreviated MR, have become ever more popular as new forms of MR are developed. The basic definition of magnetoresistance is the change in resistance due to application of a magnetic field.

10.5.1   Classical Magnetoresistance

Sections 10.1–10.3 have shown that magnetic fields acting on semiconductors produce both Hall effect and magnetoresistance. This type of magnetoresistance is shown in Table 10.1 to be in the range of a 1–10% change. This type of "classical" magnetoresistance is commonly called AMR, for anisotropic magnetoresistance, since its conductivity in (10.4) is anisotropic.

Since the same semiconducting material produces both Hall effect and magnetoresistance, either can be used to make a sensor. Since the magnetoresistance changes of Table 10.1 classical AMR are only a few percent, many engineers prefer to use the Hall voltage, even though it requires addition of the two Hall electrodes.

10.5.2   Giant Magnetoresistance

In 1988 a much larger percentage change in magnetoresistance was discovered and appropriately named *giant magnetoresistance (GMR)*. Throughout the 1990s, GMR was developed extensively, especially for sensing magnetic fields recorded on computer disk drives, magnetic stripes on credit cards, and so on [8, 9]. In 2007 the discoverers of GMR, Albert Fert and Peter Grünberg, shared the Nobel Prize in physics.

GMR is very different from classical MR. It requires a multilayer structure of very thin films. It attains change in resistance as high as 65% due to the change in magnetization of the successive magnetic layers. For example, layers of Fe can be coupled antiferromagnetically (with opposite electron spins [3]) through nonmagnetic

Cr layers. When the magnetic field to be sensed is applied, the antiferromagnetic coupling can be overcome, making the magnetic layers parallel. The resistance varies as the cosine of the angle between the magnetizations. Thus the resistance in the antiparallel (AP) case is much greater than the resistance in the parallel case.

There are several different designs for GMR sensors. One sensor has two ferro-magnetic (FM) layers sandwiching the GMR material. The two layers are often called *soft adjacent layers* (SALs). Often one of the layers has fixed magnetization, while the other is varied by the field to be sensed. Maximum current (minimum resistance) occurs when the two layers have parallel magnetism.

GMR is now very commonplace for read heads for hard disk computer drives. A later development to increase storage density is called CPP-GMR [10, 11]. CPP stands for current perpendicular to plane, in contrast with the present CIP-GMR, where CIP stands for current in plane. The last section of this chapter will discuss a recent type of GMR sensor called a spin valve sensor.

10.5.3 Newer Forms of Magnetoresistance

There are additional forms of magnetoresistance. All are being investigated to possi-bly replace GMR in the near future.

- *Tunneling Magnetoresistance (TMR)*. Again placed between two SALs, the material used here is a very thin insulator carrying current by quantum tunneling of electrons when subjected to a magnetic field. This tunneling effect was first observed in 1975 but is presently being developed to possibly replace GMR for disk drive read heads. A TMR of 8% was achieved in 2011 using a magnetic tunnel junction sensor [12] for magnetic nanoparticles of the type discussed in Example 7.7. Quantum mechanical effects such as tunneling are mostly beyond the scope of this book.

- *Colosal Magnetoresistance (CMR)*. Insulating manganese oxide crystals are changed from nonmagnetic insulators to FM conductors by the magnetic field to be sensed. However, such a change requires very low temperatures and a field greater than 1 T. Thus CMR is not very promising at the present time.

- *Extraordinary Magnetoresistance (EMR)*. This more recent form of MR is based upon classical MR, but adds conducting particles to the semiconducting material. As shown in Example 10.4, adding highly conducting materials alters the electric field and current flow patterns. In EMR, many conducting particles are placed inside the semiconductor, thus causing concentrations in the current density. The percentage change in resistance can exceed 35% [13–15].

10.6 MAGNETORESISTIVE HEADS FOR HARD DISK DRIVES

As explained above, MR material is commonly placed near an SAL. In the head of Figure 10.5, an MR layer called a *stripe* is separated from an SAL by a thin nonmagnetic spacer. The applied current in the MR stripe is in the direction out of the

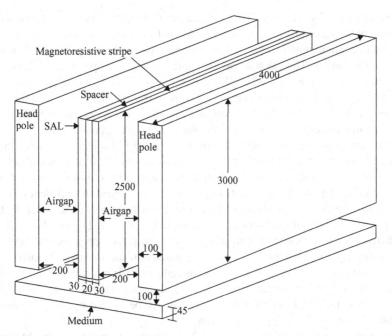

FIGURE 10.5   Geometry of typical magnetoresistive head. Dimensions in nanometers are typical of the 1990s and have since been reduced. The applied current flows out of the page.

page. The nanometer dimensions of Figure 10.5 are typical of MR heads of the mid 1990s [16]. While small, nowadays MR heads are even smaller, making them true examples of nanotechnology and microelectromechanical systems (MEMS) [17].

The head shown in Figure 10.5 is used to sense, that is, to read the magnetic bits previously recorded on the disk medium. The highly permeable head poles direct some of the flux from the disk to the MR stripe. The magnetized medium typically has a B–H curve with a slope of two or three times the permeability of air and a residual flux density B_r of approximately 0.5 T.

Assuming the MR stripe exhibits classical MR, the anisotropic conductivity of (10.4)–(10.9) has been used in 3D finite-element software [16]. The computed MR due to the magnetized media was 2.1%, and the computed voltage contours have the slant of the type shown in Figure E10.6.1.

10.7   GIANT MAGNETORESISTIVE SPIN VALVE SENSORS

GMR introduced in Section 10.5 has led to the development of *spin valve sensors*. They contain two FM layers separated by a normal metal (NM) layer. The magnetization of one FM layer is fixed, while the magnetization of the other FM layer can rotate in response to the external magnetic field that is being sensed.

As mentioned in Chapter 2, magnetization is caused by electron motion. There are two kinds of electron motion [18]. *Orbital* electron motion (along orbits such as planets orbiting the sun) can be likened to current flowing in a loop, thereby creating magnetic fields according to Ampere's law. *Spin* of electrons was postulated in 1925 to explain certain magnetic observations and later was explained theoretically using quantum mechanics. The electron can be approximated as a sphere with charge on its surface; if this sphere (or other 3D shape) rotates or spins, the resulting current creates a magnetic field. The direction of this spin magnetic field reverses if the spin reverses. According to the Pauli exclusion principle, each energy level in an atom contains a maximum of two electrons, and they must have opposite spins.

Spin valves operate using the two spin states of the FM material, which is here assumed to be cobalt (Co) [19] but could also be a compound of iron and cobalt. In the two-current model, the two spin directions along a fixed quantization axis are up (denoted by $\uparrow$) and down (denoted by $\downarrow$) and are assumed independent of each other. In Co, the two spin states are shifted in energy with density of states (DOS, the energy per unit range of energy) of $\downarrow$ electrons larger than the DOS of $\uparrow$ electrons. The differences in DOS at the Fermi energy level (a typical electron energy) between the two spin types contributes to a spin-asymmetry in the transport properties of the FM. The smaller DOS of the $\uparrow$ electrons reduces their relaxation rate and thus they have a much larger conductivity than the $\downarrow$ electrons. Thus if a small voltage is applied across the FM layer most of the current will be carried by the $\uparrow$ electrons; the current is said to have spin polarization.

The in-plane resistance of a system of NM layers sandwiched between FM layers is examined as follows. Assume the magnetization in consecutive FM layers is parallel (P) and AP (antiparallel) to the spin quantization axis. Assuming no spin-flip scattering, the two spin types are independent parallel conducting channels, and their conductances (reciprocals of resistances) simply add as in a simple DC electric circuit. Denoting the resistances as $R_\uparrow$ and $R_\downarrow$, the total parallel resistance (for the P layer) from circuit theory is the product over the sum:

$$R_P = R_\uparrow R_\downarrow / (R_\uparrow + R_\downarrow) \qquad (10.24)$$

In the AP case, the two spin layers each have the same resistance $\frac{1}{2}(R_\uparrow + R_\downarrow)$, giving the total AP resistance:

$$R_{AP} = (R_\uparrow + R_\downarrow)/4 \qquad (10.25)$$

The GMR ratio factor is thus:

$$F_{GMR} = (R_{AP} - R_P)/R_P \qquad (10.26)$$

Substitution gives:

$$F_{GMR} = (1/4)(R_\uparrow + R_\downarrow)^2 / R_\uparrow R_\downarrow \qquad (10.27)$$

Note that the larger the difference between the two spin resistances, the greater is F_{GMR}. Improving the conductance of the FM layer, which is dominated by up-spin

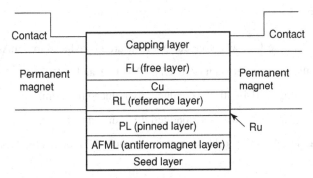

FIGURE 10.6   Schematic of cross section of a CIP spin-valve GMR sensor. The ruthenium (Ru) and copper (Cu) spacer layers are in between the PL and RL, and RL and FL, respectively. Electrical contacts are made to the biasing permanent magnets. Not shown are electrically isolated conducting shields (such as will be discussed in Chapter 13) that are above and below the sensor [19].

electrons, can be done by reducing the number of impurities or increasing the material grain size, thereby reducing $R_\uparrow$ more than $R_\downarrow$, thereby increasing F_{GMR}.

Figure 10.6 shows a way to use these two spin resistances to make a CIP (current-in-plane) spin valve GMR sensor. The purpose of the seed layer is to set the correct crystal texture and to optimize grain size. The antiferromagnet layer (AFML) helps keep the magnetization of the pinned layer (PL) in a fixed direction. The thin ruthenium (Ru) layer promotes a strong interaction between the PL and the reference layer (RL), keeping them AP. A typical thickness of the Ru layer is 0.9 nm. The thickness of the PL and RL layers is adjusted so that they have nearly equal magnetic moment (total magnetization). This makes the net moment of the PL–Ru–RL system nearly zero, forming what is called a synthetic antiferromagnet. Thus the PL and RL do not respond to the external field to be sensed, nor do they disturb the free layer (FL). A Cu spacer separates the RL and FL; on top of which is a capping layer typically made of TaN.

The GMR signal in Figure 10.6 is generated in the RL–Cu–FL layers. The FL magnetization direction rotates in response to the field to be sensed, changing the resistance from its minimum R_{min} for parallel FL and RL magnetizations, to its maximum for AP magnetizations. The resistance as a function of the angle θ between the FL and the RL magnetization directions is [19]:

$$R = R_{\text{min}}[1 + \tfrac{1}{2}F_{\text{GMR}}(1 - \cos\theta)] \qquad (10.28)$$

The FL is biased by permanent magnets in Figure 10.6 so that its magnetization is perpendicular to the RL magnetization in the absence of the external field to be sensed.

Spin valve magnetoresistive read heads such as shown in Figure 10.6 are now very popular in magnetic disk drives. The next chapter includes a discussion of other types of recording heads. It contrasts and compares them with magnetoresistive heads.

PROBLEMS

10.1 Redo Example 10.1 when the current measured is reduced to 2 mA.

10.2 Show all the steps between (P10.1.7), (P10.1.8), and (P10.1.9).

10.3 Redo Example 10.4 with the conductivity reduced to 4 S/m.

10.4 Redo Example 10.4 with the conductivity reduced to 1 S/m.

10.5 Redo Example 10.7 for the current of Example 3.2 reduced by 50%.

REFERENCES

1. Greiner RA. *Semiconductor Devices and Applications*, McGraw-Hill Book Co.; 1961, pp. 80–81.
2. Brauer JR, Ruehl JJ, MacNeal BE, Hirtenfelder F. Finite element analysis of Hall effect and magnetoresistance. *IEEE Trans Electron Devices* 1995;42:328–333.
3. Kittel Charles. *Introduction to Solid State Physics*, 3rd ed. New York: John Wiley & Sons; 1967, pp. 240–244.
4. Gauthier S. Tracking motion with multipole-magnet Hall sensing. *Motion System Design* 2012;54:34–37.
5. Simpkins A, Todorov E. Position estimation and control of compact BLDC motors based on analog Hall effect sensors. *Proceedings of the American Control Conference*, 2010, pp. 1948–1955.
6. Klaiber B, Turpin P. Sensing currents for maximum efficiency. *EE Times Magazine*, 2012, pp. 26–30.
7. Schneider PE, Horio M, Lorenz RD. Integrating GMR field detectors for high-bandwidth current sensing in power electronic modules. *IEEE Trans Ind Appl* 2012;48:1432–1439.
8. White RL. Giant magnetoresistance: a primer. *IEEE Trans Magn* 1992;28:2482–2487.
9. White RL. Giant magnetoresistance materials and their potential as read head sensors. *IEEE Trans Magn* 1994;30:346–352.
10. Takagishi M, Koi K, Yoshikawa M, Funayama T, Iwasaki H, Sahashi M. The applicability of CPP-GMR heads for magnetic recording. *IEEE Trans Magn* 2002;38:2277–2282.
11. Fukuzawa H, Yuasa H, Hashimoto S, Koi K, Iwasaki H, Takagishi M, Tanaka Y, Sahashi M. MR ratio enhancement by NOL current-confined-path structures in CPP spin valves. *IEEE Trans Magn* 2004;40:2236–2238.
12. Lei ZQ, Leung CW, Li L, Li GJ, Feng G, Castillo A, Chen PJ, Lai PT, Pong, PWT. Detection of iron-oxide magnetic nanoparticles using magnetic tunnel junction sensors with conetic alloy. *IEEE Trans Magn* 2011;47:2577–2580.
13. Moussa J, Ram-Mohan LR, Sullivan J, Zhou T, Hines DR, Solin SA. Finite element modeling of enhanced magnetoresistance in thin film semiconductors with metallic inclusions. *Phys Rev* 2001;B64:184410.
14. Moussa J, Ram-Mohan LR, Rowe ACH, Solin SA. Response of an extraordinary magnetoresistance read-head to a magnetic bit. *J Appl Phys* 2003;94:1110.
15. Solin SA. Magnetic field nanosensors. *Scientific American* 2004;291:71–77.

16. Brauer JR, Jenich MM. *Finite Element Analysis of a Magnetoresistive Head*, IEEE Magnetic Recording Conference, Minneapolis, MN, September 13–15, 1993.

17. Williams EM. *Design and Analysis of Recording Magnetoresistive Heads*, New York: Wiley-IEEE Press; 2000.

18. Cullity BD, Graham CD. *Introduction to Magnetic Materials*, 2nd ed. Hoboken, NJ: Wiley IEEE Press; 2009.

19. Heinonen OG, Singleton EW, Karr BW, Gao Z, Cho HS, Chen Y. Review of the physics of magnetoresistive readers, *IEEE Trans Magn* 2008;44:2465–2471.

CHAPTER 11

Other Magnetic Sensors

Besides the Hall effect and magnetoresistive sensors of the preceding chapter, many other types of magnetic sensors are commonly used. This chapter examines the major remaining types of magnetic sensors.

11.1 SPEED SENSORS BASED ON FARADAY'S LAW

The preceding chapter discussed how a toothed steel wheel rotating next to a Hall sensor is used in many automotive safety systems. In many cases, the speed or velocity is to be sensed, not the position, and thus a pickup coil can be used instead of the Hall sensor, as shown in Figure 11.1. Note that, as in Figure 10.3, a permanent magnet serves as the usual source of the magnetic field. The Hall sensor of Figure 10.3 is replaced in Figure 11.1 by a coil wound around a steel pole piece. A photograph of an actual automotive variable reluctance velocity sensor is shown in Figure 11.2. The cutaway in Figure 11.2 shows the coil, which is wound of hundreds of turns of very thin copper wire.

The voltage induced in the coil is determined by Faraday's law of Equation (2.33):

$$V = -N \frac{d\varphi}{dt} \qquad (11.1)$$

where N is the number of turns and φ is the flux passing through the coil. As the toothed steel wheel in Figure 11.1 rotates with angular position θ, the reluctance changes as analyzed in Example 3.2, producing a changing flux through the coil. Assume for simplicity that the flux φ is sinusoidal:

$$\varphi(\theta) = \varphi_{pk} \sin n_T \theta \qquad (11.2)$$

where n_T is the number of teeth on the wheel. Assume that the toothed wheel is rotating at constant mechanical angular velocity Ω in radians per second. Thus:

$$\theta = \Omega t \qquad (11.3)$$

Magnetic Actuators and Sensors, Second Edition. John R. Brauer.
© 2014 The Institute of Electrical and Electronics Engineers, Inc. Published 2014 by John Wiley & Sons, Inc.

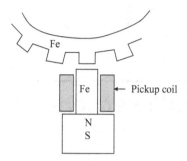

FIGURE 11.1   Toothed wheel speed sensors.

Substituting (11.2) and (11.3) into (11.1) gives:

$$V = -\Omega N n_T \varphi_{\text{pk}} \cos(n_T \Omega t) \qquad (11.4)$$

Notice that the magnitude of the induced voltage is proportional to the speed. At very low speed, the voltage is very small. Thus the sensors of Figures 11.1 and 11.2 are indeed speed sensors. While it is true that their voltages can be integrated over time to obtain position, they are inherently speed sensors, not position sensors. In most cases, the coil is open-circuited so that the voltage observed is that of (11.4).

Example 11.1   Speed Sensor Output for Toothed Wheel of Example 3.2   Example 3.2 is to be used as a speed sensor by winding an open-circuited pickup coil of N turns on top of the exciting coil in Figure E3.2.1. Assume that the magnetic flux varies sinusoidally between the values of Example 3.2 and that its armature has n_T teeth equally spaced around 360°. Find the expression for the output voltage as a function of the armature speed Ω.

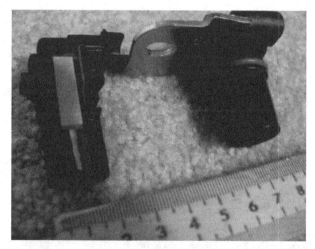

FIGURE 11.2   Actual automotive speed sensor stators, consisting of a coil wound around steel adjacent to a permanent magnet. A cutaway stator is shown to the left of a complete stator.

Solution From Example 3.2, the flux varies from 62.85E−6 to 628.5E−6 Wb. Assuming a sinusoidal variation gives:

$$\varphi(\theta) = 62.85\text{E–}6 + (628.5 - 62.85)\text{E–}6 \sin n_T \theta \qquad (\text{E}11.1.1)$$

Substituting in (11.2) and (11.4) gives the output voltage:

$$V = -\Omega N n_T (566\text{E–}6) \cos(n_T \Omega t) \qquad (\text{E}11.1.2)$$

11.2 INDUCTIVE RECORDING HEADS

The preceding chapter examined magnetoresistive heads for reading (sensing) magnetized media such as hard disks. Besides magnetoresistive heads, inductive heads are commonly used. Inductive heads have a voltage induced by Faraday's law.

Inductive heads are the most common means of writing on, that is, magnetizing, the disk medium. A typical inductive head is shown in Figure 11.3. Its coil is wound with many turns of very fine wire, and therefore, for a given current, the coil can create a much higher magnetizing field intensity than a magnetoresistive head with its single current-carrying conductor. The ampere-turns in the head coil create magnetic flux lines in the disk medium. Inductive heads can also be used for reading, that is, sensing the magnetic field pattern written on the disk. However, their reading performance is usually inferior to that of magnetoresistive heads.

The inductive head read signal is based on Faraday's law, as is the signal produced by a moving-magnet microphone. For simplicity, assume that a sinusoidal pattern is magnetized on the recording disk and produces the flux of (11.2) linking the head coil:

$$\varphi(\theta) = \varphi_{\text{pk}} \sin n_C \theta \qquad (11.5)$$

where the number of teeth n_T of (11.1) is changed to n_C, the number of cycles of magnetization along the disk track circumference. Assuming the disk is traveling at

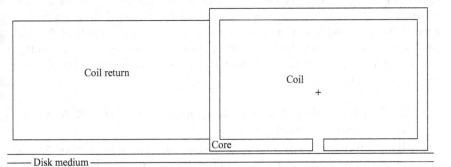

FIGURE 11.3 A typical inductive head and adjacent disk medium.

angular speed Ω in radians per second, then, from Faraday's law, the signal volts "read" at the head coil is similar to (11.3), becoming:

$$V = -\Omega n_C N \varphi_{\text{pk}} \cos(n_C \Omega t) \qquad (11.6)$$

Since φ_{pk} is produced by one magnetized bit (half cycle) on the disk, it equals B_{pk} times the area of a half cycle. Thus:

$$\varphi_{\text{pk}} = B_{\text{pk}}(2\pi r_{\text{disk}} w_{\text{track}})/(2n_c) \qquad (11.7)$$

where r_{disk} is the disk radius, and w_{track} is the disk magnetization track width. The flux density B_{pk} is limited by saturation, etc. Substituting (11.7) in (11.6) gives:

$$V = -\Omega r_{\text{disk}} \pi w_{\text{track}} N(B_{\text{pk}}) \cos(n_C \Omega t) \qquad (11.8)$$

Notice that the magnitude of the voltage sensed is proportional to the product of the angular speed Ω times the disk radius. When the radius is reduced by disk drive miniaturization, the read voltage of (11.8) is reduced. In contrast, the voltage of a magnetoresistive read head is proportional only to the change in flux density, not speed. Thus magnetoresistive read heads have almost totally replaced inductive read heads in hard disk drives.

The magnetization pattern produced on the disk medium in actual operation depends on the history of all previous write operations. Predicting all such hysteresis effects is a very challenging task. Several books and a huge number of technical papers continue to be written about magnetic recording [1,2] and magnetic hysteresis [3,4]. Also, all permanent magnets used in actuators and sensors must undergo magnetization, and modeling their *magnetizing fixtures* requires similar techniques [5].

As mentioned in Chapter 1, disk drives often utilize the read signal of the head as a position sensor for the voice coil actuator in a feedback control servo system [1]. The voice coil actuator must *seek* the desired track in minimum time for rapid information transfer to or from the disk [6,7].

Example 11.2 Writing and Reading with an Inductive Head   The inductive head of Figure E11.2.1 is to be used for both writing and reading. It has 20 turns of wire carrying 0.05 A for writing. Assume that the disk medium has a relative permeability of 3 and has been previously unmagnetized. Dimensions of the head core, of relative permeability 2000, are 80 μm in x and 50 μm in y. The head core is 4 mm deep in the z direction. The medium is 3 mm thick in the radial direction and spaced by 1 mm from the head.

(a) Find the field in the disk medium during writing using Maxwell. Assume all eddy currents are negligibly small.

(b) If the coil current is reset to zero, and the section of the disk between the head poles (below the head airgap) is magnetized with $B_r = 0.5$ T in the direction parallel to the disk, find the flux pattern throughout the medium and head.

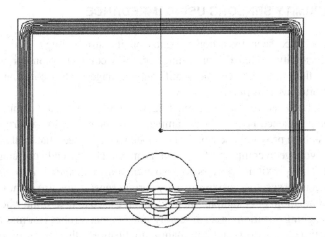

FIGURE E11.2.1   Computer display of an inductive head with computed flux lines during write operation on a previously unmagnetized disk.

Solution   Maxwell's magnetostatic solver was first used with no permanent magnet materials, and later with the bit region being permanently magnetized with $B_r = 0.5$ T in the direction parallel to the disk.

(a) Computed flux lines are shown in Figure E11.2.1. All flux densities are less than 2 T, with the highest densities in the head core.

(b) The computed flux lines are shown in Figure E11.2.2. All flux densities are less than 0.3 T, with the highest density in the bit. Most of the head core has flux density on the order of 0.15 T, less than 10% of its flux density during writing.

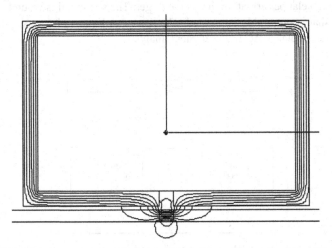

FIGURE E11.2.2   Computer display of an inductive head with computed flux lines produced by a single bit on the disk immediately below the head gap.

11.3 PROXIMITY SENSORS USING IMPEDANCE

The AC impedance seen by a coil depends on its surroundings. If the adjacent surrounding vicinity, called the proximity, contains conducting and/or permeable objects, then the impedance can be significantly changed. Thus coils carrying AC currents are often used as *proximity sensors.*

Since metal objects are always conducting, proximity sensors are commonly used to detect their presence nearby. For example, metal detectors for security systems commonly include proximity sensors. The impedance they see, that is, their ratio of complex AC voltage to complex AC current, is affected by the eddy currents induced in the metal. Thus proximity sensors are also commonly called *eddy current sensors.*

Besides security systems, proximity eddy current sensors are also commonly used in factory automation. For example, such sensors are used to determine when a metal part is approaching a station on an assembly line, so that a robot can find and work on the part. In both security and automation applications, the sensor output signal must undergo either simple or advanced signal processing to determine the next action.

The object being sensed is often called the *target.* Most eddy current proximity sensors detect the target using frequencies of 50 Hz through about 100 kHz. Both real and imaginary parts of the sensor impedance are usually monitored in order to detect the target and its location.

11.3.1 Stationary Eddy Current Sensors

The first step in understanding proximity sensors is to analyze such sensors when the target is not moving, or moving so slowly that motional eddy current effects are negligible. The effects of higher speed motion will be investigated later.

An example of a proximity sensor is shown in Figure 11.4. It is basically a solenoid actuator with a clapper armature that is the target. The sense coil is wound on a steel stator core that is preferably laminated. The sensor of Figure 11.4 was analyzed by two-dimensional planar AC finite-element analysis.

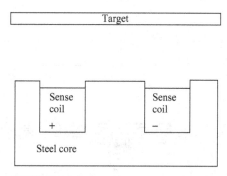

FIGURE 11.4 A typical proximity sensor.

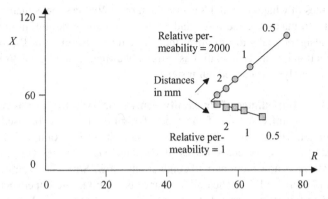

FIGURE 11.5 Impedance plane plots for proximity sensor of Figure 11.4 with various targets and positions. The target has conductivity 1.E7 S/m and a relative permeability of either 2000 or 1. The frequency is 1 kHz. The airgap distance between the stator poles and the target is 0.5 mm or greater.

The plot of complex-plane impedance for two target materials and multiple positions is shown in Figure 11.5 for frequency of 1 kHz. Ideally, the frequency should be selected to see the greatest impedance changes with variable target materials and positions. Note that the higher permeability target material gives higher inductive reactance X than the lower permeability material, as one would expect. Note also that the resistive R (lossy) component in Figure 11.5 rises as the target approaches, due to the conductive losses in the target. The target material and its distance can thus be determined from the complex-plane impedance.

Note in Figure 11.5 that as the target approaches, the impedance follows a path called a locus. The complex-plane impedance locus is thus very revealing of the target.

Complex-plane impedance loci are commonly used in *non-destructive testing (NDT) and non-destructive evaluation (NDE)*. For example, an AC current-carrying coil can be moved along a steel pipe. Any changes in the eddy current flow patterns of the pipe due to cracks, imperfect geometries, etc. will cause changes in the impedance locus. Many books and papers are devoted to electromagnetic sensors for NDT and NDE [8,9].

In NDE, the evaluation of the complex-plane impedance locus is a type of *inverse problem*. Signal processing software is employed to attempt to determine the type of any flaw and its severity. Examples include pipes in nuclear power plants and security checkpoints at airports. Certain NDE signals are caused by unacceptable metallic flaws that must be detected in order to register an alarm. Another example is eddy current sensors suspended down holes in the earth; their signals are determined by subterranean conductivities and can thus be processed to find petroleum.

Any sensor must, to some extent, perturb the object being sensed. Eddy current proximity sensors produce forces on their metallic targets. For example, the sensor of Figure 11.4 is also a magnetic solenoid actuator with the target as the armature. The armature forces of Chapters 7 and 8 act on the target. Whereas the actuator designer usually seeks to maximize the force and resultant armature motion, in most cases, the sensor designer seeks to minimize the target force and motion. Because eddy current prediction by hand calculation is extremely difficult, usually electromagnetic finite-element analysis is carried out instead.

Example 11.3   Impedance of Proximity Sensor for Two Target Positions   An eddy current proximity sensor of the type of Figure 11.4 is to be used with an aluminum target. The stator is identical to the steel stator of Example 7.1, but it is now laminated. The stator coil consists of 10 turns, each carrying 0.2-A rms. The armature material is now aluminum of conductivity 3.6E7 S/m. The target and sensor are shown in Figure E11.3.1. The coil resistance is 0.1 mΩ. The impedance seen by the coil is to be found for airgap distances of 2 mm and 4 mm using Maxwell.

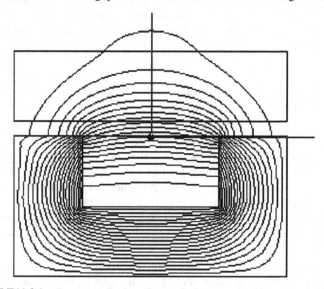

FIGURE E11.3.1   Computer display of a proximity sensor and its computed flux lines.

Solution   The easiest way to make this model is to begin with the model of Example 7.1. Changes required include making the solution type "Eddy Current" and changing the armature material. For the original 2-mm airgap, the flux linkage output for 1 turn is $\lambda = 2.375\text{E}{-}6 - j1.195\text{E}{-}7$. The series impedance for one turn is then found using (6.30):

$$Z = j\omega\lambda/I \qquad (\text{E}11.3.1)$$

where $I = 2$ A and $\omega = 377$. The resulting series impedance (for 10 turns) is:

$$Z_s = R_s + jX_s = (10)(22.53\text{E}{-}6 + j448.\text{E}{-}6) = 22.53\text{E}{-}5 + j448\text{E}{-}5$$

$$(\text{E}11.3.2)$$

The total impedance is found by adding the 0.1-mΩ coil resistance (as explained in Chapter 8), obtaining:

$$Z(0.002) = 32.53\text{E--}5 + j448\text{E--}5 \qquad (\text{E}11.3.3)$$

The corresponding flux lines are shown in Figure E11.3.1.

To obtain results at the 4-mm position, the armature is moved upward by 2 mm using the "Arrange" and "Move" commands. The flux linkage output for 1 turn is $\lambda = 2.66\text{E--}6 - j8.215\text{E--}7$, greater evidently because the aluminum is further away and repels less flux. The series impedance for one turn is then found using (6.30):

$$Z = j\omega\lambda/I \qquad (\text{E}11.3.4)$$

where $I = 2$ A and $\omega = 377$. The resulting series impedance (for 10 turns) is:

$$Z_s = R_s + jX_s = (10)(155\text{E--}6 + j501.4\text{E--}6) = 155\text{E--}5 + j501\text{E--}5 \qquad (\text{E}11.3.5)$$

The total impedance is found by adding the 0.1-mΩ coil resistance to obtain:

$$Z(0.004) = 165\text{E--}5 + j501\text{E--}5 \qquad (\text{E}11.3.6)$$

11.3.2   Moving Eddy Current Sensors

The effect of target velocity on the sensor signal is indicated by the *magnetic Reynolds number* [10]:

$$R_m = \mu\sigma Vl \qquad (11.9)$$

where V is the velocity and l is the length of the conductive armature in the direction of the velocity and motion. The magnetic Reynolds number is dimensionless. If it is less than one, then the velocity effects are negligible, and the methods of the preceding section apply.

If R_m exceeds one, then the velocity effects are significant. The larger R_m, the more the magnetic flux distribution is altered by motional induced voltage. Hence magnetic actuators and sensors with high magnetic Reynolds numbers must be analyzed including velocity effects. Usually magnetic actuators have zero armature speed at time zero, and rarely reach speeds with high Reynolds numbers. However, in the case of *magnetic brakes,* the speed is high and is reduced by eddy current braking forces [11, 12].

Magnetic sensors often sense objects with high velocities and high Reynolds numbers. An example is NDE sensors for steel heat exchanger tubes in nuclear

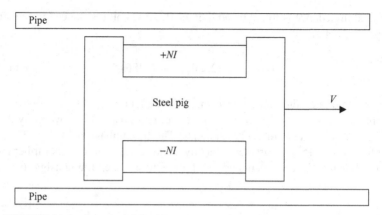

FIGURE 11.6 A typical AC eddy current NDT sensor traveling through a pipe.

power plants or for pipelines made of steel. These eddy current sensors are often pulled through pipes at fairly high velocities, and thus R_m often exceeds one. A typical traveling sensor, often called a pig, is shown in Figure 11.6. Note that if there is no defect in the pipe, then the geometry of Figure 11.6 does not change with time, even though there is a relative velocity between the pig and the pipe. A perfect or *smooth* pipe becomes an imperfect pipe if there is any roughness on its surface due to corrosion cracks, etc.

Finite-element methods have been developed for such problems with smooth conducting armatures traveling at constant high velocities [11,12]. In such formulations, the matrix equation becomes unsymmetric [12]. For more general moving armatures with imperfect non-smooth (changing) geometry with motion, finite-element methods will be discussed in Chapters 14 and 15.

11.4 LINEAR VARIABLE DIFFERENTIAL TRANSFORMERS

The variation of magnetic sensor signal with position is often nonlinear, whereas a linear correspondence between position and signal voltage is highly desirable. For example, the proximity sensor impedances of Figure 11.5 and of Example 11.3 are not linear with position.

A sensor with an extremely linear relation between position and voltage magnitude is the *LVDT, linear variable differential transformer.* A typical LVDT is shown in Figure 11.7 and is made of E-shaped steel, laminated to make the magnetic Reynolds number nearly equal to zero. The LVDT has a primary winding excited with AC voltage V_p. It also has two secondary windings with induced open-circuit voltages V_{s1} and V_{s2}.

The two secondary voltages are induced by Faraday's law transformer action, and are equal and opposite if the laminated steel armature is at zero position, that is, centered below the primary winding and its laminated steel pole. If the armature is positioned to the right (a positive position x), then the total induced voltage $V_{s2} - V_{s1}$

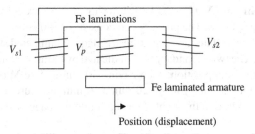

FIGURE 11.7 A typical LVDT.

is positive and proportional to x. The relation between the LVDT voltage output $V_s = V_{s2} - V_{s1}$ and x is shown in Figure 11.8.

Various LVDT designs are possible, and the frequency of operation may also vary. Typical frequencies range from 50 Hz to 25 kHz. Figure 11.8 shows that the output voltage increases with frequency, but can be extremely linear with position. Note also that the voltage is reduced as the load impedance is reduced.

LVDTs are unfortunately quite expensive. Their optimum frequency usually lies from 500 Hz to 5 kHz. To generate this frequency, electronics required includes the following.

- A 60 Hz to DC power supply
- An oscillator of the desired frequency
- A phase shifter
- AC and DC amplifiers

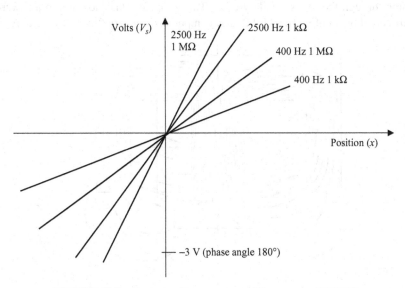

FIGURE 11.8 Typical voltage versus position curves of LVDT.

Example 11.4   Simple LVDT   An LVDT is shown in Figure E11.4.1 as drawn in Maxwell's drawing window. The primary coil consists of 1000 turns, each carrying 2-A rms 1000 Hz. Both the stator and armature are made of laminated steel with relative permeability 2000. The two secondary coils indicated by the arrows are open-circuited and connected in series opposition, each having 500 turns. Use Maxwell to find the total secondary output voltage for the 20-mm × 6-mm armature at centered (zero) position and moved 4 mm to the right. The depth into the page is 100 mm.

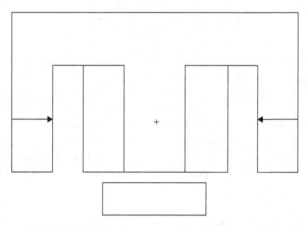

FIGURE E11.4.1   Geometry of LVDT to be modeled.

Solution   Since there are no significant eddy currents, a magnetostatic solution suffices. The model shown in Figure E11.4.1 was input into Maxwell to find the flux passing through the two secondary coils. The computed flux plot at zero armature position is shown in Figure E11.4.2. The computed secondary fluxes are $-2.93E-5$

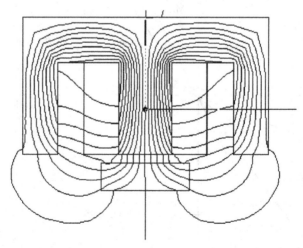

FIGURE E11.4.2   Computer display of flux lines computed in LVDT for armature position $= 0$.

and +2.94E−5 Wb, the magnitudes differing due solely to the mesh not being perfectly symmetric. The induced voltage is thus approximately zero.

Next, the "Arrange" and "Move" commands were used to move the armature 4 mm to the right. The computed secondary fluxes are −2.59E−5 and +3.58E−5 Wb for 1 m depth. For the actual 0.1 m depth, the total flux linking the secondary is thus +0.99E−6 Wb. To find the secondary voltage magnitude, use Faraday's law:

$$|V| = N d\phi/dt = N\omega\phi = 2\pi f N\phi \qquad (E11.4.1)$$

Substituting the total flux, frequency, and 500 turns, the LVDT output for this position is 3.11 V.

11.5 MAGNETOSTRICTIVE SENSORS

Magnetostriction is the phenomenon of elastic deformation that accompanies magnetization [13]. Magnetic fields cause certain materials to deform. This deformation is a form of motion in magnetostrictive and *magnetorheological* actuators. However, motions of such actuators are usually much less than 1 mm, and thus magnetostrictive actuators are not further discussed in this book.

Magnetostrictive sensors, on the other hand, are useful and fairly commonplace. They require, however, materials in which magnetic flux density and mechanical properties are strongly interdependent. While steel has small magnetostrictive effects, recently developed materials exhibit much stronger magnetostriction.

The material constitutive properties are determined by the following matrix equation for planar magnetostriction:

$$\begin{bmatrix} \sigma_{xx} \\ \sigma_{yy} \\ H_x \\ H_y \end{bmatrix} = \begin{pmatrix} K_1 & K_2 & X_1 & X_2 \\ K_3 & K_4 & X_3 & X_4 \\ X_5 & X_6 & M_1 & M_2 \\ X_7 & X_8 & M_3 & M_4 \end{pmatrix} \begin{bmatrix} \varepsilon_{xx} \\ \varepsilon_{yy} \\ B_x \\ B_y \end{bmatrix} \qquad (11.10)$$

Note that the left-hand column vector contains both planar structural stresses σ and magnetic field intensities. The right-hand column vector contains both planar structural strains ε and magnetic flux densities. The 4×4 matrix is the material tensor, containing purely structural entries K, purely magnetic entries M, and magneto-mechanical coupling cross-term entries X.

A strongly magnetostrictive material is Terfenol-D, which is made up of rare earth elements Tb (terbium), Dy (dysprosium), and also Fe. A typical Terfenol-D actuator is shown in Figure 11.9. Application of a magnetic field causes the Terfenol-D membrane to bend.

A commonly used magnetostrictive position sensor is shown in Figure 11.10. The magnetostrictive material acts as an acoustic waveguide. The linear position sensor is composed of four elements.

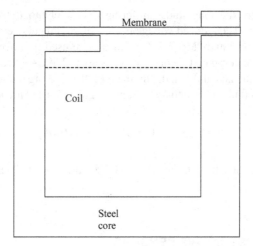

FIGURE 11.9   A magnetostrictive actuator with Terfenol-D membrane.

(1) Waveguide sensing element

(2) Sensing and output conditioning electronics module

(3) Protective housing

(4) External permanent magnet

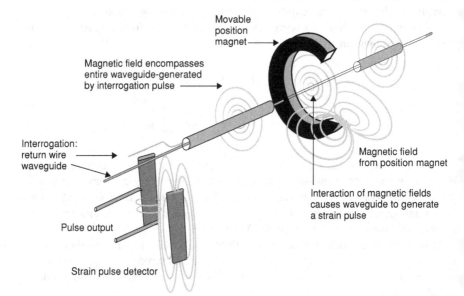

FIGURE 11.10   A magnetostrictive position sensor. This picture of a Temposonics® sensor is courtesy of MTS Sensors.

A sonic strain pulse is induced in the magnetostrictive waveguide by the momentary interaction of two magnetic fields. One field comes from a movable permanent magnet which passes along the outside of the sensor tube, while the other field comes from a current pulse or interrogation pulse applied along the waveguide. The interaction of the two magnetic fields produces a strain pulse, which travels at sonic speed along the waveguide until the pulse is detected at the head of the sensor. The position of the magnet is determined with high precision by measuring the elapsed time between the application of the interrogation pulse and the arrival of the resulting strain pulse. Hence accurate non-contact position sensing is achieved without any wear to the sensing components.

Another popular use of magnetostrictive sensors is for torque sensing. For control of rotating electric motors or internal combustion engines, accurate sensing of shaft torque is often required. Since the shaft is rotating, connecting it directly to electrical sensing wires is highly impractical. Instead, the strain in the shaft can be measured with a non-contact magnetostrictive sensor [14], and the strain is proportional to torque.

11.6 FLUXGATE SENSORS

Materials such as steel with nonlinear *B–H* curves can be used to make *fluxgate sensors*. The sensor material is driven between the constant-permeability and the saturable regions of its *B–H* curve. A typical fluxgate sensor is shown in Figure 11.11. The inductance seen by its coil is a function both of its coil current and the external magnetic field to be sensed.

Fluxgate sensors are used as proximity sensors, for navigational and geomagnetic field measurement instruments, compasses, and for position and speed sensing. A

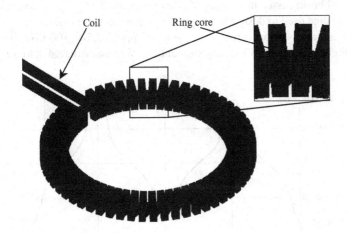

FIGURE 11.11 Computer display of a typical fluxgate sensor with a toroidal coil wound on a ring core.

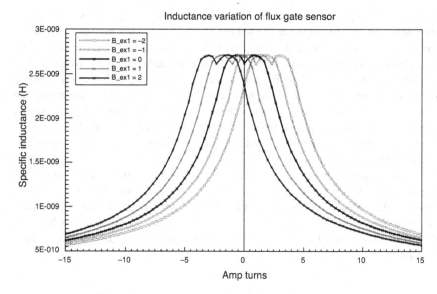

FIGURE 11.12 Typical fluxgate sensor inductance variation.

properly designed fluxgate sensor has an output linearly proportional to the magnetic field being sensed.

Finite-element analysis aids the design of fluxgate sensors [15]. The inductance seen by the coil for various external magnetic flux densities computed for a typical fluxgate sensor is plotted in Figure 11.12. Note how the curves shift with the external field. The curves can then be exported from the Maxwell finite-element software to a circuit analysis program such as SPICE or Simplorer®. SPICE, which stands for Simulation Program with Integrated Circuit Emphasis, is widely available in several forms, as will be discussed in Chapter 15. Simplorer, available from Ansys, Inc., is a software package used to design and analyze complex technical systems.

Typical saturable sensor current response to an applied 2.5-V, 100-kHz sinusoid is shown in Figure 11.13 for no applied external field. When an external field is applied,

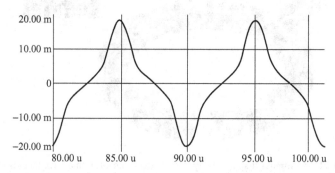

FIGURE 11.13 Typical fluxgate sensor current response to an applied 2.5-V 100-kHz sinusoid with no external magnetic field. Note that the positive and negative areas are equal.

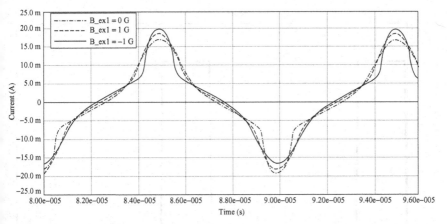

FIGURE 11.14   Typical fluxgate sensor current response to an applied 2.5-V 100-kHz sinusoid with applied external magnetic fields. Note that the positive and negative areas are unequal when an external field is applied.

the current response is shifted as shown in Figure 11.14. The sensor behaves similarly when excited with a square wave voltage, as shown by the current in Figure 11.15. Thus fluxgate sensors are often driven by a square wave voltage such as shown in Figure 11.16. Such a voltage can be created by two alternately fired MOSFETs as shown in Figure 11.17. The sensing circuit Simplorer model, including SPICE models of op-amp integrated circuits, is shown in Figure 11.18. The output voltage of Figure 11.18 is shown in Figure 11.19 for various fields to be sensed [15].

In many cases, there are two external fields to be sensed, and the differential flux density between the two is to be found. The final fluxgate sensor model for such a system is shown in Figure 11.20. The MOSFETs are fired by state machines. The

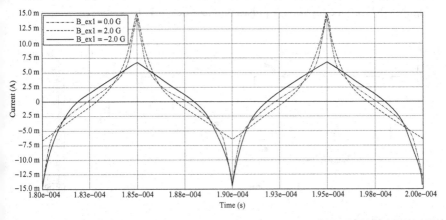

FIGURE 11.15   Typical fluxgate sensor current response to an applied 1.5-V 100-kHz square wave with applied external magnetic fields. Note that the positive and negative areas are unequal when an external field is applied.

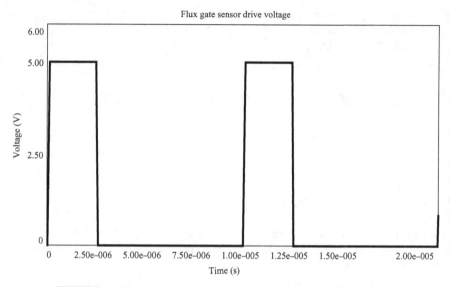

FIGURE 11.16 Typical square wave voltage applied to fluxgate sensor.

system model transient response to a change in the external field from 0 to −0.2 mT is shown in Figure 11.21. Thus the response time of the fluxgate sensor is shown to be in the millisecond range [15]. Fluxgate sensors are able to detect even smaller flux densities, down to the 0.1 nT range.

Somewhat similar to fluxgate sensors in use of nonlinear *B–H* materials is the *Wiegand wire* sensor patented in 1974 [16]. It has an outer coating or shell with high coercive force that surrounds a core of soft magnetic material. At the instant when the outer shell is fully magnetized and the core begins to carry flux, both the core and shell switch magnetization polarity. This flux switch generates a significant voltage. Thus Wiegand wire sensors are often used in place of the Hall sensors in the encoders of Figure 10.4, producing voltage pulses on the order of 1–10 V.

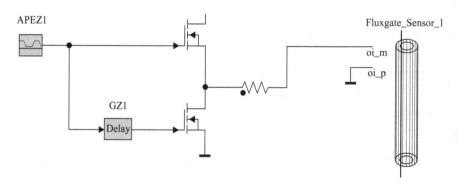

FIGURE 11.17 A typical fluxgate sensor drive circuit containing two MOSFETs, modeled in Simplorer.

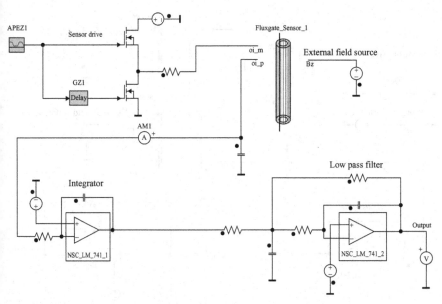

FIGURE 11.18   A typical Simplorer model of fluxgate sensor, including op-amp integrated circuit models originally developed for SPICE.

11.7   CHATTOCK COIL FIELD AND CURRENT SENSOR

A sensor for AC magnetic fields and currents near iron is the *Chattock coil* shown in Figure 11.22. Note that the Chattock coil has a small-diameter winding of long arc length *l*, which is often semicircular as shown. Because the Chattock coil has an air

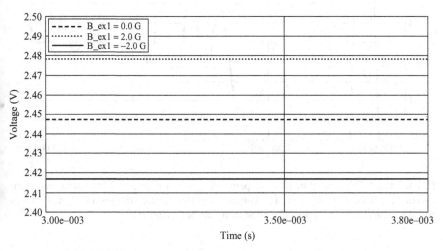

FIGURE 11.19   Computed output voltages of Figure 11.18 model.

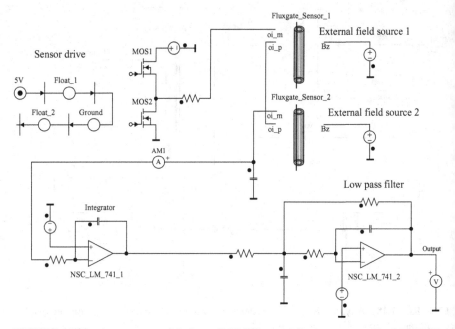

FIGURE 11.20 A systems model of two-field differential fluxgate sensor. The MOSFETs are fired by a state machine model, commonly used in control systems.

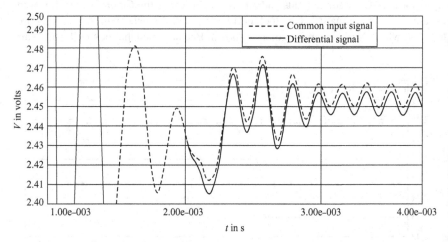

FIGURE 11.21 Computed transient response of sensor of Figure 11.20 to a step change in external field from 0 to −0.2 mT. The common and differential voltages are plotted on the vertical axis, and time in seconds is plotted along the horizontal axis.

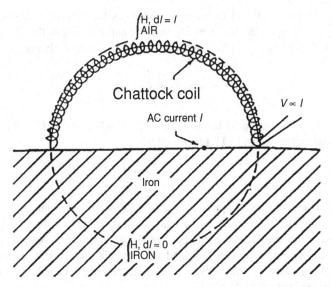

FIGURE 11.22   Chattock coil and its use to determine the line integral of $\mathbf{H} = \mathrm{MMF} = \mathrm{AC}$ current I enclosed.

core and thus is lightweight, it can be easily moved over iron surfaces to detect AC currents. Both impressed currents and induced eddy currents can be detected using their magnetic fields [17]. Very similar to the Chattock coil is the Rogowski coil, which usually has an arc closer to 360°; both are named after their inventors.

The AC voltage induced in the Chattock coil is derived by applying Faraday's law:

$$V = -N\frac{d\varphi}{dt} \qquad (11.11)$$

where N is the number of turns and the magnetic flux φ is sinusoidal with $\varphi = \varphi_{peak}$ $e^{j\omega t}$. The flux is $\varphi = B A$ where B is the flux density and A is the cross-sectional area of the coil. Since the Chattock coil has an air core and μ_0 is the permeability of free space, we obtain:

$$V = -jNA\omega\mu_0 H \qquad (11.12)$$

In Figure 11.22 with one coil at one location:

$$V = -jNA\omega\mu_0 H_{ave} \qquad (11.13)$$

which can be rewritten as:

$$V = -jA\omega\mu_o(NH_{ave}) \qquad (11.14)$$

For each turn of the Chattock coil, this becomes:

$$V = -jA\omega\mu_o(L_T N_L H_{ave}) \qquad (11.15)$$

where L_T is the Chattock coil turn length, and N_L is the number of turns per unit length. If the average of the flux intensity over the entire length of the Chattock coil of Figure 11.22 is examined, a final substitution can be made to obtain:

$$V = \mu_o \omega N_L A \int H \cdot dl \qquad (11.16)$$

Applying Ampere's circuital law to Figure 11.22 and assuming $H = 0$ in the iron (due to its high permeability), (11.14) becomes:

$$V = \mu_o \omega N_L A I \qquad (11.17)$$

Thus the Chattock coil output voltage is directly proportional to the current enclosed I, which in many cases is the eddy current induced in the iron [17]. For example, steel laminations in the stators of large generators may be damaged such that their interlaminar insulation fails and thus contains large eddy currents. The magnetic field produced by the eddy currents is commonly detected and located by scanning with a Chattock coil [17].

11.8 SQUID MAGNETOMETERS

Magnetometer is a term commonly used for very sensitive magnetic field sensors. Fluxgate sensors can be used in magnetometers.

SQUID stands for *s*uperconducting *qu*antum *i*nterference *d*evice, and is the most sensitive way of measuring magnetic fields at frequencies as low as 1 Hz (essentially DC). Unfortunately, the superconducting (zero resistance) coils require cooling to very low temperatures.

If magnetic flux passes through a ring made of a superconducting material as shown in Figure 11.23, a current is induced in the ring. Without any disturbances the

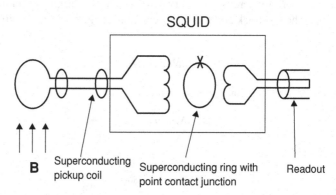

FIGURE 11.23 A schematic of SQUID. The current in the readout coil is proportional to the magnetic flux density being sensed.

current would continue flowing forever. The magnitude of the induced current is an extremely sensitive indicator of the flux density. It turns out that the flux through the ring is quantized in that the flux can only take on values that are integer multiples of a basic flux quantum. To measure the current in the superconducting ring, a Josephson junction is used. It is a thin layer of insulation or a narrowing of the superconductor to form a weak link. In 1962, Josephson predicted that a supercurrent could flow through this weak link that would be an oscillating function of the magnetic field intensity in the weak link. The periodic variations in this supercurrent produced by the quantized flux can accurately measure the magnetic field to be sensed. The periodic variations in the current have the same pattern as the interference fringes produced by the diffraction of light, and are related to the flux quantization. Usually, the ring is inductively coupled to a radio-frequency circuit that supplies a known bias field and yields the detector output current. Sensitivity is increased by coupling the ring with the weak link to a larger superconducting ring without a weak link, as shown in Figure 11.23 [18].

The sensitivity of SQUIDs is of the order of 10 fT or $1.E-14$ T, approximately the most sensitive level in Figure 10.1. The ability to set a null level by adjusting the bias field in the radio-frequency circuit makes the device particularly useful for differential field measurements. For example, if the null level is set to the average flux density of the earth, the SQUID will readily detect anomalies in the field. The sensing loop can be configured to be sensitive to a gradient in the measuring field by cross-connecting two parallel turns. Because the sense loop is superconducting, it has an essentially DC response to magnetic fields. By orienting the coils properly, the gradient of a component of the external field in any direction can be sensed, which is important in magnetic separators as discussed in Chapters 5 and 7. The ability to fabricate a high sensitivity gradiometer that can measure the partial derivatives of (5.26) is a unique advantage of this technology over other magnetic sensors [18].

Quantum mechanics is also involved in Magnetic Resonance Imaging (MRI), which has been analyzed elsewhere [19]. Sometimes, magnetic ferrite nanoparticles are used to provide contrast in MRI, and such particles are also being investigated for DNA testing [20].

11.9 MAGNETOIMPEDANCE AND MINIATURE SENSORS

Some of the newest developments in sensors are for *magnetoimpedance (MI)* sensors. MI sensors were developed in Japan on micromachined chips in 2001 and 2002 [21]. At that time, they were made of amorphous FeCoSiB wire that had been tension-annealed for surface anisotropy. Their impedance changes due to skin effects and is proportional to the square root of the externally applied magnetic field. More recently, other materials have been used and the response has been made linear with the applied field [22–24].

Various magnetic sensor types are compared in Table 11.1. Note that MI sensors make good magnetometers, for their resolution is down to the same nanotesla range as fluxgate sensors. The speeds in Table 11.1 can be considered to be bandwidths;

TABLE 11.1 Comparison of Sensor Types

Type	Length	Sensitivity	Speed	Wattage
Hall	100 μm	0.5 E−4 T	1 MHz	10 mW
MR	100 μm	0.1 E−4 T	1 MHz	10 mW
GMR	100 μm	0.01 E−4 T	1 MHz	10 mW
Fluxgate	20 mm	0.1 E−9 T	5 kHz	1 W
MI	2 mm	0.1 E−9 T	1 MHz	10 mW

all are 1 MHz except for fluxgate sensors. Not in the table is a new type of magnetic sensor called a *magnetic microwire* [25] which uses magnetostriction in a glass-coated permeable wire to remotely detect stresses in materials.

The lengths in the table are minimum dimensions; note that fluxgate sensors require the most space, followed by MI sensors. Fluxgate sensors also require more input power; the other types only require about 10 mW. Miniaturization to sizes well below 1 mm not only reduces size and weight, but also reduces power requirements. Reduced input power for battery-powered sensors enables reduced battery size and may even enable sensors to be powered by new *energy-harvesters*. These devices generate power from motion or vibration of mobile platforms such as automobiles.

The sensors in Table 11.1 with lengths of 100 μm can be used to make *motes*, which are remote wireless sensors. Of very recent development, motes are sometimes called *smart dust*. If inexpensive enough, hundreds or more could be sprinkled around a region to monitor its detailed magnetic field distribution. IEEE standard 802.15.4 has been enhanced to cover wireless mesh networks called "ZigBee." IEEE standard P1451.4 enables *smart sensors* to *plug and play* into computers, including note book and smaller computers. In automobiles, trucks, and construction equipment, sensors are commonly interfaced using CAN BUS standard technology.

11.10 MEMS SENSORS

Microelectromechanical systems (MEMS) include miniature magnetic sensors and actuators. With all dimensions much less than 1 mm, MEMS actuators have limited output force and are therefore of limited use. MEMS magnetic sensors, however, can output useful voltage or current and thus are increasingly being developed. MEMS sensors are usually made by special techniques used in semiconductor chip manufacture, including *micromachining*.

Many MEMS magnetic sensors use Lorentz force. For example, a miniature bar-shaped permanent magnet moves in a magnetic field without requiring input power. The motion can be detected optically or by other means to sense magnetic fields as small as 200 nT [18]. Such MEMS Lorentz force sensors are linear over a huge range to as high as 50 T [26].

Another MEMS sensor is called the xylophone resonator [18]. A MEMS beam is supplied an AC current with frequency equal to its mechanical resonant frequency. A

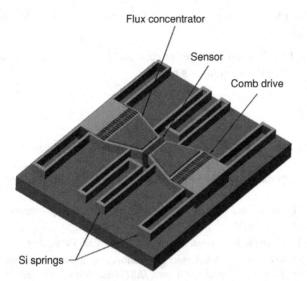

FIGURE 11.24 A MEMS flux concentrator [17]. There is a space between the substrate and the flux concentrators on the MEMS permeable flaps.

DC magnetic field applied perpendicular to the beam axis then vibrate the beam with amplitude proportional to the magnetic field. The vibration amplitude is detected optically.

A MEMS device that concentrates magnetic fields and thus improves the sensitivity of magnetic sensors is the flux concentrator shown in Figure 11.24. It is made of a silicon substrate and soft magnetic flux concentrator flaps. According to the reluctance concepts of Chapter 3, decreasing the gap between the flaps increases the flux density at the sensor. The flaps are forced to oscillate by the applied AC voltage to the two electrostatic comb drives. Typically the MEMS structure resonates mechanically at about 10 kHz, modulating the field at the sensor to increase the operating frequency and the signal-to-noise ratio. Thus the flux concentrators should increase the sensitivity by one to three orders of magnitude [18]. Chapter 13 will further discuss signal-to-noise ratio.

PROBLEMS

11.1 Redo Example 11.1 when the ampere-turns are reduced from 1000 to 500.

11.2 Redo Example 11.2 when the head core relative permeability is reduced to 200.

11.3 Redo Example 11.3 when the frequency is changed to: (a) 100 Hz, (b) 200 Hz, (c) 400 Hz.

11.4 Redo Example 11.3 when the target material is changed to steel with relative permeability 100 and conductivity 1.E6 S/m.

11.5 Find the LVDT output voltage for Example 11.4 when the frequency is increased to 5 kHz.

11.6 Find the LVDT output voltage for Example 11.4 when the armature position is: (a) –4 mm, (b) +2 mm, c) +6 mm.

REFERENCES

1. Jorgensen F. *The Complete Book of Magnetic Recording*, Blue Ridge Summit, PA: TAB Books Inc.; 1980.
2. Comstock RL. *Introduction to Magnetism and Magnetic Recording*, New York: John Wiley & Sons; 1999.
3. Mayergoyz ID. *Mathematical Models of Hysteresis and Their Applications*, 2nd ed. New York: Academic Press; 2003.
4. Della Torre E. *Magnetic Hysteresis*, New York: Wiley-IEEE Press; 2000.
5. VanderHeiden RH, Arkadan AA, Brauer JR. Nonlinear transient finite element modeling of a capacitor discharge magnetizing fixture. *IEEE Trans Magn* 1993; 29: 2051–2054.
6. Kobayashi M, Horowitz R. Track seek control for hard disk dual-stage servo systems. *IEEE Trans Magn* 2001; 37: 949–954.
7. Ashar KG. *Magnetic Disk Drive Technology: Heads, Media, Channel, Interfaces, and Integration*, New York: Wiley-IEEE Press; 1996.
8. Ida N. *Numerical Modeling for Electromagnetic Non-Destructive Evaluation*, London: Chapman & Hall; 1995.
9. Gotoh Y, Takahashi N. Proposal of detecting method of plural cracks and their depth by alternating flux leakage testing: 3D nonlinear eddy current analysis and experiment. *IEEE Trans Magn* 2004; 40: 655–658.
10. Woodson HH, Melcher JR. *Electromechanical Dynamics*, New York: John Wiley & Sons; 1968.
11. Rodger D, Leonard PJ, Karaguler T. An optimal formulation for 3D moving conductor eddy current problems with smooth rotors. *IEEE Trans Magn* 1990; 26: 2370–2372.
12. Ida N, Bastos JPA. *Electromagnetics and Calculation of Fields*, 2nd ed. New York: Springer-Verlag; 1997. pp 420–422.
13. Jay F (ed.). *IEEE Standard Dictionary of Electrical and Electronic Terms*, 2nd ed. New York: Wiley-Interscience; 1977.
14. Chen Y, Snyder JE, Schwichtenberg CR, Dennis KW, McCallum RW, Jiles DC. Metal bonded Co-ferrite composites for magnetostrictive torque sensor applications. *IEEE Trans Magn* 1999; 35: 3652–3654.
15. Steward D. *Fluxgate Sensor Analysis*, Pittsburgh, PA: Power Point presentation prepared for Ansoft Corporation; 2002.
16. Davis ML, Wiegand: the man, the effect, the wire. *Motion Systems Design* 2011; 24–28.
17. Rettler RE, Brauer JR, Ravenstahl M. Finite element eddy current computations in turbo-generator rotors and imperfect stators. *IEEE Int Electric Machines & Drives Conf* 2005; 1598–1605.
18. Lenz J, Edelstein AS. Magnetic sensors and their applications. *IEEE Sensors J* 2006; 6: 631–649.

19. Jin J. *Electromagnetic Analysis and Design in Magnetic Resonance Imaging*, Boca Raton, FL: CRC Press; 1998.

20. Hoffman A. Magnetic viruses for biological and medical applications. *Magnetics Business & Technology* 2005; 24: 1–3.

21. Mohri K, Uchiyama T, Shen LP, Cai CM, Honkura Y, Aoyama H. Amorphous wire and CMOS IC-based sensitive micromagnetic sensors utilizing magnetoimpedance (MI) and stress-impedance (SI) effects. *IEEE Trans Magn* 2002; 38: 3063–3068.

22. Delooze P, Panina L, Mapps D, Ueno K, Sano, H. Sub-nano tesla resolution differential magnetic field sensor utilizing axymmetrical magnetoimpedance in multilayer films. *IEEE Trans Magn* 2004; 40: 2664–2666.

23. Sandacci S, Makhnovskiy D, Panina L, Mohri K, Honkura Y. Off-diagonal impedance in amorphous wires and its application to linear magnetic sensors. *IEEE Trans Magn* 2004; 40: 3505–3511.

24. Wang X, Yuan W, Zhao Z, Li X, Ruan J, Yang X. Giant magnetoimpedance effect in CuBe/NiFeB and CuBe/Insulator/NiFeB electroless-deposited composite wires. *IEEE Trans Magn* 2005; 41: 113–115.

25. Praslička D, Blazek J, Smelko M, Hudak J, Cverha A, Mikita I, Varga R, Zhukov A. Possibilities of measuring stress and health monitoring in materials using contactless sensor based on magnetic microwires. *IEEE Trans Magn* 2013; 49: 128–131.

26. Ripka P, Janosek M. Advances in magnetic field sensors. *IEEE Sensors J* 2010; 10: 1108–1116.

PART IV

SYSTEMS

CHAPTER 12

Coil Design and Temperature Calculations

Magnetic actuators and sensors are often components of large systems, such as automobiles. The actuators and sensors must perform reliably in the system environment and interface properly with the system.

To interface properly with system power supplies, the coils of the magnetic components must be properly designed, as described in this chapter. As part of the coil design, the temperatures developed by the coils and the components must be predicted. The relationship between electromagnetics (EM) and temperatures T is shown in Figure 12.1 as a block diagram.

12.1 WIRE SIZE DETERMINATION FOR DC CURRENTS

Many DC power supplies consist of a fixed voltage. For example, automobiles have a standard 12 V supply, although the voltage under no load is actually 14 V, and other higher voltages are being added to accommodate ever-growing electrical demands.

As discussed in Chapter 7 and other previous chapters, the magnetic field and force are determined by ampere-turns NI. Ampere's law states that NI produces the field, not voltage. Thus the engineer designing a magnetic actuator or sensor must ensure that the coil carries the proper NI.

For DC voltages V, Ohm's law gives the relation between V and current I:

$$I = V/R \tag{12.1}$$

where R is the coil resistance in ohms. Similar to reluctance of Chapter 3 and as depicted in Figure 3.2, resistance is related to geometry:

$$R = l/(\sigma S_c) \tag{12.2}$$

where l is length of the conducting coil wire, σ is its conductivity, and S_c is the cross-sectional area of the wire.

Magnetic Actuators and Sensors, Second Edition. John R. Brauer.
© 2014 The Institute of Electrical and Electronics Engineers, Inc. Published 2014 by John Wiley & Sons, Inc.

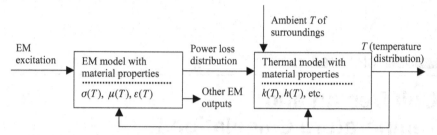

FIGURE 12.1 Modeling temperature and its effects in magnetic devices, to be discussed in this chapter.

For copper at room temperature (20°C), σ is 5.8E7 S/m. The other common coil material, aluminum, has a room temperature conductivity of approximately 3.6E7 S/m. Both materials, however, change their conductivity with temperature T, as indicated in Figure 12.1. For $0 < T < 100$°C, the conductivity of copper obeys:

$$\sigma = \frac{5.8\text{E7 S/m}}{1 + 0.00393(T - 20°\text{C})} \tag{12.3}$$

That is, the conductivity of copper decreases by approximately 0.4% for each degree Celsius rise in temperature. The resistivity of copper, being the reciprocal of conductivity, increases with temperature, having a positive temperature coefficient.

While square or rectangular wire is sometimes used, usually the wire used to wind coils has a circular cross section. Thus its cross-sectional area S_c is related to wire diameter d:

$$S_c = \pi r^2 = \pi d^2/4 \tag{12.4}$$

Instead of choosing any wire diameter, one chooses an available *wire gauge*. Different countries can use different gauges. The American Wire Gauge (AWG) is a number, usually an integer, that gives the bare wire diameter [1]:

$$d = (0.00826)(1.123^{-\text{AWG}}) \text{ m} \tag{12.5}$$

In AWG, every 6-gauge decrease gives a doubling of the wire diameter, and every 3-gauge decrease gives a doubling of the cross-sectional area of the wire. Also common is the metric wire gauge, which is simply 10 times the diameter in millimeters. Thus increasing wire size is accomplished by decreasing AWG or by increasing metric gauge.

However, all wires require insulation. The insulation can be a very thin enamel or other coating, consisting of one or more layers called *builds*. One can consult wire tables to determine insulated wire diameters.

The coil occupies a given coil winding area, as shown for example in any of the DC actuators of Chapter 7. The relation between the coil winding area S_w and the copper (or other wire material) area is the *packing factor*:

$$F_p = NS_c/S_w \qquad (12.6)$$

For circular wires, even when they are tightly wound, the packing factor is typically only about 75%. Conservative designs assume a maximum $F_p = 0.70$ or less [1].

Based on the above equations, various coil design procedures can be followed. Given a certain winding area and voltage, the desired NI can be obtained. However, if I is too high a value, then the coil temperature may become too high [2], as will be discussed later in this chapter. Maximum permissible current densities vary with cooling and other factors, but a reasonable value in some cases is 40 A/mm². All coil insulation is only capable of withstanding a certain temperature rise, for example, 40°C.

Example 12.1   Simple DC Coil Design at a Given Temperature   An axisymmetric copper coil is to be designed to operate at a maximum temperature of 60°C. The available winding area S_w is 1.E−3 m² and the average coil radius is 5 cm. Assume a packing factor of 70% and a DC voltage of 12 V. If $NI = 1000$ ampere-turns, find the bare wire diameter, the number of turns N, and the current density.

Solution   The length of a turn l_T is 2π times the 0.05-m radius, and thus $l_T = 0.31416$ m. The conductivity σ of the copper at 60°C is found using (12.3) to be 5.012E7 S/m. This value can be used in the resistance equation from (12.2) and (12.4):

$$R = 4\frac{Nl_T}{\sigma \pi d^2} \qquad (E12.1.1)$$

The other key equation, based on (12.1), is:

$$NI = NV/R \qquad (E12.1.2)$$

Substituting (E12.1.1) and solving for the bare diameter gives:

$$d = \sqrt{\frac{4NIl_T}{\sigma F_p \pi V}} \qquad (E12.1.3)$$

Inserting the known values of σ, F_P, V, l_T, and NI gives $d = 0.975$ mm. Then using (12.6):

$$N = F_p S_w/S_c = 4F_p S_w/(\pi d^2) \qquad (E12.1.4)$$

Inserting the known values gives $N = 938$ turns. Thus each turn carries $1000/938 = 1.066$ A. The current density is the current per unit area of copper, and here is 1.43 A/mm², a reasonably small value.

12.2 COIL TIME CONSTANT AND IMPEDANCE

In all DC and AC magnetic actuators and sensors, the time constant is important. When a DC voltage V is applied (as a step), the current rises exponentially according to:

$$I = \frac{V}{R}(1 - e^{-t/\tau_e})\qquad(12.7)$$

where the electric time constant $\tau_e = L/R$. When an AC voltage V_{AC} is applied, then (ignoring capacitive effects), the steady-state AC current obeys the complex phasor relation:

$$I_{AC} = \frac{V_{AC}}{Z} = \frac{V_{AC}}{R + j\omega L}\qquad(12.8)$$

where Z is the complex AC impedance. Thus the current has the phase angle:

$$\theta_I = -\arctan(\omega L/R)\qquad(12.9)$$

Hence the L/R ratio is important for both DC and AC magnetic devices.

One expects that the coil design would affect the L/R time constant and the impedance. Thus this section investigates the design of a coil in a given coil winding area S_w. We seek to choose N for minimum time constant L/R and minimum energy W lost in ohmic power in the coil. Inductance L, resistance R, and the related τ_e and W can be calculated for a coil of given area S_w and ampere-turns NI. The derivation that follows ignores velocity effects.

From Chapter 3, L is proportional to the square of N and to permeance P of (3.14):

$$L = N^2\mathcal{P} = (NI/I)^2\mathcal{P} = (NI)^2\mathcal{P}/(I^2)\qquad(12.10)$$

Resistance R of the coil is proportional to length of wire (here N times the length of each turn l_T) divided by the product of conductivity and wire area times packing factor:

$$R = Nl_T/(\sigma F_p S_c/N)\qquad(12.11)$$

$$R = N^2 l_T/(\sigma F_p S_c)\qquad(12.12)$$

The related electric circuit time constant $\tau_e = L/R$ is then:

$$\tau_e = N^2\mathcal{P}/(\sigma F_p S_c)/(N^2 l_T)\qquad(12.13)$$

$$\tau_e = \mathcal{P}(\sigma F_p S_c)/l_T\qquad(12.14)$$

Thus the time constant L/R is independent of turns N.

Energy loss W (power loss integrated over time) is proportional to the square of I and to resistance:

$$W = kI^2 R = k\,(NI/N)^2 N^2 l_T/(\sigma\, S_c\, F_p) \qquad (12.15)$$

$$W = k(NI)^2 l_T/(\sigma\, S_c\, F_p) \qquad (12.16)$$

Hence W depends (as would be expected) on excitation NI, but does not depend on whether the coil designer chooses a high N and low I or a low N and high I.

Thus the engineer designing a high speed actuator cannot vary N over a given area and expect the coil current rise time to be minimized. Instead, what must be done to reduce the time for the current to rise to a given amperage is to increase the voltage applied to the coil. This voltage increase is usually temporary to avoid excessive coil power loss and overheating.

For the Bessho actuator of Figure 5.2, the time constant τ_e is 20 ms [3]. The inductance computed by magnetostatic finite-element software [4] for a 10-mm airgap = 3.15 H, which then leads to a resistance $R = L/\tau_e = \sim150\ \Omega$, which is believed to be close to the actual R [4]. Even if the number of turns were changed from the 3300 of Figure 5.2, the initial time constant would remain 20 ms.

12.3 SKIN EFFECTS AND PROXIMITY EFFECTS FOR AC CURRENTS

The above DC resistance R can be increased significantly under AC operation. From Faraday's law, the magnetic fields produced by the AC current induce additional AC currents in the coil conductors. These additional currents produce additional losses, and sometimes change the current density distribution from uniform to nonuniform.

The skin depth equation (8.1) of Chapter 8 applies not only to steel but to coil conductors, and is repeated here:

$$\delta = \frac{1}{\sqrt{\pi f \mu \sigma}} \qquad (12.17)$$

where f is the AC frequency in Hz and σ is electrical conductivity. Coils of copper and aluminum have permeability $\mu = 12.57\mathrm{E}{-7}$, the permeability of air. The derivation of the above skin depth assumes one-dimensional conducting slabs, but (12.17) is an approximate indicator of the depth of currents and magnetic fields for many practical problems.

When the skin depth is much greater than the wire radius, then AC magnetic flux density can be shown to produce losses similar to those of (8.2):

$$\frac{P_e}{v} = \frac{\omega^2 \sigma}{24}\left(w_y^2 B_x^2 + w_x^2 B_y^2\right) \qquad (12.18)$$

where B_x and B_y are components of the peak AC magnetic flux density passing through the z-directed wire, $\omega = 2\pi f$, P_e is power in watts, v is the volume of the conducting material in cubic meters, w_x is the wire width in the x direction, and w_y is the wire width in the y direction. The losses of (12.18) are sometimes a significant addition to the I^2R losses of the DC resistance.

However, when the skin depth is smaller than the wire radius, then *skin effect losses* are said to occur. The current density distribution is no longer uniform. While the average of the current density J over the wire cross section remains the same, the average power loss density increases. The total power loss is:

$$P = \int \frac{J^2}{\sigma} dv \qquad (12.19)$$

Thus skin effect increases power loss, and occurs whenever conductor radius exceeds skin depth. A way to avoid skin effect loss is to replace large conductors by many conductors of much smaller radius, as is done in *Litz wire*. Unfortunately, Litz wire is much more expensive than solid copper or aluminum wire.

Skin effect is usually considered to be associated with the magnetic field of a single isolated wire in air. Because other wires add contributions to the magnetic field, other wires in the proximity cause changes in the power losses. These effects of other wires are included in the *proximity effect*.

Since the magnetic fields are affected by all wires and all magnetic materials, the best way to analyze skin and proximity effects is to use finite-element analysis. The software must include eddy currents and must constrain the total current in each conductor to the same value.

Example 12.2   Skin Effect in an Isolated Conductor   A circular copper wire carrying 100 A peak is placed in air, far from any other materials. If its radius is 10 mm, find its current density and magnetic flux density distributions at 400 Hz and 1 Hz using Maxwell. Also find the power loss and resistance per meter.

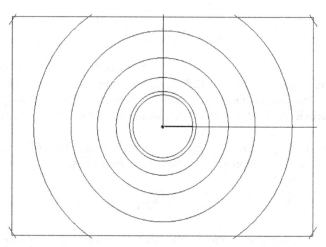

FIGURE E12.2.1   Computer display of computed flux lines in isolated wire at 400 Hz.

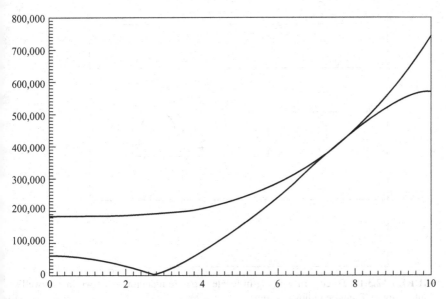

FIGURE E12.2.2   Computed current density at 400 Hz versus radius from center of wire. The upper curve is the total magnitude, and the lower curve is the magnitude at phase angle zero.

Solution   Use Maxwell's "Eddy Current" solver with the wire centered in its default model area, placing "balloon" boundaries around the area. At 400 Hz the power loss computed is 0.49036 W/m. Thus the 400-Hz resistance $R = P/I^2$, where I is rms, is 0.98072E−4 Ω/m. Figure E12.2.1 shows the computed flux lines; note that they appear to be concentrated near the wire surface. Figure E12.2.2 graphs the computed current density versus wire radius. Note that the current density decays exponentially from the surface, as expected for an isolated wire. Since the skin depth at 400 Hz from (12.17) is 3.304 mm, the decay over the 10-mm radius appears reasonable.

At 1 Hz, the computed power loss is reduced to 0.2758 W/m. Thus the 1-Hz resistance is 0.5516E−4 Ω/m, which agrees closely with the DC resistance of 0.5488 Ω/m obtained by (12.2). Thus the resistance at 400 Hz is approximately 1.8 times the DC resistance. Graphing the current density versus radius shows an essentially uniform current density throughout the wire at 1 Hz.

Example 12.3   Skin and Proximity Effects in Stator Coil with Clapper   An axisymmetric copper coil is placed in a ferrite "cup core" inductor shown in Figure E12.3.1 as created in Maxwell's default drawing area. The area is assumed to confine the flux. The ferrite in the lower stator and upper clapper armature has a relative permeability of 1000 and conductivity of 0.01 S/m. The coil has three copper conductors, each of radius 4 mm and carrying 50 A 400 Hz. Use Maxwell to find the power loss, flux line plot, and current density distribution, showing skin and proximity effects in the three wires.

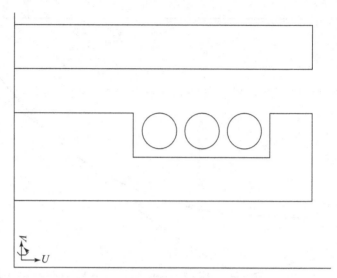

FIGURE E12.3.1   Three-turn winding in ferrite cup core inductor, as drawn in Maxwell's default window. The wire radius is 4 mm.

Solution   The model of Figure E12.3.1 was solved by Maxwell's "Eddy Current" solver. The computed power loss is 0.635 W. The flux line distribution is shown in Figure E12.3.2 at phase angle zero. The current density distribution can be plotted in color, but is instead graphed in Figure E12.3.3 along a plane cutting all three wires. Note that each of the three wires has a different current density profile (because each has a different flux density), but all that appear to have the same average current density as they should.

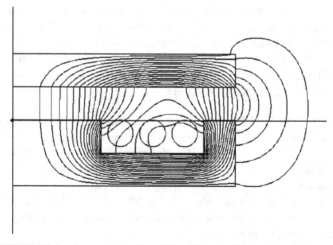

FIGURE E12.3.2   Computer display of flux line plot obtained for Figure E12.3.1.

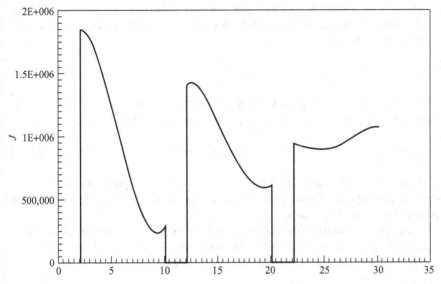

FIGURE E12.3.3   Graph of computed current density J in amperes per square meter versus position along plane cutting through the centers of the three wires in Figure E12.3.1.

12.4   FINITE-ELEMENT COMPUTATION OF TEMPERATURES

As indicated in Figure 12.1, the power losses in magnetic actuators and sensors determine their operating temperature. Since high temperatures can cause device failure and/or fire, it is important to be able to compute temperature distributions.

Thermal computations must in general include three means of heat transfer: conduction, convection, and radiation. This book will examine steady-state thermal computations. Time variation of temperature is studied elsewhere [5, 6]. Appendix A includes the symbols, dimensions, and units of heat.

12.4.1   Thermal Conduction

Steady-state thermal conduction is governed by Poisson's differential equation [5, 6]:

$$\nabla \cdot k \nabla T = -\frac{P}{v} \tag{12.20}$$

where T is temperature, k is thermal conductivity, and the right-hand side is the negative power per unit volume. Thermal conductivities are similar to electrical conductivities in that they vary from high in metals to low in insulators and also vary somewhat with temperature. Approximate room temperature (20°C) values include 400 W/(m °C) for copper and 80 W/(m °C) for typical steel. Heat conduction is compared with electric and magnetic circuits in Figure 3.2.

There is an analogy between (12.20) and one of Maxwell's equations of Chapter 2. Replacing the electric flux density of (2.49) by permittivity times electric field intensity E gives:

$$\nabla \cdot \varepsilon E = \rho_v \qquad (12.21)$$

where the right-hand side from Chapter 2 is volume charge density. Replacing E by the negative gradient of electric scalar potential ϕ_v of (2.39) gives:

$$\nabla \cdot \varepsilon \nabla \phi_v = -\rho_v \qquad (12.22)$$

Note that (12.22) is analogous to (12.20). Permittivity ε replaces thermal conductivity k, volume charge density replaces power per unit volume, and electric scalar potential replaces temperature.

Finite-element analysis of thermal conduction problems is commonplace [6,7]. Since Maxwell is used elsewhere in this book, its electrostatic solution of (12.22) will be used to solve thermal conduction by analogy.

Example 12.4   Steady Thermal Conduction Computation Using Analogy to Electrostatics   An axisymmetric copper coil is surrounded by glastic insulation and a cylindrical steel core as shown in Figure E12.4.1. The thermal conductivities are

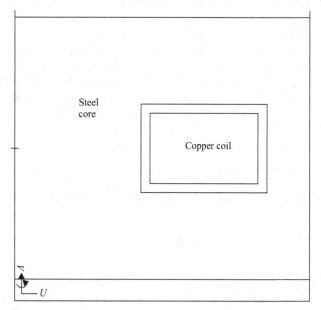

FIGURE E12.4.1   Axisymmetric inductor of radius 70 mm and height 60 mm. The temperature distribution is to be computed when there are power losses in the steel and in the copper coil of size 26 mm radially and 16 mm high, surrounded by glastic insulation 2 mm thick on all sides.

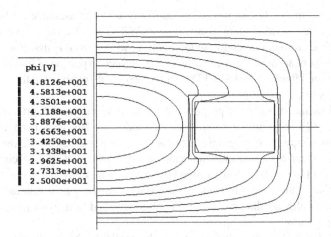

FIGURE E12.4.2   Computer display in black and white of computed voltage contours in inductor of Figure E12.4.1, also available in color. Since voltage is here analogous to temperature, the maximum temperature is approximately 48°C.

400, 10, and 79 W/(m °C), respectively. The copper power loss density is 8000 W/m^3 and the steel core loss density is 4000 W/m^3. Use Maxwell's electrostatic solver to compute the temperature distribution, assuming that the outermost steel of the cylinder is at 25°C.

Solution   The relative permittivities to be entered are the above thermal conductivities divided by free space permittivity 8.854E−12. Under "Sources and Boundary Conditions" in Maxwell SV (or under "Excitations" in Maxwell version 16) enter the charge densities of 8000 C/m^3 in the copper coil and 4000 C/m^3 in the steel core; then enter 25 V for the upper, lower, and outer steel boundaries. The computed voltage contours are shown in Figure E12.4.2. Note that the maximum contour value is approximately 48 V in the steel along the axis of symmetry. Thus the maximum temperature is 48°C. Alternatively, thermal finite-element software can be used to also obtain 48°C.

12.4.2 Thermal Convection and Thermal Radiation

Besides thermal conduction, thermal convection and thermal radiation are usually very important in determining temperatures of magnetic actuators and sensors. The easiest way to analyze them is to use the film coefficient h defined by:

$$Q = hA\Delta T \qquad (12.23)$$

where Q is heat flow, A is the area of the surface, and ΔT is the temperature difference across the convective and/or radiative region. In many cases ΔT is the temperature rise above the temperature of the surroundings, called the ambient temperature. The coefficient h has SI units of W/(°C m^2).

While the film coefficient varies tremendously with the available airflow paths and temperatures, a value often assumed is 10 W/(°C m^2) for *free convection and radiation*. Free convection is without fan cooling, while *forced convection* has a fan or wind. For more accurate determination of the convection portion of film coefficient h, computational fluid dynamics (CFD) can be carried out, often using finite-element fluid analysis [6].

Thermal radiation is usually proportional to the fourth power of surface absolute temperature [6] in SI units of kelvin (see Appendix A). However, even at surface temperatures common to magnetic actuators and sensors, say 60°C, the portion of film coefficient h due to radiation can be more than that of free convection. Radiation heat transfer is also determined by surface emissivity and geometric view factors [6]. Since both convection and radiation are commonly nonlinear functions of temperature, nonlinear film coefficient h is often required in thermal finite-element computations [7].

Thermal finite-element software is widely available. Maxwell automatically includes electromagnetic power losses in its thermal computations when linking to compatible thermal or computational fluid dynamics software.

12.4.3 AC Magnetic Device Cooled by Conduction, Convection, and Radiation

As an example of temperature computation in laminated magnetic devices, consider the three-phase inductor shown in Figure 12.2. It has a core made of laminations of Armco M6 grade steel of thickness 0.356 mm. To maintain inductance at frequencies above 60 Hz, the laminations are considerably thinner than the 0.635 mm thickness commonly used for 60-Hz laminations. The laminations are stacked to form a core of height of 190.5 mm.

The laminated core material has a *B–H* curve and core loss curves provided by Armco. To model the core loss using Maxwell's thermal finite-element solver, the Armco core loss curves are fitted to the core loss equation [8, 9]:

$$p = K_h B^2 f + K_c (Bf)^2 + K_e (Bf)^{1.5} \qquad (12.24)$$

where p = power loss density in watts per cubic meter, K_h = hysteresis coefficient, K_c = classical eddy current coefficient, K_e = excess (or anomalous) eddy current coefficient due to domains, B = peak magnetic flux density in teslas, and f = frequency in Hz. For the Armco steel laminations used, least-squares curve fitting gives K_h = 66.9, K_c = 0.210, and K_e = 1.92.

To reduce high frequency losses in the copper windings of the three-phase inductor, the winding turns are made of multiple strands. Figure 12.3 shows the three-phase windings, one on each arm of the laminated core. Each phase has seven turns made of eight strands of octagonal wire. The overall lamination width is 302 mm and the total height (including the top "I" core segment) is 361 mm. The geometry of Figure 12.3 was developed as an Autocad® DXF file, which was then directly input into Maxwell.

FIGURE 12.2   Photograph of laminated three-phase 425-A rms 88-μH inductor for which temperatures are to be found.

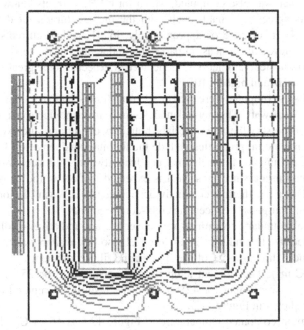

FIGURE 12.3   Computer display of geometry of laminated three-phase inductor, also showing computed magnetic flux lines at time zero, when 14.14 A (10-A rms) of frequency 100 Hz flows in the leftmost winding. Note that there are eight-core segments.

Besides the core and copper losses, losses may occur in the bolts holding the laminations together. The bolts are made of solid steel with a B–H curve and electrical conductivity (2.E6 S/m) assumed for SAE 1010 steel.

The three-phase laminated inductor of Figures 12.2 and 12.3 was input to Maxwell, which was used to compute its electromagnetic performance followed by its thermal performance. The magnetic flux density distribution and related electrical parameters and losses were computed over a wide range of current amplitudes and frequencies. Figure 12.3 shows a typical computed magnetic flux line pattern.

The loss distributions in the various copper windings, the various segments of the steel core, and the steel bolts, were computed for frequencies ranging from 1 Hz to 20 kHz. Skin effects and proximity effects in the stranded copper windings were included in the computations. To check the copper loss computation of 0.0507 W at 1 Hz and 10 A rms, the simple DC resistance formula (12.2) was used where length $l = 0.1905$ m, electrical conductivity $\sigma = 5.8E7$ S/m for Cu, and S_c is the area of the octagonal-shaped strand. Substituting the area and multiplying by I^2 gives a loss of 0.050 W, agreeing closely with the 0.0507 W computed by finite elements.

At higher frequencies for the same 10-A current, finite-element computations show that the copper loss increases from 1.184 W at 100 Hz to 32.266 W at 20 kHz. The increase is due to skin and proximity effects, which also cause changes in the magnetic field flux pattern shown in Figure 12.4 at 4 kHz, which differs in the winding area from Figure 12.3.

The inductance was also computed, which for 60-Hz currents below 750 A was 89.8 µH. This agrees reasonably well with the manufacturer's rated 88 µH and measured 60-Hz inductance of approximately 96 µH at low currents; the small difference is believed due at least in part to the inductance of the leads. Because of skin effects within steel laminations, inductance is expected to be greatly reduced at frequencies 20 kHz and higher [10].

Next, the losses computed were used in a thermal finite-element model, again using the Maxwell SV software. The thermal model is made from the electromagnetic model by adding the necessary thermal materials, boundary conditions, and excitations.

The thermal material properties include the thermal conductivities. Even though the steel used is laminated, the thermal conductivity for solid steel was used; this is believed appropriate since most heat is expected to flow in the xy plane of the laminations. Other key thermal conductivities include 10.2 W/(m °C) for the glastic insulation in the "airgaps" between the eight-core segments.

The thermal boundary conditions are convective and radiative, involving a film coefficient. Because the inductor of Figure 12.2 also has available air flow paths at its ends (in the z direction out of the figure), the film coefficient assumed here is $h = 15$ W/(°C m^2). The ambient temperature is assumed to be 25°C. The 60-Hz loss distributions were input for 425 A rms. The resulting computed temperature distribution is shown in Figure 12.5.

The maximum computed temperature in Figure 12.5 is 163.5°C, which agrees reasonably well with the measured maximum temperature of 156°C. The computed temperatures are significantly affected by the assumed film coefficient; for example, changing it from 15 W/(°C m^2) to 20 W/(°C m^2) causes the maximum temperature to fall to 145°C.

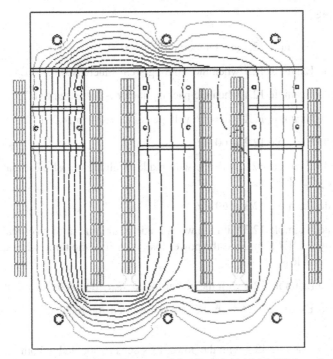

FIGURE 12.4   Computer display of computed flux lines for 14.14 A (10 A rms) of frequency 4 kHz in leftmost winding. Note that due to skin and proximity effects the winding leakage flux lines differ from the low frequency flux lines in Figure 12.3.

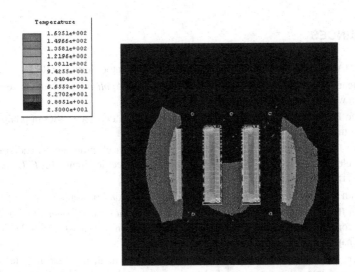

FIGURE 12.5   Computer display in black and white of computed temperature distribution in laminated inductor, also available in color.

PROBLEMS

12.1 Redo Example 12.1 at a temperature of 100°C.

12.2 Redo Example 12.1 when the winding area is increased to 1.5E−3 m².

12.3 Redo Example 12.1 when the packing factor is increased to 0.75.

12.4 Redo Example 12.1 when the voltage is increased to 14-V DC.

12.5 Redo Example 12.2 at 60 Hz.

12.6 Redo Example 12.2 with the wire made of aluminum of electrical conductivity 3.54E7 S/m.

12.7 Redo Example 12.3 at 60 Hz.

12.8 Redo Example 12.3 with the wires made of aluminum of electrical conductivity 3.54E7 S/m. Find the power lost in each of the three wires as well as the total loss.

12.9 Redo Example 12.4 with the power loss density doubled in the coil.

12.10 Redo Example 12.4 with the power loss density doubled in the steel core only.

12.11 Redo Example 12.4 with the power loss density doubled in both the coil and the core.

12.12 Redo Example 12.4 with the thermal conductivity of the insulation reduced to 4 W/(m °C).

REFERENCES

1. Juds MA. *Notes on Solenoid Design*, Milwaukee, WI: Eaton Corporate R&D; 2012.
2. Flanagan WM. *Handbook of Transformer Design and Applications*, 2nd ed. New York: McGraw-Hill, Inc.; 1993.
3. Bessho K, Yamada S, Kanamura Y. Analysis of transient characteristics of plunger type electromagnets. *Electr Eng Jpn* 1978;98:56–62.
4. Brauer JR, Ruehl JJ, Hirtenfelder F. Coupled nonlinear electromagnetic and structural finite element analysis of an actuator excited by an electric circuit. *IEEE Trans Magn* 1995;31:1861–1864.
5. Holman JP. *Heat Transfer*. New York: McGraw-Hill Book Co.; 1963.
6. Brauer JR (ed.). *What Every Engineer Should Know About Finite Element Analysis*, 2nd ed. New York: Marcel Dekker, Inc.; 1993, Chapter 4 "Thermal Analysis" by V. D. Overbye and Chapter 5 "Fluid Analysis" by N. J. Lambert.
7. Brauer JR, Wallen P. Coupled 3D electromagnetic, structural, and thermal finite element analysis as integral components of electronic product design, *Proceedings of the IEEE Wescon Conf.*, October 1996, pp. 358–364.

8. Skibinski GL, Schram BG, Brauer JR, Badics Z. Finite element prediction of losses and temperatures of laminated and composite inductors for AC drives, *Proceedings of the IEEE Int. Electric Machines and Drives Conf.*, June 2003.

9. Fitzgerald AE, Kingsley C, Umans SD. *Electric Machinery*, 6th ed. New York: McGraw-Hill Book Co.; 2003.

10. Brauer JR, Beihoff BC, Cendes ZJ, Phillips KP. Laminated steel eddy loss vs. frequency computed using finite element analysis. *IEEE Trans Ind Appl* 2000;36:1132–1137.

CHAPTER 13

Electromagnetic Compatibility

Just as in the preceding chapter, where the magnetic device design must be compatible with system voltages and temperatures, the device must also possess *electromagnetic compatibility*. EMC, the common abbreviated term, is related to EMI, *electromagnetic interference*. Magnetic actuators and sensors must not produce unacceptable EMI, nor have their performance affected adversely by EMI.

13.1 SIGNAL-TO-NOISE RATIO

EMI is electromagnetic interference or noise, where noise is any undesirable signal [1]. There are four general types of EMI.

- **Conducted emissions** are noises produced by the magnetic device that travel to other devices in any system via conductors such as power supply wires.
- **Radiated emissions** are noises produced by the magnetic device that travel to other devices in any system via electromagnetic radiation in air or free space.
- **Conducted susceptance** refers to noises produced in the magnetic device by conduction through wires from other devices.
- **Radiated susceptance** refers to noises produced in the magnetic device by electromagnetic radiation in air or free space.

Whether noise is produced by conduction or radiation, the important factor is whether it dominates the desired signal. For example, if a radio signal, such as a voice and/or music, is large compared with noise, such as static due to lightning, then performance is generally satisfactory. If the noise becomes almost as large as, or larger than, the signal, then performance is usually unsatisfactory. The signal-to-noise ratio, abbreviated as S/N or SNR, is thus a key parameter.

For best S/N, the noise should be reduced and/or the signal increased. If the noise has a different frequency spectrum than the signal, filtering circuits can be employed to reduce the noise. Hence most EMI and EMC specifications by regulatory agencies

Magnetic Actuators and Sensors, Second Edition. John R. Brauer.
© 2014 The Institute of Electrical and Electronics Engineers, Inc. Published 2014 by John Wiley & Sons, Inc.

(such as the Federal Communications Commission in USA) are in terms of frequency. However, if noise occupies the same frequency band as the signal, then other methods must be used such as digital signal processing and adaptive filters.

Conducted noise can usually be reduced by proper circuit design. For example, conducted noise due to IR voltage drops caused by current pulses in an actuator coil can be reduced if the wires from the power supply to the actuator and other devices are large enough to reduce the resistance and if proper grounding is employed [1,2]. For automotive conducted susceptibility the biggest challenge often is alternator *load dump*, but careful filtering and shielding can prevent it from causing malfunction of proximity sensors [3].

Radiated noise is usually a more difficult problem, whether the magnetic device is emitting it or subjected to it. Magnetic actuators, which often have large ampere-turns, tend to emit EMI. Magnetic sensors, which often are very sensitive to magnetic fields, tend to be very susceptible to EMI.

Inductive magnetic sensors are especially susceptible to radiated EMI. For example, inductive pickup coils of the velocity sensors of Section 11.1 induce voltages proportional to frequency according to Faraday's law. Thus high frequency EMI can induce large noise voltages, larger than the desired signal at low speeds. In automotive EMI engineering [4] involving such velocity sensors on all four wheels, one method of noise suppression is to connect the four sensors so that *common-mode noise* voltages induced in all four wheels essentially cancel out, leaving mostly the desired signal voltage [5].

13.2 SHIELDS AND APERTURES

The best way to reduce radiated emissions and radiated susceptance is to use a *shield*. A shield is a highly conductive enclosure, almost always made of metal. Ideally, it would *completely* enclose and seal off the magnetic actuator or sensor without interfering with its operation.

From Chapter 8, conductive materials possess the skin depth:

$$\delta = \frac{1}{\sqrt{\pi f \mu \sigma}} \tag{13.1}$$

where f is frequency in Hz, μ is permeability, and σ is electrical conductivity. The key parameter in a shield is the ratio of its thickness T to skin depth. The larger the ratio, the lower is the transmissibility of electromagnetic fields from one side of the shield to the other. Thus the larger the ratio of T to skin depth, the better the shield is at preventing EMI.

Shield thickness T equals skin depth at a frequency f_{min} found from (13.1):

$$f_{min} = \frac{1}{\pi \mu \sigma T^2} \tag{13.2}$$

For noise frequencies below the minimum frequency f_{min} the shield is ineffective. Note that for lowest f_{min}, the conductivity, permeability, and thickness should all be as high as possible.

Shield thickness is limited by weight, size, and cost considerations. To reduce the weight, aluminum is a very common shield material. However, the higher permeability of steel causes it to be chosen sometimes to shield low frequencies. In fact, it will be shown that steel shields even reduce DC magnetic field transmission from one side to the other.

For magnetic actuators and sensors, magnetic field EMI is usually the main concern, rather than electric field EMI. However, above a certain frequency range *coupled electromagnetic* radiation, mentioned in Chapter 2, becomes the most significant form of EMI. Thus electric fields are also important, and in fact are usually measured during EMC tests. The frequency at which coupled electromagnetic fields become significant is when the component size, including the shield size, has a dimension of approximately 10% of a wavelength. Wavelength λ is given by:

$$\lambda = c/f \qquad (13.3)$$

where c is the speed of light that equals 3.E8 m/s in air or free space.

Unfortunately, almost all shields must contain at least one *aperture*, a term meaning opening or hole. For magnetic sensors, the object or field to be sensed cannot be entirely separated from the sensor, so an aperture is usually required. An example of magnetic sensor with shields is the spin-valve GMR sensor in Figure 10.6, where the shields are above and below, leaving apertures around all the sides. For magnetic actuators, the object to be moved must usually be accessible and thus an aperture is required. For both magnetic actuators and sensors, electrical voltages or currents must be accessed through some sort of terminal.

A typical shield and aperture are shown in Figure 13.1. The shield is a rectangular box with a top piece that is attached above a gasket [6, 7]. To reduce computer modeling time, only one-fourth of the box is needed. The shield has a conductivity

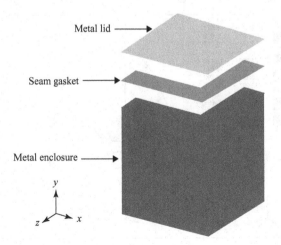

FIGURE 13.1 One-quarter of a rectangular shielding box that is 1 mm thick. The quadrant size is 110 × 110 mm of the 440-mm² box. The box height is 115 mm, on top of which is the 5-mm thick gasket and the 1-mm thick lid.

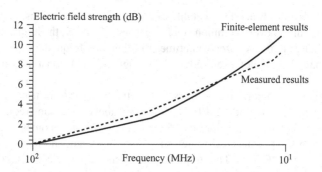

FIGURE 13.2 Computed and measured leakage electric field from shield of Figure 13.1.

of 4.8E7 S/m and the gasket has a complex relative permeability of $100-j\,100$ along with a relative permittivity of 10. Inside the box is a vertically oriented half-loop current source lying in the $x = 0$ plane. The top of the loop is located 50 mm down from the top of the lid. The electric field outside the box is to be computed over the frequency range 100 kHz–100 MHz. At 100 MHz, the wavelength from (13.3) is 3 m. Since the box has maximum dimension of almost 0.5 m, coupled electromagnetic fields can be significant in the range of about 60 MHz and higher. Thus a coupled 3D electromagnetic finite-element formulation, to be presented in the next chapter, was used in all computations for Figure 13.1.

The computed electric field intensity versus frequency is plotted in Figure 13.2, where it is shown to compare fairly closely with measurements available to 1 MHz [6, 7]. Note that the field is plotted in decibels (dB) because of the large amplitude range of electric and magnetic fields (as in Figure 10.1). The computed 3D electric field distribution in three cut planes around the box is displayed in Figure 13.3.

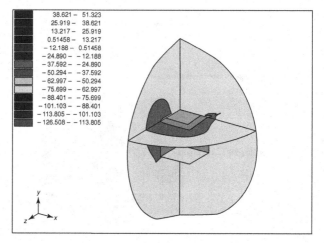

FIGURE 13.3 Computer display of computed electric field distribution in dB emitted by shield of Figure 13.1 at 1 MHz. This display appears in color on the computer screen but is here shown in black and white.

Similar finite-element computations have been carried out for over 100 apertures at frequencies as high as approximately 10 GHz [8].

Some magnetic actuators and sensors inherently possess some shielding. For example, axisymmetric plunger actuators in Chapter 7 almost completely surround their coils by steel, and thus usually emit less EMI than planar actuators with clapper armatures.

Example 13.1   Cylindrical Shield without and with an Aperture   A circular copper wire loop carrying 1000 ampere-turns AC is shielded by a cylindrical metal can as shown in Figure E13.1.1. The coil is stranded such that its skin and proximity effects are negligible; its inner radius is 1 mm, outer radius 2 mm, and its height is 1 mm. The coil is centered in the can of inner radius 3 mm, inner height 3 mm, and thickness 0.4 mm. The can has an optional 1-mm airgap below its top. Use finite elements to find the magnetic flux density just inside and outside the top corner of the can for the following cases.

(a) Aluminum can at 60 Hz and 10 kHz with no airgap. The aluminum electrical conductivity is 3.6E7 S/m.

(b) Aluminum can at 10 kHz with 1-mm airgap.

(c) Steel can at 0.1 Hz (essentially DC) and 10 kHz. The steel has relative permeability of 1000 and electrical conductivity 2.E6 S/m.

(d) Steel can at 10 kHz with 1-mm airgap.

Solution   Since the overall shield size at the frequencies to be analyzed is much less than 10% of a wavelength, Maxwell's "Eddy Current" solver will be used. Under

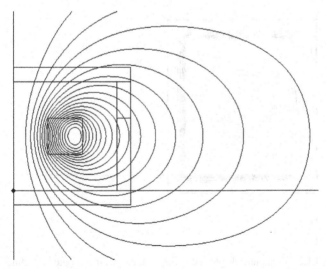

FIGURE E13.1.1   Computer display of cylindrical can with computed flux lines at 60 Hz with complete aluminum shield.

"Sources and Boundary Conditions" use a balloon outer boundary and stranded source. Results are as follows.

(a) At 60 Hz the aluminum is not a good shield, as shown by its flux lines in Figure E13.1.1. Note that they pass through the aluminum as if it were not present, because the skin depth in aluminum at 60 Hz using (13.1) is 10.8 mm, much greater than the shield thickness. Flux densities at the top right inner edge of the can are 0.029 T inside and 0.021 T outside, essentially natural decay with distance. At 10 kHz, some shielding occurs, because the skin depth is now 0.84 mm. The flux densities are 0.031 T inside and 0.015 T outside.

(b) When the aperture is cut in the aluminum, there is more flux density outside. The flux density decays from 0.035 T inside the aperture to 0.020 T outside.

(c) The skin depth in the steel is 0.113 mm at 10 kHz, less than the 0.4-mm thick shield. At 0.1 Hz, B is 1.1 mT just inside the shield and only 0.577 mT outside the shield; the shield itself has flux densities on the order of 0.6 T. At 10 kHz, B is 3.77 mT just inside the shield and only 0.163 mT outside the shield; the shield itself has inner flux densities of 1.6 T. The flux line plot is shown in Figure E13.1.2, and shows how the shield confines most of the magnetic flux.

(d) Cutting the aperture produces 97 mT in the same outside location that previously had only 0.163 mT. As shown in Figure E13.1.3, the aperture allows a lot of flux to pass, although the shield still is helpful.

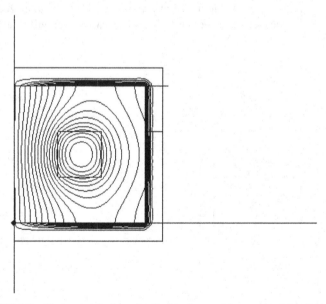

FIGURE E13.1.2 Computer display of cylindrical can with computed flux lines at 10 kHz with complete steel shield.

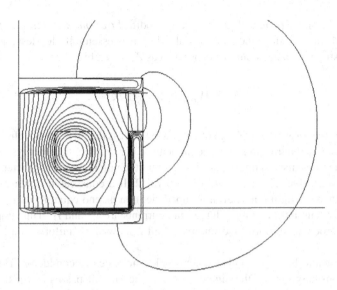

FIGURE E13.1.3   Computer display of cylindrical can with computed flux lines at 10 kHz with steel sheet with an aperture.

13.3   TEST CHAMBERS

Test chambers are commonly used to test radiated EMC behavior of magnetic actuators and sensors. To evaluate emissions, electromagnetic *absorption chambers* are often used. They are quite expensive and large, because they must contain a great deal of absorptive (also called electromagnetic anechoic) material, usually wedge-shaped pieces facing the emitting device. They absorb rather than reflect EMI, and thus simulate the "open air" environment in which many magnetic devices are placed. Smaller, less expensive chambers called transverse electromagnetic (*TEM*) *cells* and *triplate cells* are often used to determine radiated susceptibility.

13.3.1   TEM Transmission Lines

A TEM field has **E** and **H** vectors both lying in the plane transverse (perpendicular or normal to) the direction of energy propagation. **E** and **H** are perpendicular to each other, and the ratio of their magnitudes is called the wave impedance:

$$Z_w = |E_F|/|H_F| = \sqrt{\mu/\varepsilon} \tag{13.4}$$

where the energy is propagating in the forward (*F*) direction in a lossless material with permeability μ and permittivity ε. Note that the subscript *F* on *E* and *H* indicates the forward-traveling "waves" of *E* and *H*. Substituting the material properties of air, its $Z_w = 377\ \Omega$. Assuming no reflections or losses, the only components of **E** and **H** are those in (13.4), and thus the **E** and **H** fields are uniform and known.

TEM cells and triplate cells are basically modified *transmission lines*. A common basic TEM transmission line is a coaxial cable. It is essentially lossless, and has a characteristic impedance commonly denoted as Z_o, given by:

$$Z_o = |V_F|/|I_F| = \sqrt{L/C} \tag{13.5}$$

where V_F is the forward-traveling voltage wave and I_F is the forward-traveling current wave. Also, L is the transmission line inductance per unit length in henrys per meter and C is the transmission line capacitance per unit length in farads per meter. Typical coax (coaxial cable) Z_o values are 50 and 75 Ω. If a coax cable is matched with an impedance (including a Thevenin impedance) on its end equal to its Z_o, then no reflections occur and the only voltages and currents are V_F and I_F. Thus impedance matching produces uniform and known V and I, as well as uniform and known $\mathbf{E}$ and $\mathbf{H}$.

A coax cable has a grounded outer shield and a center conductor. The center conductor usually carries the signal voltage. The current passes down the center conductor and returns via the outer shield. The characteristic impedance can be altered by changing L and C. They are changed by the following.

(a) Changing the material between the shield and the center conductor. For example, different plastics have different permittivities.

(b) Changing the geometry of the center conductor and outer shield. For circular coax, varying these two radii changes Z_o. For other shapes, such as elliptical or rectangular, other values of Z_o can be obtained. Changing the geometry changes both L and C. It is possible, however, to vary the geometry along a transmission line and yet obtain the same L/C ratio, thereby maintaining a constant Z_o. Maintaining constant Z_o prevents reflections, thereby helping to maintain the desired uniform and known V, I, and $\mathbf{E}$, $\mathbf{H}$.

Besides the above lossless transmission line, lossy (with loss) transmission lines also occur. Their characteristic impedance is given by [9]:

$$Z_o = |V_F|/|I_F| = \sqrt{(R + j\omega L)/(G + j\omega C)} \tag{13.6}$$

where R is the series resistance per unit length in ohms per meter, and G is the parallel conductance per unit length in siemens per meter. Because j is the square root of minus 1, the characteristic impedance in the lossy case can be a complex number. The angular frequency $\omega = 2\pi f$, where f is the frequency in hertz.

In TEM transmission lines, the electric and magnetic fields are at right angles to each other and can be computed separately. If lossless, electrostatic and magnetostatic solutions apply. If lossy, the magnetic field computations should include eddy current losses at the frequency f as described in the preceding chapter.

FIGURE 13.4   Typical small TEM cell (33 × 33 × 64 cm).

13.3.2 TEM Cells

A TEM cell is a TEM transmission line that has a big "bulge" in it that serves as the cell or chamber for the device under test. There are many ways to make the "bulge" and thus many varieties and sizes of TEM cells exist. Figure 13.4 shows a typical small TEM cell. Note that the coax connectors on both ends are visible.

Note also in Figure 13.4 that the TEM cell is completely enclosed. However, Figure 13.4 shows that doors are placed in the enclosure walls to allow the device being tested to be inserted. Also couplings through the doors and walls enable measurements of the fields and the device performance.

Interiors of TEM cells can be of various designs. The field is only approximately uniform throughout the entire interior.

13.3.3 Triplate Cells

A triplate cell is a simplification of a TEM cell in which some walls are removed to create three metal "plates." The device to be tested can thus be easily inserted and altered. The triplate in Figure 13.5 has the coax connectors visible on both ends.

Both triplate cells and TEM cells bombard the device under test with both electric and magnetic fields. To find the ratio E/H for an arbitrary-shaped triplate or TEM cell, assume that the load impedance on the cell end matches the characteristic impedance, giving $V/I = Z_o$. Then from (13.5):

$$V/I = (L/C)^{1/2} \qquad (13.7)$$

$$V^2/I^2 = L/C \qquad (13.8)$$

FIGURE 13.5 Typical triplate cell.

Substituting energy expressions (6.17) and (6.23) gives:

$$V^2/I^2 = (2W_{\text{mag}}/I^2)/(2W_{\text{el}}/V^2) \tag{13.9}$$

Thus

$$W_{\text{mag}} = W_{\text{el}} \tag{13.10}$$

This equality has been verified using finite-element computations. Thus a feature of a matched triplate or TEM cell is that its *electric and magnetic fields have equal energies.*

Now the magnetic energy W_{mag} is the volume integral of magnetic energy density, or the average magnetic energy density times the volume:

$$W_{\text{mag}} = (1/2)\mu_o\, H_{\text{ave}}^2 v \tag{13.11}$$

where H_{ave} is the average H and v is volume. The electric energy is the volume integral of electric energy density, or the average electric energy density times the volume:

$$W_{\text{el}} = (1/2)\varepsilon_o E_{\text{ave}}^2\, v \tag{13.12}$$

where E_{ave} is the average E.

Substituting (13.11) and (13.12) into (13.10) gives:

$$(1/2)\mu_o H_{\text{ave}}^2 = (1/2)\varepsilon_o E_{\text{ave}}^2 \tag{13.13}$$

Hence

$$E_{\text{ave}}/H_{\text{ave}} = (\mu_o/\varepsilon_o)^{1/2} = 377 \tag{13.14}$$

Thus in any triplate cell or TEM cell that has a load equal to its characteristic impedance, that is $V/I = Z_o$, the average ratio of E/H is 377. Since the voltage contours (for the **E** field) and the magnetic flux lines are identical as discussed in the example below, **E** and **H** fields obey this ratio everywhere in matched triplate cells and TEM cells. Most commercial cells produce a specified **E** field, but (13.14) obtains the cell **B** field magnitude as:

$$B = \mu_o E / 377 \qquad (13.15)$$

Since magnetic sensors and actuators are usually more susceptible to **B** field EMI than to **E** field EMI, (13.15) is often useful.

Example 13.2   Characteristic Impedance of Triplate   The cross section of the main center section of a triplate cell is shown in Figure E13.2.1. Its dimensions are 60 × 60 cm. Its three plates are all 5 mm thick. Its center plate is 50.8 cm wide. Find its characteristic impedance using Maxwell software. Also find its voltage contours and magnetic flux lines, assuming it is matched, thereby indicating the variation of the fields over the cell cross section.

Solution   To find characteristic impedance using the above formulas, L and C must be found. First, Maxwell's "Electrostatic" solver is used to find C and the electric field. Figure E13.2.1 shows computed voltage contours for $V = 10$ V between the adjacent plates. In most regions $E = V/d = 33.3$ V/m in the vertical direction. However, near the corners of the plates E is much higher, and away from the plates E decays and changes direction ("fringes"). Maxwell also computes the electric energy stored W_E as 2.2564E–9 J/m for $V = 10$ V between the plates. Then from the capacitance energy formula (6.23), solving for capacitance gives:

$$C = 2 W_{el} / V^2 \qquad (E13.2.1)$$

the capacitance C is found to be 45.13 pF/m.

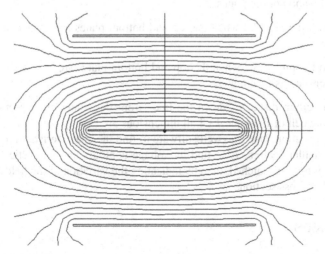

FIGURE E13.2.1   Computer display of triplate cell center cross section and its computed voltage contours or magnetic flux lines.

Next, the same triplate was analyzed for its magnetic fields and inductance using Maxwell's "Eddy Current" solver. To include skin effects expected at audio frequencies (20– 20 kHz) and higher, the plates are assumed to have their currents on their skin, and thus the plate permeability is assumed to be 0.1% that of air. The computed magnetic flux lines appear identical to Figure E13.1.1. For 10 A, the computed magnetic energy is 1.2513E-5 J/m depth. Then from the well-known magnetic energy formula (6.17), solving for inductance gives:

$$L = 2W_{\text{mag}}/I^2 \qquad \text{(E13.2.2)}$$

the inductance L is found as 250.3 nH/m. Then from (13.5), the characteristic impedance of the triplate is 74.5 Ω. This agrees reasonably well with the 75 Ω specified by the triplate manufacturer.

The magnetic flux lines and voltage contours of Figure E13.1.1 show that a large area of the region between the plates has an approximately uniform field. The field is uniform in most locations away from plate corners. Thus, in using triplate cells, the magnetic device under test must be considerably smaller than the cell and should be placed near the center of the cell. If desired, finite-element models of the device under test can be developed [10] and included in the model of the cell.

PROBLEMS

13.1 Redo Example 13.1 at a frequency of 20 kHz.

13.2 Redo Example 13.1 for steel with a relative permeability of 500 and conductivity 1.5E6 S/m.

13.3 Redo Example 13.2 when the center plate is expanded to the same width as the top and bottom plates.

13.4 Redo Example 13.2 when the top and bottom plates are expanded to 75 cm wide.

13.5 Redo Example 13.2 when the top and bottom plates are 75 cm wide and the center plate is 60 cm wide.

13.6 At a certain distance "down the line" from the center cross section of the triplate cell of Example 13.2 all dimensions are exactly one-half those of the center section. Use Maxwell software to find the characteristic impedance of the smaller section. Also find its voltage contours and magnetic flux lines, assuming it is matched, thereby indicating the variation of the fields over this smaller cross section.

REFERENCES

1. Ott HW. *Noise Reduction Techniques in Electronic Systems*. New York: John Wiley & Sons; 1976.

2. Paul CR. *Introduction to Electromagnetic Compatibility.* New York: John Wiley & Sons; 1992.

3. Repas R. Prox sensors for vehicles. *Machine Design* 2005;37.

4. Markel MS, Brauer JR, Brown BS. Using finite element software to predict automotive EMC. *Proceedings of the EMC/ESD International Conference,* Denver, CO, April 1993.

5. Bauer H (ed.). *Automotive Handbook,* 4th ed. Stuttgart, Germany: Robert Bosch GmbH; 1996, also available from SAE, Warrendale, PA, p 638.

6. Brauer JR, Brown BS. Mixed-dimensional finite elements for models of electromagnetic coupling and shielding. *IEEE Trans Electromagn Compat* 1993;35:235–241.

7. Harada T, Hatakeyama K, Inomata M, Masuda N, Fujihara N. Suppressing electromagnetic field leakage through gaps in metal enclosures using magnetic materials. *Proceedings of the 9th International Zurich Symposium and Technical Exhibition on EMC, Zurich, Switzerland,* March 1991, pp 725–730.

8. Brauer JR. Homogenized finite element model of a beam waveguide resonator antenna with over one hundred coupling holes. *Proceedings of the Applied Computational Electr.omagnetic Society Annual Review of Progress,* Monterey, CA, 1998.

9. Brauer JR. EMC of automotive wiring analyzed by finite elements. Paper presented at *SAE Passenger Car Meeting,* Dearborn, MI, September, 1985; also in *SAE Transactions.*

10. Bracken JE, Bardi I, Mathis A, Polstyanko S, Cendes ZJ. Analysis of system level electromagnetic interference from electronic packages and boards. *Proceedings of the IEEE Meeting on Electrical Performance of Electronic Packaging,* Austin, TX, October, 2005.

CHAPTER 14

Electromechanical Finite Elements

For magnetic actuators and sensors, mechanical parameters usually influence performance more than any other nonelectromagnetic parameters. For example, armature mass directly affects the speed of response of a magnetic actuator. Mechanical and electromagnetic behaviors are coupled together and described as *electromechanical* behavior.

This chapter and the next are devoted to electromechanical analysis. The present chapter uses electromechanical finite elements to analyze the behavior of magnetic actuators and sensors.

14.1 ELECTROMAGNETIC FINITE-ELEMENT MATRIX EQUATION

The magnetic finite elements described in Chapter 4 are not sufficient by themselves to completely analyze magnetic actuators and sensors. Subsequent chapters have described the use of finite elements to analyze electric fields and coupled electromagnetic fields. In addition, the electric circuit exciting the actuator or sensor must be included in many analyses.

An electromagnetic finite-element formulation that is capable of analyzing all types of electromagnetic fields is described here. While many other formulations exist, each with its own advantages, the formulation given here is easy to understand and is directly related to the structural finite-element formulation to follow.

As Chapter 4 showed, an energy functional can be used to derive a finite-element matrix equation. In Chapter 4 the functional consisted of magnetic energy and input electrical energy for magnetostatic finite elements. Here for electromagnetic finite elements that include not only magnetostatic energy, but also eddy current power loss, electrostatic energy, and coupled electromagnetic field energy, additional terms must be added. Here instead of minimizing an energy functional w by setting its derivative with respect to an unknown potential p to zero, we set the differential

Magnetic Actuators and Sensors, Second Edition. John R. Brauer.
© 2014 The Institute of Electrical and Electronics Engineers, Inc. Published 2014 by John Wiley & Sons, Inc.

energy variational δw to zero. The complete electromagnetic energy variational is [1]:

$$
\delta w = \int\limits_{\text{vol}} dv \int\limits_{t_o} dt \left\{ \left[\delta \left(\nabla \frac{\partial \psi}{\partial t} \right) + \delta \frac{\partial \mathbf{A}}{\partial t} \right] \cdot [\varepsilon] \left(\nabla \frac{\partial \psi}{\partial t} + \frac{\partial \mathbf{A}}{\partial t} \right) \right.
$$

$$
- \left[\delta \left(\nabla \psi \right) + \delta \mathbf{A} \right] \cdot [\sigma] \left(\nabla \frac{\partial \psi}{\partial t} + \frac{\partial \mathbf{A}}{\partial t} \right) - \delta \left(\nabla \times \mathbf{A} \right) \cdot [\upsilon] \left(\nabla \times \mathbf{A} \right)
$$

$$
\left. - \alpha \delta [\upsilon](\nabla \cdot \mathbf{A}) \cdot (\nabla \cdot \mathbf{A}) - \delta \psi(\rho_v) + \delta \mathbf{A} \cdot \mathbf{J} \right\}
$$

$$
+ \int\limits_{\text{surf}} ds \int\limits_{t_o} dt \left[\delta \mathbf{A} \cdot (\mathbf{H} \times \mathbf{u}_n) - \delta \psi \left(\mathbf{u}_n \cdot \left(\mathbf{J} + \frac{\partial \mathbf{D}}{\partial t} \right) \right) \right]
$$

$$
+ \int\limits_{\text{surf}} ds \delta \Psi (\mathbf{u}_n \cdot \mathbf{D})|_{t_o} \tag{14.1}
$$

where the potential component Ψ is the time integral of the electric scalar potential ϕ_v of (2.39), that is:

$$
\Psi = \int dt \phi_v \tag{14.2}
$$

The first right-hand side energy term in (14.1) is due to stored electric energy, the second term is energy lost due to eddy currents and other conduction currents, and the third is due to stored magnetic energy. The fourth term imposes uniqueness on 3D magnetic vector potential $\mathbf{A}$ by penalizing its divergence by a factor α, and is needed only for 3D nodal finite elements; the 3D tangential vector (edge) elements mentioned in Chapter 4 do not require this term (nor do 2D or axisymmetric elements). The fifth and sixth terms represent volume energy excitations on potential components Ψ and $\mathbf{A}$, respectively. The final surface integral terms (with unit vector normal to the surface) allow energy to be input through the boundaries of the finite-element model.

The 3D electromagnetic finite-element formulation derived by setting the energy variational of (14.1) to zero contains three matrices: $[K_E]$, $[B_E]$, and $[M_E]$. They appear in the electromagnetic finite-element equation [1,2]:

$$
[M_E]\{\ddot{u}_E\} + [B_E]\{\dot{u}_E\} + [K_E]\{u_E\} = \{F_E(t)\} \tag{14.3}
$$

where $[K_E]$ is the reluctance stiffness matrix of (4.13) and is here defined in terms of shape functions N by the finite-element volume integral:

$$[K_E] = \int dv [\nabla \times N]^T [v][\nabla \times N] \qquad (14.4)$$

where the central factor, the reluctivity tensor matrix, is the inverse of the magnetic permeability tensor matrix $[\mu]$. The $[B_E]$ matrix of (14.3) is called the conductance matrix and is proportional to the conductivity tensor according to:

$$[B_E] = \int dv [\nabla N]^T [\sigma][\nabla N] \qquad (14.5)$$

and accounts for energy (power) loss necessary to solve eddy current problems. Finally, to solve problems where electric field energy storage is included, the permittance matrix proportional to permittivity is defined as:

$$[M_E] = \int dv [\nabla N]^T [\varepsilon][\nabla N] \qquad (14.6)$$

and is needed in coupled electromagnetic problems.

The unknown electromagnetic vector $\{u_E\}$ in (14.3) contains in general four components at each finite-element node:

$$\{u_E\} = \begin{bmatrix} A_1 \\ A_2 \\ A_3 \\ \Psi \end{bmatrix} \qquad (14.7)$$

where the three components of magnetic vector potential $\mathbf{A}$ in many cases are x, y, and z components, but may also be in other coordinate systems (global or elemental).

The right-hand side of (14.3) is the electromagnetic excitation vector:

$$\{F_E\} = \begin{bmatrix} J_{\text{vector}} \\ 0 \end{bmatrix}\Bigg|_{\text{node}} + \begin{bmatrix} H_{\text{tan}} \\ J_{\text{norm}} \end{bmatrix}\Bigg|_{\text{surface}} + \begin{bmatrix} \nabla \times \mathbf{H}_c \\ 0 \end{bmatrix}\Bigg|_{\text{volume}} \qquad (14.8)$$

where the vector J_{vector} includes the current densities at the finite-element nodes as described in two dimensions in (4.12), but here allowing for all 3D components. The second term in (14.8) allows for 3D $\mathbf{H}$ fields to be imposed on surfaces of finite-element models, as do the TEM cells and triplate cells of the preceding chapter. Finally, to allow permanent magnet excitations, the third term is added that involves the distribution of the coercive field intensity $\mathbf{H}_c$ of Chapter 5 as detailed elsewhere [1].

Note that (14.3) includes the permittance matrix times the second time derivative (indicated by the double dot) of $\{u_E\}$ and the conductance matrix times the first time derivative (indicated by the single dot) of $\{u_E\}$. It also includes the reluctance matrix times $\{u_E\}$ as expected from Chapter 4. Because of its time derivatives, (14.3) can solve both static and dynamic (time-varying) electromagnetic problems.

The derivation of the general electromagnetic matrix equation (14.3) is based on electromagnetic energy. Instead of minimizing an energy functional as in Chapter 4, an energy variational including all forms of energy is set to zero [1, 2].

Because (14.3) includes all forms of electromagnetic energy as well as time derivatives, it can solve many types of electromagnetic problems. They are the following.

(1) Static problems: electrostatics, magnetostatics, current flow.

(2) AC problems, in which each dot is replaced by $j2\pi f$, where f is any frequency.

(3) Transient problems, in which the dots involve time steps (as in Chapter 9).

(4) Nonlinear magnetostatic problems (with nonlinear B–H curves).

(5) Nonlinear transient problems, using time steps and nonlinear B–H curves.

(6) Modal solutions of resonance problems, usually at high frequencies only encountered by magnetic actuators and sensors when undergoing EMC tests.

14.2   0D AND 1D FINITE ELEMENTS FOR COUPLING ELECTRIC CIRCUITS

The general formulation of (14.3) allows for 3D finite elements, 2D finite elements, and also one dimensional (1D) and zero dimensional (0D) finite elements. The 1D and 0D finite elements allow electric circuits to be included in the finite-element model. While other methods of including electric circuits are available [3], the 0D technique described here is especially easy to understand.

Zero dimensional (0D) finite elements connect nodes independent of their geometric dimensional distances [2]. Since conventional electric circuit elements R, C, and L connect circuit nodes independent of their distance and only dependent on topology, 0D finite elements can model circuit elements. Thus the general formulation (14.3) includes the three 0D finite elements labeled as res, cap, and ind in Figure 14.1. Because circuit nodes have voltage as their potential, the res, cap, and ind elements have the time integral of voltage (14.7) as their potential.

The cap 0D finite element of C farads contributes to the permittance $[M_E]$ matrix the terms:

$$[M_c] = C \begin{bmatrix} 1 & -1 \\ -1 & 1 \end{bmatrix} \tag{14.9}$$

The res 0D finite element of R ohms contributes to the conductance $[B_E]$ matrix the terms:

$$[B_R] = \frac{1}{R} \begin{bmatrix} 1 & -1 \\ -1 & 1 \end{bmatrix} \tag{14.10}$$

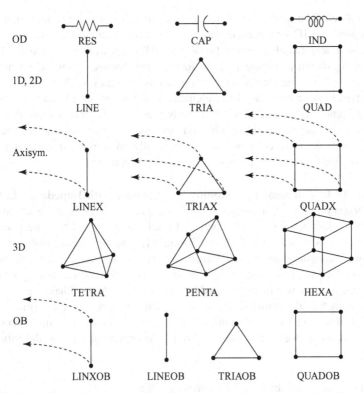

FIGURE 14.1   Finite elements of various dimensions used in the general formulation (14.3).

Finally, the ind 0D finite element of L henrys contributes to the reluctance $[K_E]$ matrix the terms:

$$[K_L] = \frac{1}{L} \begin{bmatrix} 1 & -1 \\ -1 & 1 \end{bmatrix}$$ (14.11)

Inserting the above three terms in (14.3) yields the equation:

$$C\frac{\partial V}{\partial t} + \frac{V}{R} + \frac{1}{L} \int V dt = I$$ (14.12)

where V is the voltage difference between the two nodes and I is their current. Because (14.12) agrees exactly with Kirchhoff's current law of conventional electric circuit theory for a parallel RLC circuit, the 0D finite-element method of analyzing circuits is validated.

To couple the above 0D circuit finite elements into a 3D or 2D finite-element model, 1D finite elements can be used. As shown in Figure 14.1, there are two types of 1D elements, the line and the linex. The linex is used only for axisymmetric problems, while the line is more general.

The line element usually is made highly conductive in order to represent wires and/or coils. The 1D line element can then connect 0D circuit elements into 2D or 3D finite-element models of magnetic devices. All of the available 2D and 3D finite elements are shown in Figure 14.1, including *open boundary* finite elements. The balloon boundary finite elements used in previous examples, such as Example 13.1, are one type of open boundary finite element. The types of open boundary elements shown in Figure 14.1 are called ABC elements, which stands for *asymptotic boundary condition* or (for high frequency problems) *absorbing boundary condition*. The ABC elements have the features of producing essentially no increase in computer times and serving well for static [4], transient, and AC problems [5].

Example 14.1   Constant Permeability Transformer with Impedance Loading its Secondary   The transformer shown in Figure E14.1.1 is made of constant permeability laminated (zero conductivity) steel. As discussed for linear variable differential transformers (LVDTs) in Chapter 11, placing a load impedance on a transformer secondary affects transformer currents and voltages. The transformer in Figure E14.1.1 has a 1-Ω resistive load impedance on its four turn secondary winding. Its one turn primary winding is excited by 120-V 60-Hz AC and has a resistance of 0.1 Ω. The rectangular primary and secondary windings are separated from the core by 0.25 m. The core has dimensions shown in Figure E14.1.1. Use the formulation described in the above sections to find both transformer currents for the following cases.

(a) The core permeability is 1.E5 times that of air.
(b) The core permeability is 500 times that of air.

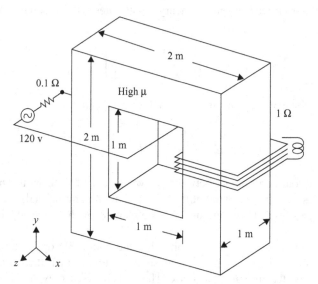

FIGURE E14.1.1   Transformer of Example 14.1.

(c) Same as (b) but the secondary winding is made nonrectangular by reducing its separation from the primary from 0.5 to 0.25 m over one-half of its depth in the z direction.

Solution   The AC methods of this chapter were used [6] to obtain the results listed in Table E14.1.1. For case (a), the currents in the row with relative permeability 1.E5 were computed, along with the flux line plot shown in Figure E14.1.2 with the visible 3D, 1D, and 0D finite elements. Note that the computed primary current I_1 and secondary current I_2 agree closely with those of an ideal transformer listed in the top row of Table E14.1.1, which have a ratio exactly equal to their turns ratio.

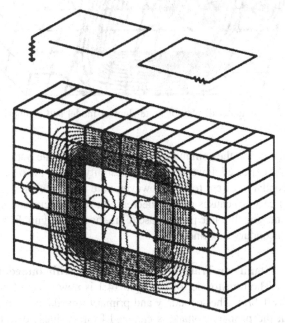

FIGURE E14.1.2   Computer display of finite-element model and computed flux lines for Example 14.1 case (a).

TABLE E14.1.1   Calculated Transformer Current Magnitudes

Core μ_r	Secondary Coil	I_1 (A)	I_2 (A)
Infinity	Ideal	738.4	184.6
1.E5	Rectangle	738.5	184.6
500	Rectangle	1099.1	93.5
500	3D, closer	1098.9	93.6

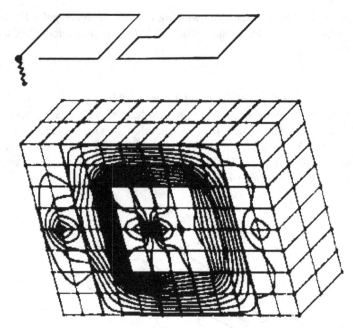

FIGURE E14.1.3   Computer display of finite-element model and computed flux lines for Example 14.1 case (c).

For cases (b) and (c), the computed currents are listed in the bottom two rows of Table E14.1.1. As expected, lowering the permeability causes the primary current to rise and the secondary current to fall. However, note that the closer coupling of the two windings of case (c) causes a slight decrease in primary current and increase in secondary current. The flux line plot of case (c) is shown in Figure E14.1.3, showing the shift in the secondary flux lines.

Example 14.2   Nonlinear Transformer Waveforms with Impedance Loading its Secondary   The transformer of Example 14.1 is now made of nonlinear steel with conductivity 1 S/m. The secondary and primary impedances remain as in Figure E14.1.1, but the primary voltage is changed from a sinusoid to the triangular waveform shown in Figure E14.2.1 of period 16 ms with 1200 V peak. This large voltage produces saturation, and the steel is assumed to have the B–H curve listed elsewhere [2]. The waveforms of steel flux density and secondary current are to be computed during two periods after the primary voltage is applied.

Solution   The nonlinear transient methods of this chapter were used [6] to obtain the waveforms shown in Figures E14.2.2 and E14.2.3. Note that the flux density waveform is "clipped" at approximately 1.5 T due to saturation, as one might expect. Because the secondary voltage and current are proportional to the time derivative of flux density according to Faraday's law, the current waveform is "peaked." Thus saturation effects are fully computed, including the losses in the steel [6].

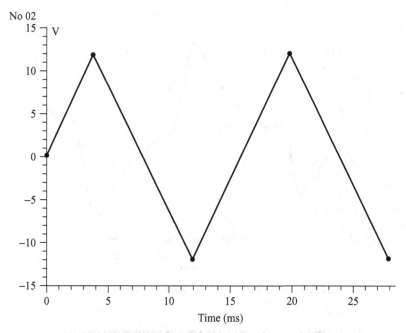

FIGURE E14.2.1   Applied primary voltage waveform for Example 14.2.

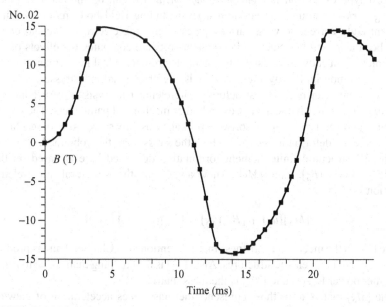

FIGURE E14.2.2   Computed flux density in transformer steel core of Example 14.2.

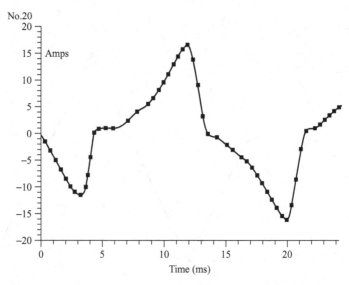

FIGURE E14.2.3 Computed secondary current waveform for Example 14.2.

14.3 STRUCTURAL FINITE-ELEMENT MATRIX EQUATION

To allow motion of the armatures of magnetic actuators and sensors, some of the electromagnetic finite elements shown in Figure 14.1 must be allowed to move. The simplest type of motion is *rigid body motion*, and it can be modeled by placing *moving surfaces* around 3D armatures. By including rigid body motion in finite-element eddy current solvers, various types of moving electromechanical devices have been analyzed [7]. Such finite-element software computes the effects of rigid body motion on eddy currents and transient electromechanical performance.

To allow nonrigid body motion, that is, deformations and stresses, structural finite elements can be used. Structural finite-element analysis is very commonly used to compute mechanical stresses and other mechanical parameters. The method, available in several commercial software packages, usually solves first for mechanical displacements (deformations), from which the stresses are then obtained.

The 3D structural finite-element formulation described here is based on three matrices: $[K_S]$, $[B_S]$, and $[M_S]$. They appear in the structural finite-element equation [2]:

$$[M_S]\{\ddot{u}_S\} + [B_S]\{\dot{u}_S\} + [K_S]\{u_S\} = \{F_S(t)\} \tag{14.13}$$

where $[K_S]$ is the mechanical stiffness matrix mentioned in Chapter 4 and is produced by springs and other elastic structures. Also, $[B_S]$ is the damping matrix due to friction and other power losses, and $[M_S]$ is the mass matrix.

The $[M_S]$ matrix term thus represents the mass times acceleration of Newton's second law of motion. The $[K_S]$ matrix term represents Hooke's law of spring force, while the $[B_S]$ matrix term represents forces observed in dashpots. All three matrices are common to magnetic actuators and sensors. The matrices are assembled element by element, where various structural elements are available. Detailed matrix derivations for structural finite-element types such as truss, rod, bar, beam, triangular plate, quadrilateral plate, and hexahedron appear in books [2, 8].

Appendix A includes symbols, dimensions, and units of mechanics. Also, Figure 3.2 shows the key relations for a simple mechanical structure and compares them with those for magnetic circuits, electric circuits, and heat conduction.

The unknown structural vector $\{u_S\}$ in (14.13) here contains components at each finite-element node:

$$\{u_S\} = \begin{bmatrix} u_1 \\ u_2 \\ u_3 \end{bmatrix} \tag{14.14}$$

where the three components of translation u in many cases are x, y, and z components, but may also be in other coordinate systems.

The right-hand side of (14.13) is the structural excitation vector, consisting of the applied forces at nodes. The forces may be produced by electromagnetic fields as described in Chapter 5.

Note that the structural equation (14.13) is directly analogous to the electromagnetic equation (14.3). Both include an $[M]$ matrix times the second time derivative (indicated by the double dot) of $\{u\}$ and a $[B]$ matrix times the first time derivative (indicated by the single dot) of $\{u\}$. They both also include a $[K]$ matrix times $\{u\}$. Because of their time derivatives, (14.3) and (14.13) can solve both static and dynamic (time-varying) problems.

14.4   FORCE AND MOTION COMPUTATION BY TIME STEPPING

Coupled electromechanical problems can be solved in the time domain by solving both (14.3) and (14.13) at each time step [9]. The two equations are coupled at each time step through their right-hand side excitation vectors. The electromagnetic excitation vector has additional terms due to motion. All finite elements can have both electromagnetic and structural material properties.

The Maxwell stress tensor is used to compute the distributed structural force produced by electromagnetic fields at surface nodes. Chapter 5 presented the component of Maxwell stress normal to a surface; the complete nonlinear stress tensor $[\tau]$ for all pressure components relates pressure and direction n according to [10]:

$$\{P\} = [\tau]\{n\} \tag{14.15}$$

$$\begin{bmatrix} P_x \\ P_y \\ P_z \end{bmatrix} = \begin{pmatrix} B_x H_x - w_m & B_x H_y & B_x H_z \\ B_y H_x & B_y H_y - w_m & B_y H_z \\ B_z H_x & B_z H_y & B_z H_z - w_m \end{pmatrix} \begin{bmatrix} n_x \\ n_y \\ n_z \end{bmatrix} \tag{14.16}$$

where w_m is linear or nonlinear magnetic energy density as in Chapter 5 and n_x, n_y, and n_z are components of the unit vector $\boldsymbol{n}$ normal to the surface. Integrating the above pressure over a closed surface s gives the force component F_{Si} in the direction i [10]:

$$F_{Si} = \oint (\tau_{ij} n_j)ds \qquad (14.17)$$

Thus

$$\frac{F_{Si}}{s} = P_i = \tau_{ij} n_j \qquad (14.18)$$

Substituting (14.16) gives the pressure in terms of normal N and tangent T components of the magnetic fields [10]:

$$P_N = B_N H_N - w_m; \quad P_T = B_T |H_N| \qquad (14.19)$$

As shown in Figure 14.2 [10], there are two ways of computing magnetic force on an object. The first method is to completely surround the object with the surface, such as the armature of a magnetic actuator. This one-sided integration method

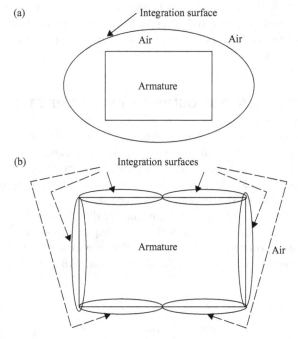

FIGURE 14.2   Integration surfaces for Maxwell stress tensor computation. (a) one-sided integration, (b) two-sided integration.

obtains the total force on the armature, but not its distribution. The second method is to place multiple integration surfaces encompassing both steel and air sides of all finite elements on the armature surface. As these integration regions are collapsed to the armature surface, they perform two-sided integrations along the armature face. Because of the higher **H** field and magnetic energy density in air, the force on steel and other high permeability materials is outward toward the higher energy density air region. Also, if steel has regions of varying saturation, the two-sided integrations can be carried out at finite-element interfaces within the steel. Thus the two-sided method computes the spatial *distribution* of force, which is helpful in understanding actuator behavior and is essential in obtaining mechanical stresses.

As the motion proceeds, nodes can move. In the examples given below, air is modeled with a very low modulus of elasticity of 1 Pa, which produces very small contributions to the $[K_S]$ matrix. Thus the air elements easily collapse or expand as motion occurs. If the motion is very large, then some air elements may become very distorted, which reduces their accuracy. For large motion, it may be necessary to not only move nodes but also reconnect the nodes through new finite elements.

14.5 TYPICAL ELECTROMECHANICAL APPLICATIONS

In the examples below the motion is small enough that no element reconnection is required. All results are obtained by time stepping using both nonlinear electromagnetic and structural finite elements. All eddy current effects, including motional EMFs, are included in the solutions. While in structural finite-element analysis air is rarely modeled, air finite elements are here required for electromagnetic fields and also appear in the structural finite-element model. The key parameter usually needed for magnetic actuators and sensors is response time, which is computed as follows by electromagnetic and structural finite elements.

14.5.1 DC Solenoid with Slowly Rising Current Input

The Bessho DC axisymmetric plunger solenoid actuator of Figure 5.2 has been previously studied in Chapters 5, 7, 9, and 12. As mentioned in Chapter 12, its time constant is 20 ms according to the original paper by Bessho et al. [11]. The current waveform applied in the Bessho experimental tests obeys:

$$I(t) = I_{DC}(1 - e^{-t/0.020}) \qquad (14.20)$$

where the final current I_{DC} is 0.5 A during the experimental tests.

To compute the motion of this actuator, the finite-element model shown in Figure 14.3 is made with all elements having both electromagnetic and structural properties [9]. The steel finite elements are given a modulus of elasticity of 200.E9 Pa and Poisson's ratio of 0.265, as typical for many steels [2]. As mentioned previously, the electromagnetic steel elements have nonlinear *B–H* curves and electrical conductivity of 1.7E6 S/m. The air finite elements are given a structural modulus of

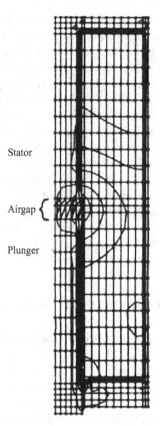

Stator

Airgap {

Plunger

FIGURE 14.3 Computer display of coupled electromagnetic and structural finite-element model of Bessho magnetic actuator with applied current. The computed flux lines are shown at 5 ms for the 15-mm airgap.

elasticity of 1 Pa. Initial airgaps are set to match the experimental tests at 10, 15, and 20 mm. As set in the experiments, the total moveable mass is 6 kg.

The computed position of the armature versus time is shown in Figure 14.4 for all three initial airgaps [9]. Also shown in Figure 14.4 are the corresponding measured position curves. Note that computations and measurements agree quite well. The largest disagreement is for the 20-mm airgap, which produces the greatest node motion and element distortion. Note that the time to close, often the most important parameter for magnetic actuators, is accurately predicted by the coupled finite-element analysis.

14.5.2 DC Solenoid with Step Voltage Input

Like most DC actuators, the Bessho solenoid is most likely to be excited by a DC voltage step. As mentioned in Chapter 12, the time constant of the Bessho solenoid of 20 ms is unchanged even if its number of turns is changed. However, it is important

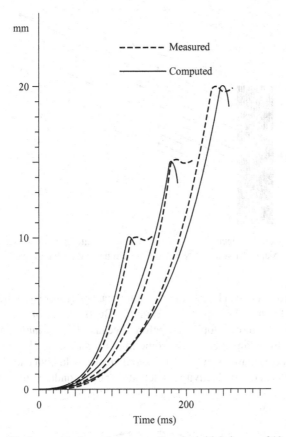

FIGURE 14.4   Displacement of armature versus time for initial airgaps of 10, 15, and 20 mm. All curves are for the applied current waveform of (14.19) with 0.5-A DC and total moving mass of 6 kg.

to note that the current waveform of (14.19) is based upon conventional LR circuit theory and does not include the effects of motional EMF. As we shall see, when a voltage is applied, *motion effects can change the current waveform and thus change the actuator response time.*

To apply a voltage, 0D and 1D finite elements are added to Figure 14.3, producing the finite-element model shown in Figure 14.5 [12]. The 0D element is a resistor of value 200 Ω, and it is driven by a step voltage source of 100 V. The final DC current is 100 V/200 Ω = 0.5 A, the desired value. The initial airgap is assumed to be 10 mm.

The computed current versus time is shown in Figure 14.6 [12]. Note that for times less than 40 ms, the current increases exponentially with a time constant of approximately 20 ms as in Section 12.2. However, from approximately 40 to 120 ms, the current *decreases* due to the motional voltage produced by the moving armature. The computed armature position versus time is approximately the same as the 10-mm curve in Figure 14.4, which shows substantial motion beginning at approximately

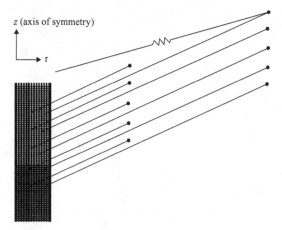

FIGURE 14.5 Coupled electromagnetic and structural finite-element model of Bessho magnetic actuator with applied voltage. Note that the circuit is distributed over 10 points within the coil area.

40 ms. At approximately 117 ms, the armature reaches closure and thus at that time the current of Figure 14.6 dips to a minimum. Beyond that time, the current rises again, but with a different slope than before, because L has changed with armature position. The current dip at closure is sometimes used to sense closure.

The above computations are for measured steel conductivity of 1.7E6 S/m. To examine the effects of conductivity on actuator speed, conductivity σ was varied as

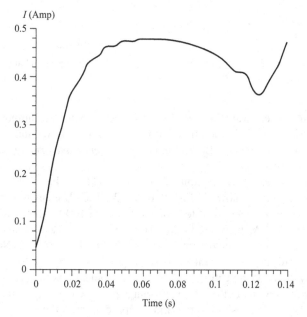

FIGURE 14.6 Computed current versus time for Bessho actuator with applied step voltage.

TABLE 14.1 Closing Times (ms)^a Computed Using Electromechanical Finite Elements for Various Step Excitations of *V* or *I*, Conductivities σ, and Mechanical Parameters

Case	$\sigma = 0$	%	$\sigma = 1.7\text{E}6$	%	$\sigma = 5\text{E}6$	%	$\sigma = 10\text{E}6$	%
1. *V*, 6 kg	115	0	<u>117</u>	1.7	121	5.2	126	9.6
2. *V*, 1.4 kg	67	0	72	7.5	75	11.9	80	19.4
3. *I*, 1.4 kg	46	0	51	10.9	56	21.7	62	34.8
4. *I*, 1.4 kg, spring	49	0	54	10.2	61	24.5	67	36.7

^{a}The percentage changes in closing times due to eddy currents are computed, with Bessho's baseline case underlined.

listed in Table 14.1. Besides varying σ from 0 to 10.E6 S/m, four different cases are examined in the table. Case 1 is the voltage step examined in Figure 14.6; it shows that varying conductivity only varies closing time by about 11 ms. In case 2, the moving mass is lowered to the armature mass only, which is 1.4 kg. In case 3, the 0.5-A current is input as a step at time zero, producing the fastest closing times, especially for zero conductivity. Finally, in case 4 a small opposing spring of 1500 N/m is added, typical of relay spring forces [13].

14.5.3 AC Clapper Solenoid Motion and Stress

The Eaton 60-Hz AC clapper solenoid actuator of Figures 4.3 and 8.4 can be analyzed using coupled nonlinear electromagnetic and structural finite elements. The spring in Figure 4.3 is made up of multiple steel coils and relay leaves, which interact to produce the nonlinear spring force versus position curve shown in Figure 14.7. When the armature is in the open position with the 5.97-mm airgap of Figure 4.3, the spring is partially compressed to produce a preload of 2.55 N. The slope of Figure 14.7, called K_{Snl}, varies from as low as 443 N/m to as high as 3.E4 N/m [14]. When the

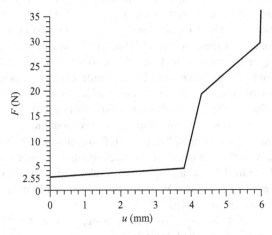

FIGURE 14.7   Measured nonlinear spring curve of actuator spring and relay of Eaton AC actuator.

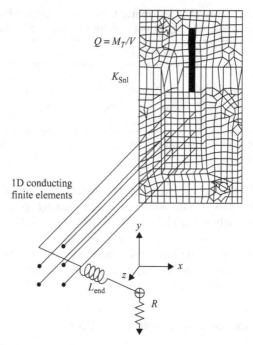

FIGURE 14.8   Coupled finite-element model of right-half of Eaton AC actuator.

airgap is closed at 5.97 mm, the slope in Figure 14.7 becomes very high to prevent further motion.

The nonlinear spring K_{Snl} is included in the coupled finite-element model shown in Figure 14.8. Just as for the Bessho actuator, the total moving mass M_T is somewhat greater than simply the mass of the magnetic armature. Thus the mass density ρ of the armature structural finite elements is set to M_T divided by the armature volume. The steel 2D finite elements have the same structural properties as in the Bessho actuator, with modulus of elasticity 2.E11 Pa and Poisson's ratio of 0.265.

The electromagnetic finite-element model is also shown in Figure 14.8, and consists of 0D, 1D, and 2D finite elements. The 2D finite-elements model the nonlinear steel, the air, the coil area, and the aluminum shading ring. To account for 3D end resistance of the shading ring, the aluminum conductivity is reduced somewhat [14]. Figure 14.8 also contains six 1D line finite elements modeling the coil, which are attached to 0D finite elements modeling the coil R and end turn inductance L_{end}. The end turn inductance computed using 3D finite elements is 23 mH [14]; adding it to the 2D model of Figure 14.8 ensures accurate results with minimal computer time. The model of Figure 14.8 ignores AC hysteresis losses in the steel, but if they are significant they can be included [15].

The time required for closure of AC actuators depends on the instant within the AC cycle that the voltage is applied. Here the voltage applied at time zero is assumed to be zero and to rise thereafter as a sine wave. The resulting computed current

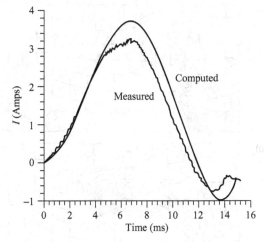

FIGURE 14.9   Coil current waveform of Eaton AC actuator.

versus time is shown in Figure 14.9. The measured current is also shown and agrees reasonably well.

The computed armature displacement versus time is shown in Figure 14.10. The leftmost curve is for an armature density ρ equal to 0.22 kg divided by the model armature volume v. However, it was realized later that the potentiometer used to trace the measured curve in Figure 14.10 added to the moving mass and produces $M_T = 0.267$ kg. Increasing the armature density accordingly yields the rightmost curve in Figure 14.10, which agrees very closely with the measured curve. It should be noted that the force of gravity is also included in Figure 14.10 because the actuator

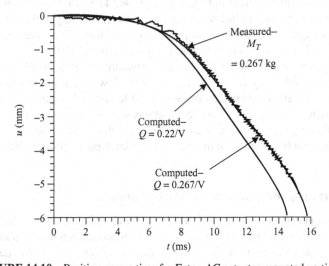

FIGURE 14.10   Position versus time for Eaton AC actuator mounted vertically.

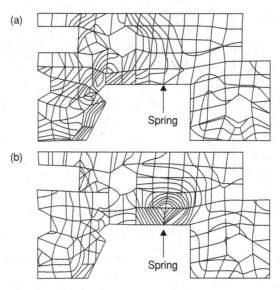

FIGURE 14.11 Eaton armature finite elements and stress contours (also available in color). (a) at $t = 15.6$ ms, maximum = 943 kPa, (b) at $t = 12$ ms, maximum = 26 kPa.

is mounted vertically. Both the computed and measured closure times in Figure 14.10 are 16.0 ms.

Besides computing the electrical and mechanical responses of Figures 14.9 and 14.10, the coupled finite-element analysis also obtains mechanical stresses. A common parameter for stress is called Von Mises equivalent stress [2], and it is plotted for the armature in Figure 14.11. The stress varies with time and with assumed spring position. The maximum stress in Figure 14.11 is 943 kPa, which is too low to cause yield or other mechanical failure of most steels. However, improper placement of bolt holes might increase the stresses and produce mechanical failure. Hence computed stress distributions can be very valuable to design engineers [16].

Since the stresses and strains of the steel armature are quite small, its deformations are also small, only a few microns at most. However, armatures of magnetic actuators are sometimes purposely made of flexible material which deforms greatly during actuation. Examples of flexible armatures are common in *magnetic switches,* including flexible reeds in *reed switches* used as relays and sensors [17, 18], and flexible ferrite rubber permanent magnets used to latch push-button switches [19].

14.5.4 Transformers with Switches or Sensors

Magnetic actuators and sensors are often driven by electronic circuits, which usually require nonlinear modeling. The simplest type of nonlinear behavior is a switch. An ideal open switch has zero current, but its voltage drop is unknown. An ideal closed switch has zero voltage, but its current is unknown.

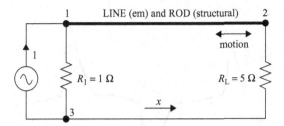

FIGURE 14.12   Time-varying resistor.

A switch is thus a time-varying resistor, which can be modeled using 1D coupled electromagnetic and structural finite elements [20, 21]. The 1D finite element in Figure 14.12 is a typical variable resistor. It behaves as a line electromagnetic finite element of Figure 14.1 and also as a 1D structural rod element. To vary the resistance, the line is stretched by applying a mechanical force that varies with time. For example, if the line has a conductivity of 1 S/m and initial length of 1 cm, then its resistance is only 0.01 Ω initially. Applying a large force to stretch the line length to 1 km, its resistance increases to 1 kΩ.

Such a switch is attached to the transformer secondary shown in Figure 14.13. Its primary is excited by a 50-Hz triangular wave. If the switch is always closed, the 86 mΩ secondary resistor has the computed voltage shown in Figure 14.14 over time from 0 to 25 ms. If the switch is opened, however, from time 15 to 20 ms, the secondary current is zero during that period but shows an inductive "kick" when reclosed in Figure 14.15.

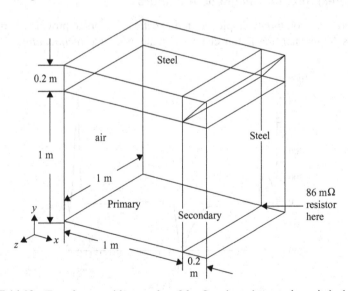

FIGURE 14.13   Transformer with secondary 86 mΩ resistor that may be switched in or out. One-quarter of the transformer is shown, with its primary and secondary coils modeled by 1D line finite elements.

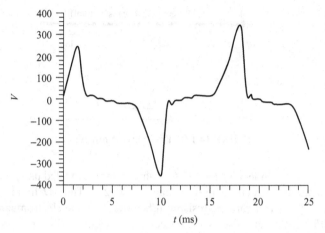

FIGURE 14.14   Computed secondary resistor voltage for transformer of Figure 14.13 when secondary switch is always closed.

Such variable resistance models can be useful for modeling electronically controlled actuators and sensors. For example, LVDTs of Chapter 11 can have varying load impedances, and magnetoresistive sensors of Chapter 10 typically behave as time-varying resistors.

14.5.5   Reciprocating Magnetic Actuators

Springs are very commonly employed in magnetic actuators to provide return force. Examples (other than the Eaton clapper actuator) include *reciprocating magnetic*

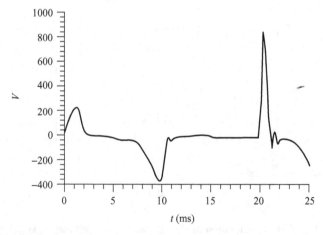

FIGURE 14.15   Computed secondary voltage for transformer of Figure 14.13 when secondary switch is closed except during time from 15 to 20 ms.

actuators. They produce repeated back-and-forth linear motion by alternately sup-plying coil current to provide magnetic force and motion in one direction, and remov-ing (or in some designs reversing) the coil current to allow the spring to move the armature back. Door bells have long been a common example.

Nowadays new types of reciprocating magnetic actuators are being developed for applications such as shakers [22] and piston compressors [23]. Ordinarily most compressor pistons are driven by rotary motors via crankshafts, connecting rods, and wrist pins. These three mechanical parts are eliminated by use of a reciprocating actu-ator (also often called a reciprocating motor). A new reciprocating magnetic actuator designed for operating compressors for refrigeration is shown in Figure 14.16. Its outer stator has coil windings that are fed with AC voltage and current. Preferably the frequency of the current is the readily available 60 Hz or 50 Hz. The armature in Figure 14.16 has NdFeB permanent magnets mounted on a nonpermeable tube. The actuator design is somewhat similar to a much older design [24] which had difficulty controlling its stroke and thus the compression ratio. Accurate control of stroke is much easier nowadays using modern magnetic sensors and feedback control electronics.

The mechanical resonant frequency of the simple spring-mass system of the arma-ture of Figure 14.16 is derived from (14.13) to be:

$$f_{res} = (1/(2\pi))(K/M)^{1/2} \tag{14.21}$$

where K is the armature spring constant and M is the total reciprocating mass. This resonant frequency is designed to match the AC drive frequency for maximum efficiency of the compressor system [23].

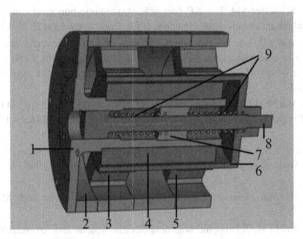

FIGURE 14.16   Reciprocating magnetic actuator for a compressor [23]. Parts are: (1) com-pression chamber, (2) outer shell, (3) left outer steel stator with copper winding (not shown), (4) inner steel stator, (5) right outer stator, (6) permanent magnet armature, (7) linear bearing, (8) piston in compression chamber, (9) spring with spring constant K.

PROBLEMS

14.1 The transformer of Example 14.1 can also be analyzed using its equivalent electrical circuit. As discussed in Chapter 8, a coil with losses is equivalent to a resistor in parallel with an inductor. From Chapter 5, the parallel magnetizing inductor equals the square of the primary turns divided by the reluctance. The secondary resistor must be referred to the primary according to ideal transformer theory. Calculate the reluctance for relative steel permeability of 1.E5 and the resulting magnetizing inductor. Then use the inductor to calculate the currents and compare them with the finite element computations of case (a).

14.2 Extend Problem 14.1 for case (b) of Example 14.1. That is, let the steel relative permeability be 500 and compare your calculated currents with the finite-element computations of case (b).

REFERENCES

1. MacNeal BE, Brauer JR, Coppolino RN. A general finite element vector potential formulation of electromagnetics using a time-integrated electric scalar potential. *IEEE Trans Magn* 1990;26:1768–1770.

2. Brauer JR (ed.). *What Every Engineer Should Know About Finite Element Analysis*, 2nd ed. New York: Marcel Dekker, Inc.; 1993.

3. Zhou P, Fu WN, Lin D, Stanton S, Cendes ZJ. Numerical modeling of magnetic devices. *IEEE Trans Magn* 2004;40:1803–1809.

4. Brauer JR, Schaeffer SM, Lee J-F, Mittra R. Asymptotic boundary condition for three dimensional magnetostatic finite elements. *IEEE Trans Magn* 1991;27:5013–5015.

5. Brauer JR, Mittra R, Lee J-F. Absorbing boundary condition for vector and scalar potentials arising in electromagnetic finite element analysis in frequency and time domains. *Proceedings of the IEEE Antennas and Propagation Society Symposium*, July 1992, pp 1224–1227.

6. Brauer JR, MacNeal BE, Hirtenfelder F. New constraint technique for 3D finite element analysis of multiturn windings with attached circuits. *IEEE Trans Magn* 1993;29:2446–2448.

7. Zhou P, Stanton S, Cendes ZJ. Dynamic modeling of electric machines. *Proceedings of the Naval Symposium on Electrical Machines*, October 26–29, 1998, Annapolis, MD, pp 43–49.

8. MacNeal RH. *Finite Elements: Their Design and Performance*. New York: Marcel Dekker, Inc.; 1994.

9. Brauer JR, Ruehl JJ. 3D coupled nonlinear electromagnetic and structural finite element analysis of motional eddy current problems. *IEEE Trans Magn* 1994;30:3288–3291.

10. Brauer JR, Ruehl JJ, Juds MA, Van der Heiden MJ, Arkadan AA. Dynamic stress in magnetic actuator computed by coupled structural and electromagnetic finite elements. *IEEE Trans Magn* 1996;32:1046–1049.

11. Bessho K, Yamada S, Kanamura Y. Analysis of transient characteristics of plunger type electromagnets. *Electr Eng Jpn* 1978;98:56–62.

12. Brauer JR, Ruehl JJ, Hirtenfelder F. Coupled nonlinear electromagnetic and structural finite element analysis of an actuator excited by an electric circuit. *IEEE Trans Magn* 1995;31:1861–1864.

13. Brauer JR, Chen QM. Alternative dynamic electromechanical models of magnetic actuators containing eddy currents. *IEEE Trans Magn* 2000;36:1333–1336.

14. Juds MA, Brauer JR. AC contactor motion computed with coupled electromagnetic and structural finite elements. *IEEE Trans Magn* 1995;31:3575–3577.

15. Bottauscio O, Chiampi M, Manzin A. Advanced model for dynamic analysis of electromechanical devices. *IEEE Trans Magn* 2005;41:36–46.

16. Overbye V, Brauer J, Bodine P. *Blazing Trails—Finite Element Pioneers in Milwaukee*. Lincoln, NE: Writers Club Press of iUniverse, Inc.; 2002.

17. Reed switch basics. *Machine Design* 2004;48.

18. Ota T, Hirata K, Yamaguchi T, Kawase Y, Watanabe K, Nakase A. Dynamic response analysis of opening and closing sensor for windows. *IEEE Trans Magn* 2005;41:1604–1607.

19. Van Zeeland T. Magnets in switch design. *Magn Bus Technol* 2004;22–23.

20. Brauer JR, Ruehl JJ. Finite element modeling of power electronic circuits containing switches attached to saturable magnetic components. *IEEE Trans Energy Convers* 1999;14:589–594.

21. Brauer JR. Time-varying resistors, capacitors, and inductors in nonlinear transient finite element models. *IEEE Trans Magn* 1998;34:3086–3089.

22. Peng M-T, Flack TJ. Numerical analysis of the coupled circuit and cooling holes for an electromagnetic shaker. *IEEE Trans Magn* 2005;41:47–54.

23. Zhang Y, Lu Q, Yu M, Ye Y. A novel transverse-flux moving-magnet linear oscillatory actuator. *IEEE Trans Magn* 2012;48:1856–1862.

24. Brauer JR. Reciprocating linear motor. U.S. Patent 4002935, 1977.

CHAPTER 15

Electromechanical Analysis Using Systems Models

While the finite-element techniques of the preceding chapter accurately predict electromechanical behavior of magnetic devices, many engineers need *systems models* instead. Magnetic actuators and sensors are often components in large electromechanical systems such as automobiles, airplanes, power systems, and computers. Design of such systems is carried out nowadays using systems software such as SPICE, MATLAB®, and Simplorer®. Hence this chapter discusses how magnetic actuators and sensors can be modeled with such systems software.

15.1 ELECTRIC CIRCUIT MODELS OF MAGNETIC DEVICES

15.1.1 Electric Circuit Software Including SPICE

Since most magnetic devices are connected to electric circuits, electric circuit simulation software is appropriate. If an equivalent electric circuit can be found for the magnetic device, then it can be inserted in the circuit software. Popular software for modeling electric and electronic circuits includes SPICE, Simplorer, and Multisim. Electric circuits can be used via analogies to also model the magnetic circuits of Chapter 3, heat transfer of Chapter 12, and mechanical motion as described below.

SPICE stands for Simulation Program with Integrated Circuit Emphasis. It is one of the most commonly used electric and electronic circuit simulators. SPICE was originally developed in the early 1970s [1] at the University of California, Berkeley, which has published several SPICE versions in the public domain. Commercial versions include PSPICE, HSPICE, and Maxwell SPICE. At the present time a version of SPICE called TINA-TI™ is available at www.ti.com, and another free version called LT spice is available at www.linear.com. SPICE models have been developed over the years for many electronic circuits and magnetic devices.

Magnetic Actuators and Sensors, Second Edition. John R. Brauer.
© 2014 The Institute of Electrical and Electronics Engineers, Inc. Published 2014 by John Wiley & Sons, Inc.

15.1.2 Simple *LR* Circuits

As discussed in earlier chapters such as Chapter 9, magnetic devices have an equivalent electric circuit that contains at least one inductance *L*. Any coil has a resistance *R*, usually placed in series with *L*. However, the inductance may be nonlinear, and thus represented as flux linkage versus current. In addition, as the armature moves, the inductance and flux linkage may vary significantly with position.

Example 15.1   SPICE *LR* Model of Axisymmetric Clapper Actuator   The DC axisymmetric clapper armature actuator of Figure E7.2.1 is to be modeled in a very simple fashion as an *LR* circuit at the modeled airgap of 2 mm. The coil resistance is 10 Ω. Eddy current losses are to be ignored. The number of turns is 2000. Prepare a SPICE circuit model and use it to find the current as a function of time for 1 s after a 12 V step is applied.

Solution   The total energy stored *W* in Example 7.2 is 0.8604 J for 2000 ampere-turns. Since the number of turns is 2000, the current $I = 1$ A. The inductance is then:

$$L = 2W/I^2 = 1.7208 \text{ H} \tag{E15.1.1}$$

The SPICE circuit model consists of the 1.7208-H inductor in series with the 10-Ω coil resistance, along with a 12-V source turned on at time zero. The circuit developed in Maxwell SPICE is shown in Figure E15.1.1. The computed current versus time is shown in Figure E15.1.2, and shows the expected exponential rise with time constant $L/R = 0.17208$ s.

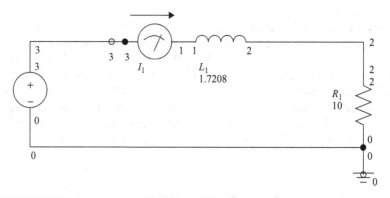

FIGURE E15.1.1   SPICE circuit series *LR* circuit for Example 7.2 at given position.

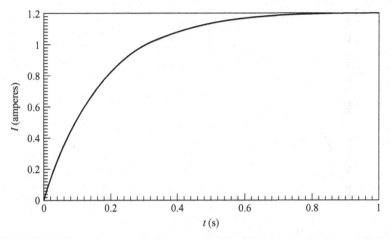

FIGURE E15.1.2   Current waveform computed by SPICE model of Figure E15.1.1.

Example 15.2   Including a Speed Voltage in Example 15.1   The above LR circuit is to be modified by including a speed voltage (of polarity to oppose the current) for $t = 0$ to $t = 0.8$ s:

$$V_{\text{speed}} = 10t^2 \qquad (E15.2.1)$$

After $t = 0.8$ s the armature stops moving, and thus the above voltage becomes zero. The inductance is assumed to be unchanged from Example 15.1 even though the armature position is changing.

Solution   A voltage source is added to the SPICE circuit of Example 15.1, obtaining the circuit shown in Figure E15.2.1. The new voltage source varies with time according to (E15.2.1) and has polarity opposing the original applied voltage. The

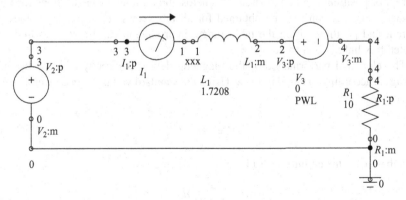

FIGURE E15.2.1   SPICE circuit series LR circuit of Example 15.1 with added speed voltage.

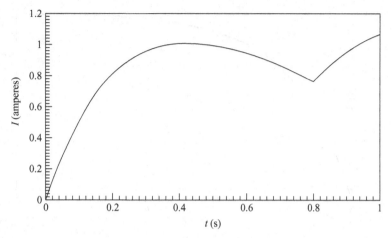

FIGURE E15.2.2 Current waveform computed by SPICE model of Figure E15.2.1.

computed current waveform is shown in Figure E15.2.2. Note that the speed voltage has produced a dip in the current at closure ($t = 0.8$ s), agreeing with the preceding chapter.

15.1.3 Tables of Nonlinear Flux Linkage and Force

In many magnetic actuators and sensors, armature motion causes large changes in flux linkage and inductance. Besides varying with armature position, flux linkage and magnetic force often vary nonlinearly with current because of nonlinear *B–H* curves. Thus current and position are typically two independent parameters which produce variations in flux linkage and force.

To account for such parametric variations, parametric finite-element analysis can be performed. A table of perhaps 20 to 100 different rows, each with a different current or position, can be prepared. Magnetostatic finite-element solutions for flux linkage and force can then be obtained for all table rows. Then for any values of current and position within the table range, interpolation can be done to find the related flux linkage and force.

The electrical performance of the magnetic device is directly affected by flux linkage λ according to Faraday's law. Using the standard voltage sign convention for inductors:

$$V = d\lambda/dt \tag{15.1}$$

Substituting the definition (6.11) for inductance L gives:

$$V = \frac{d}{dt}(LI) = L\frac{dI}{dt} + I\frac{dL}{dt} \tag{15.2}$$

The first term can be considered to correspond with Example 15.1, and the second term is the speed voltage included in Example 15.2. Thus the time variation of flux linkage determines the voltage and/or current waveforms versus time.

15.1.4 Analogies for Rigid Armature Motion

The speed and other mechanical performance parameters of the magnetic device are directly affected by the magnetic forces in the parametric table. However, as discussed in Chapter 12, mass and other mechanical parameters also affect the mechanical performance.

The same electric circuit software used to model the electromagnetic behavior of magnetic devices can also be used by analogy to model the mechanical behavior of rigid armatures. Several analogies are possible. One analogy is based on the structural equation (14.13) simplified for a rigid body with one degree of translational freedom x:

$$M_S \frac{d^2x}{dt^2} + B_S \frac{dx}{dt} + K_S x = F_S \qquad (15.3)$$

The analogous equation is for an RLC circuit of (14.12) with an additional time derivative taken on both sides:

$$C \frac{d^2V}{dt^2} + \frac{1}{R}\frac{dV}{dt} + \frac{1}{L}V = \frac{dI}{dt} \qquad (15.4)$$

Comparing the two analogous equations, the analogy for mass M_S is capacitance C, etc [2]. Thus SPICE and other electric circuit programs can be used to model rigid body motion of armatures of magnetic actuators and sensors [2]. Since the armature, stator, and other parts are all assumed to be rigid bodies, their deformations, strains, and stresses are all assumed to be zero.

15.1.5 Maxwell SPICE Model of Bessho Actuator

The Bessho actuator of Chapter 14 and previous chapters is modeled in Maxwell SPICE as shown in Figure 15.1. The SPICE model was made by exporting parametric tables of flux linkage and force from Maxwell magnetostatic finite-element analyses, represented by the small center icon of the Bessho actuator. The icon produces a speed voltage and a magnetic force. A graph of force versus position for various currents, called pull curves in Chapter 7, is shown in Figure 15.2. Maxwell SPICE automatically interpolates to obtain the force for any airgaps from 0 to 12 mm and for any currents from 0 to 1.21 A.

To the left of the Bessho icon is the electric drive circuit, in this case a 0.5-A current source rising with time constant 20 ms as discussed in previous chapters. To

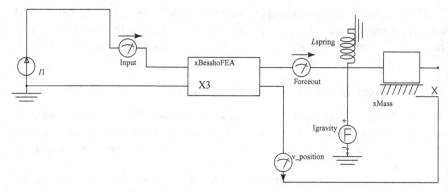

FIGURE 15.1   SPICE model of Bessho actuator. The electrical drive circuit is at the left. The actuator icon represents nonlinear magnetic flux linkage and force tables. The mechanical circuit is at the right.

the right of the Bessho icon is the SPICE analog of the mechanical system, including a mass and optional spring [3].

The circuit of Figure 15.1 does not include any eddy current effects in the conducting steel. Thus its results should be comparable to those of Table 14.1 with zero conductivity. The results of Figure 15.1 computed by SPICE are shown in Figure 15.3. There are four curves, numbered to correspond with cases 1–4 of Table 14.1. The computed closing times of Figure 15.3 differ somewhat from Table 14.1. They agree most closely with the Table 14.1 results for zero conductivity, as they should. The closing times of Figure 15.3 are reduced by 10–15% from those of Table 14.1, possibly due to different meshes and force computation methods. Also, the SPICE model

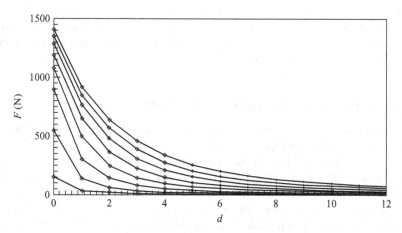

FIGURE 15.2   Pull curves computed by Maxwell parametric finite-element analysis for airgap d ranging from 0 to 12 mm and current I ranging from 0 (the bottom zero pull curve along the horizontal axis) to 1.212 A (the top pull curve).

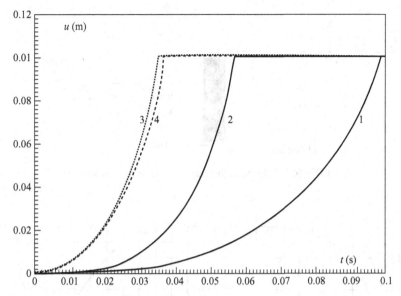

FIGURE 15.3   Computed SPICE results for Bessho armature position versus time. The four cases are those of Table 14.1 with zero conductivity.

of Figure 15.1 does not include any damping B_S or other frictional forces, whereas friction can perceptibly slow down actuator motion.

15.1.6   Simplorer Model of Bessho Actuator

Simplorer is a software package used to design and analyze complex technical systems. Its models can include electric circuit models, models of other physical domains, block diagrams, and state machine structures. One of its languages is the Simplorer Modeling Language (SML). SML is often used to model electromechanical systems, and is here used for the Bessho actuator. Simplorer is a much newer program than SPICE, and its time-stepping algorithms are more likely to converge properly. Also, Simplorer includes a large number of model libraries, including a sensor library. Simplorer instructional material with detailed worked examples is available at www.ansys.com.

The Simplorer SML model of the Bessho actuator is shown in Figure 15.4. Similar to the Maxwell SPICE model of Figure 15.1, it includes an icon of the actuator that represents the Maxwell finite-element model and its flux linkages and forces. To the left of the icon is the electric drive circuit, and to the right is the Simplorer mechanical model.

The Simplorer model was used to predict performance with the 6 kg mass and the 0.5 A input current with 20 ms rise time. The graphs in Figure 15.5 show that the closure time is 97 ms, almost identical to the SPICE results.

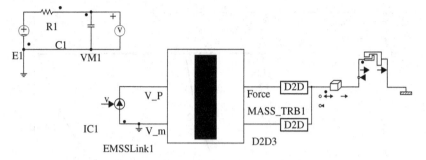

FIGURE 15.4 Simplorer model of Bessho actuator.

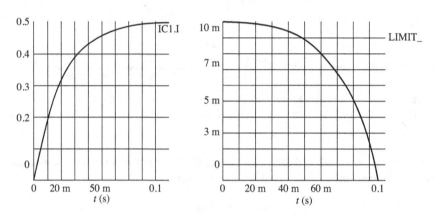

FIGURE 15.5 Computed Simplorer input and output for Bessho actuator with 6-kg moving mass. The input on the left is the current (in amperes) waveform, and the output on the right is the airgap (in meters) as a function of time (in seconds).

15.2 VHDL–AMS/SIMPLORER MODELS

15.2.1 VHDL–AMS Standard IEEE Language

VHDL–AMS stands for Very high speed integrated circuit Hardware Description Language—Analog Mixed Signal. It is a standard language used for describing digital, analog, and mixed-signal systems. It is especially popular for automotive electronic system design.

The IEEE standardized the VHDL-1076 language as a Hardware Description Language for digital models. The VHDL standard from 1993 was extended in 1999 for the description of analog and mixed-signal models in the form of the IEEE 1076.1 standard for VHDL–AMS.

A VHDL–AMS model consists of five basic elements: *entity, architecture, package, package body,* and *configuration.* Multidomain systems, such as the electrical and mechanical domains of magnetic actuators and sensors, are specified with the appropriate package entries.

The development and simulation of VHDL–AMS models is supported by Simplorer. It contains wizards, importers, exporters, schematics, netlists, digital stimuli, and other VHDL–AMS modeling capabilities. The Simplorer graphical interface contains common basic VHDL–AMS circuit components and blocks. The Simplorer VHDL–AMS solver contains an analog solver, a digital solver, and a controller that controls their interaction [4].

15.2.2 Model of Solenoid Actuator

The standard VHDL–AMS solenoid model is based upon the assumed inductance versus position relation [5]:

$$L(x) = \frac{N^2 L_o}{1 + kx} \tag{15.5}$$

where L_o is the inductance for one turn when airgap $x = 0$, N is number of turns, and k is a dimensionless coefficient. The voltage across the above inductance is:

$$V = \frac{d\lambda}{dt} = \frac{d}{dt}(LI) = N^2 L_o \frac{d}{dt}\left(\frac{I}{1 + kx}\right) \tag{15.6}$$

Since both x and I may vary with time, the above derivative gives:

$$V = N^2 L_o \left[\frac{I}{1 + kx}\frac{dI}{dt} - \frac{Ik}{(1 + kx)^2}\frac{dx}{dt}\right] \tag{15.7}$$

where the second term is the speed voltage.

The magnetic force is given by (5.13), (6.9), and (6.10) as:

$$F = \frac{\partial W_{co}}{\partial x}, \text{ where coenergy } W_{co} = \int \lambda dI \tag{15.8}$$

Since the inductance $L(x)$ of (15.5) is independent of current, the flux linkage $\lambda = LI$ becomes:

$$\lambda(I, x) = \frac{N^2 L_o I}{1 + kx} \tag{15.9}$$

and the force of (15.8) becomes:

$$F = \frac{N^2 L_o I^2}{2} \frac{\partial}{\partial x} \left(\frac{1}{1 + kx} \right) = -\frac{N^2 L_o k I^2}{2(1 + kx)^2} \tag{15.10}$$

The minus sign indicates that the magnetic force tends to close the airgap x. In terms of the flux linkage λ, the force becomes:

$$F = -\frac{k\lambda^2}{2L_o N^2} \tag{15.11}$$

Defining a new variable, the maximum inductance:

$$L_{\max} = N^2 L_o \tag{15.12}$$

results in the following VHDL–AMS model equations:

$$\lambda = \frac{L_{\max} I}{1 + kx}, \quad F_k = \frac{k}{2L_{\max}}, \quad F = -\lambda^2 F_k, \quad V = \frac{d\lambda}{dt} \tag{15.13}$$

The VHDL–AMS model includes two external electrical pins (nodes) and two mechanical pins. Given a current I and armature position x, the model equations compute λ and F. The value of F is passed to the mechanical pins as a variable. The voltage, the time derivative of λ, appears across the electrical pins.

Similar methods can be used to derive VHDL–AMS models for other magnetic components. First, the equations for stored energy, in terms of "across" and "through" variables, must be obtained. Then the forces, voltages, and losses in lossy cases, can be computed.

The VHDL–AMS solenoid model accepts a current value from electrical pins p and m, and a position value from mechanical pins $pos1$ and $pos2$. The model entity parameters are listed in Table 15.1, and the VHDL–AMS entity description is in Table 15.2. The VHDL–AMS architecture code is listed in Table 15.3.

TABLE 15.1   VHDL–AMS Solenoid Model Entity Parameters

Interface	Name	Type	Default Value	Description
GENERIC	10	Inductance	1.25E-7	Maximum L for 1 turn
"	K	Real	197.735	L coefficient
"	N	Real	1.0	Number of coil turns
TERMINAL	p, m	Electrical	—	Electrical pins
"	$pos1$	Translational	—	Mechanical pin
"	$pos2$	Translational	—	Mechanical pin

TABLE 15.2 VHDL–AMS Solenoid Entity Description for Table 15.1

```
LIBRARY IEEE;
USE IEEE.ELECTRICAL_SYSTEMS.ALL;
USE IEEE.MECHANICAL_SYSTEMS.ALL;
ENTITY solenoid IS
  GENERIC (
    L0 : INDUCTANCE := 1.25e-7;
    K : REAL :=197.735;
    N : REAL :=1.0;
  PORT (
    TERMINAL p,m : ELECTRICAL;
    TERMINAL pos1, pos2 : TRANSLATIONAL);
    QUANTITY force_out : OUT REAL := 0.0);
END ENTITY solenoid;
```

TABLE 15.3 VHDL–AMS Solenoid Model Architecture Description

```
ARCHITECTURE behav OF solenoid IS
  CONSTANT Lmax : INDUCTANCE :=L0 * N * N;
  CONSTANT Fk : FORCE :=K / (2.0 * Lmax;
  QUANTITY v ACROSS i THROUGH p TO m;
  QUANTITY position ACROSS force THROUGH pos1 TO pos2;
  QUANTITY L : INDUCTANCE;
  QUANTITY flux : FLUX;
BEGIN
  IF (position > 0.0) USE
    L == Lmax / (1.0 + K * position);
  ELSE
    L == Lmax;
  END USE;
  flux == L * i;
  v == flux'DOT;
  force == flux * flux * Fk;
  force_out == -force;
END ARCHITECTURE behav;
```

Example 15.3   Simplorer VHDL–AMS Voltage Step Response of Solenoid
The VHDL–AMS solenoid model described above is to be used to find the response to a step application of a 12-V automobile battery for 1 s. As shown in Figure E15.3.1, the total moving mass is 0.545 kg, and the force of gravity is included over the 0.0127-m stroke.

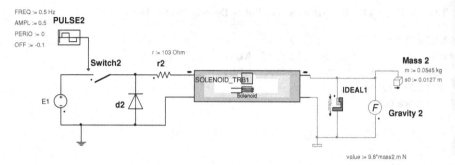

FIGURE E15.3.1 VHDL–AMS solenoid model in Simplorer with DC voltage switch.

The solenoid has 12,500 turns, a coil resistance of 103 Ω, and a diode connected as shown in Figure E15.3.1 to eliminate arcing at its switch upon turnoff. The solenoid constant k is not really a constant, and can be calculated in several ways. It can be found using the reluctance method of Chapter 3, the finite-element method of Chapter 4, and/or from measurements if the solenoid is available. In this example, finite-element solutions at two values of x are used. For $I = 0.084$ A and $x = 0$, $L = 1.251E{-}7$ H and $F = -27$ N. At the same current and $x = 0.0127$ m, $L = 3.56E{-}8$ H and $F = -0.84$ N. Find the current and force versus time from 0 to 2 s.

Solution The first approach to find k is to choose it to match L at both values of x. This should model the electrical side accurately at all times. Substituting the two L values in (15.5) gives $k = 197.735$.

Another approach is to match both L and F at time zero. Using (15.12) and (15.13) gives $L_{max} = 19.547$ and then $\lambda = 1.642$. Substituting λ in (15.13) gives $F_k = 10$ and then $k = 391.52$.

The above two values of k are quite different. It is not possible to match both L and F over the whole solenoid operating range because of its nonlinear B–H steel, flux fringing, etc. The VHDL–AMS solenoid model thus is not as accurate as the preceding models using tables from finite-element analyses at far more than two positions. Thus the Simplorer SML modeling method of Section 15.1.6 can be more accurate than a Simplorer lumped parameter VHDL–AMS model.

To choose k, one might pick a compromise value of perhaps 300. If the mechanical behavior were more important, the 391.52 value might be chosen. Here the electrical behavior must be accurately modeled, so the value 197.735 is chosen.

The resulting current waveform computed by the VHDL–AMS capability of Simplorer is shown in Figure E15.3.2, along with the computed position versus time. The computed force versus time is shown in Figure E15.3.3. Note that the force takes time to build up after the voltage is applied at time zero, and takes time to decay after the voltage is switched off at time equal to 1 s.

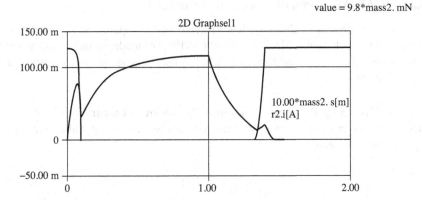

FIGURE E15.3.2   Current (slowly rising) and position versus time computed by VHDL–AMS model.

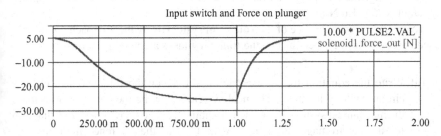

FIGURE E15.3.3   Force versus time computed by VHDL–AMS model.

15.3 MATLAB/SIMULINK MODELS

15.3.1 Software

MATLAB stands for matrix laboratory and was developed by Mathworks, Inc. A related software simulation package called Simulink is also popular.

MATLAB performs engineering calculations with and without matrices [6]. It is especially well suited for analyzing linear systems [7]. It offers a number of *toolboxes*, which are extensions for certain application areas. Its Control System Toolbox is commonly used by control systems engineers, and includes models in the Laplace transform s domain [8]. Alternative software exists such as Scilab, which is free.

Simulink provides an efficient graphical way for MATLAB users to model and simulate control systems in block diagram form. It includes many linear blocks and also some simple nonlinear blocks.

15.3.2 MATLAB Model of Voice Coil Actuator

The voice coil actuator of Chapter 7 has a force linearly proportional to current and thus is easily modeled using MATLAB as follows [9]. The model includes mechanical parameters such as M_S, B_S, and K_S of (15.3), which may be properties of a loudspeaker voice coil actuator.

Example 15.4   Step Response Times and Overshoot of Linear Voice Coil   A voice coil actuator has the proportionality constant between voice coil current and magnetic force K_a, and thus:

$$F(t) = K_a I(t) \tag{E15.4.1}$$

The actuator is shown schematically in Figure E15.4.1, and has $K_a = 0.8$ N/A. The inductance of the coil is $L = 0.02$ H and its resistance shown externally is $R = 0.5 \, \Omega$. Note that in this example $x(t)$ is the displacement and $u(t)$ is the velocity. The mass $M = 0.08$ kg, the damping friction is $B = 0.32$ N s/m, and the elasticity spring constant is $K = 1.6$ N/m.

To determine performance of the voice coil system using the approach commonly used by control systems engineers, the following tasks are to be performed:

(a) Write the equation for the electric circuit. Draw the force–voltage electric circuit analogy for the mechanical system and write the equation describing the mechanical system. Draw the s-domain (Laplace transform) block diagram for the electromagnetic actuator with $V(s)$ as the reference input and the displacement $X(s)$ as the output. Mark all the s-domain variables $V(s)$, $I(s)$, $F(s)$, velocity $U(s)$, and position $X(s)$ on the block diagram and determine the transfer function $X(s)/V(s)$.

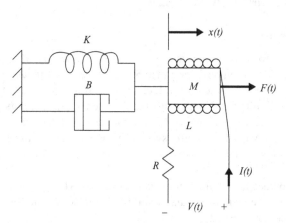

FIGURE E15.4.1   Voice coil actuator schematic diagram.

(b) Use the MATLAB *roots* function to find the roots of the system transfer function and determine the time constants. If one time constant is small relative to the other time constants, suggest a reduced-order model. Obtain the step response for the actual model and the reduced-order model on the same graph. After you have defined the numerators and denominators for both systems, you can use sys3rd = tf(num1, den1), and sys2nd = tf(num2, den2) to define the transfer function models. Then you can use:

```
ltiview('step', sys3rd, 'b', sys2nd, 'r')
```

to obtain the step response for both systems on the same linear time invariant (LTI) viewer. Right click on the plot area and determine the time domain characteristics, rise time t_r, peak time t_p, percent overshoot $P.O.$, and settling time t_s.

(c) The step response peak time and the percent overshoot of a second-order system is given by:

$$t_p = \frac{\pi}{\omega\sqrt{1 - \zeta^2}} \quad \text{and} \quad P.O. = e^{\frac{-\zeta\pi}{\sqrt{1-\zeta^2}}} \times 100 \qquad \text{(E15.4.2)}$$

where ζ is the (dimensionless) damping ratio [8]. Using the above relations, compute t_p, $P.O.$, and the settling time $t_s = 4\tau$ for the reduced-order model and compare it to the value for the actual response found in (b).

Solution

(a) The time domain and the *s*-domain equivalent circuits for the electric circuit of the actuator and the force–voltage electric circuit analogy for the mechanical system is shown in Figure E15.4.2. Note that the motional voltage is assumed negligible.

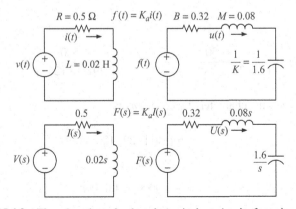

FIGURE E15.4.2   Time domain and *s*-domain equivalent circuits for voice coil actuator.

FIGURE E15.4.3 Block diagram of voice coil actuator.

From the s-domain equivalent circuit we have:

$$I(s) = \frac{1}{0.02\,s + 0.5} V(s) \quad \text{and} \quad U(s) = \frac{1}{0.08\,s + 0.32 + \dfrac{1.6}{s}} F(s)$$

$$(E15.4.3)$$

Also the given transducer equation:

$$F(s) = 0.8 I(s) \qquad (E15.4.4)$$

And the displacement in terms of velocity is:

$$X(s) = \frac{1}{s} U(s) \qquad (E15.4.5)$$

The block diagram representing the above four equations is shown in Figure E15.4.3.

Thus the transfer function relating the s-domain displacement to the s-domain voltage is:

$$G(s) = \frac{X(s)}{V(s)} = \frac{0.8}{(0.02s + 0.5)\left(0.08s + 0.32 + \dfrac{1.6}{s}\right)s} \qquad (E15.4.6)$$

or

$$G(s) = \frac{50}{(s + 25)(s^2 + 4s + 20)} = \frac{500}{s^3 + 29s^2 + 120s + 500} \qquad (E15.4.7)$$

(b) The roots of the characteristic equation are:

$$r = \text{roots}([1 \quad 29 \quad 120 \quad 500])$$

$$r = -25.0000$$

$$-2.0000 + 4.0000i$$

$$-2.0000 - 4.0000i \qquad (E15.4.8)$$

The time constants are:

$$\tau_1 = \frac{1}{25} = 0.04 \quad \text{and} \quad \tau_2 = \frac{1}{2} = 0.5 \qquad \text{(E15.4.9)}$$

$$G(s) = \frac{500}{25(0.04s + 1)(s^2 + 4s + 20)} = \frac{20}{(0.04s + 1)(s^2 + 4s + 20)}$$

$$\text{(E15.4.10)}$$

Note that $\tau_1 \ll \tau_2$. Neglecting the smaller time constant, the approximate second-order model becomes:

$$G(s) = \frac{20}{s^2 + 4s + 20} \qquad \text{(E15.4.11)}$$

To obtain the step response of the third-order and second-order systems, we use the following MATLAB commands.

```
num1 = 500; % third-order system num
den1 = [1 29 120 500]; % third-order system den
num2 = 20; % reduced-order system num
den2 = [1 4 20]; % reduced-order system den
sys3rd = tf(num1, den1) % third-order system
sys2nd = tf(num2, den2) % reduced-order system
ltiview('step', sys3rd, 'b', sys2nd, 'r')
```

The results are plotted in Figure E15.4.4.

(c) The second-order system characteristic equation is:

$$s^2 + 4s + 20 = 0 \qquad \text{(E15.4.12)}$$

$$\omega_n^2 = 20 \Rightarrow \omega_n = 4.472 \qquad \text{(E15.4.13)}$$

$$2\zeta\omega_n = 4 \Rightarrow \zeta = \frac{4}{2(4.472)} = 0.4472 \qquad \text{(E15.4.14)}$$

$$t_p = \frac{\pi}{4.472\sqrt{1 - 0.4472^2}} = 0.785 \qquad \text{(E15.4.15)}$$

$$P.O. = e^{\frac{-0.4472\pi}{\sqrt{1 - 0.4472^2}}} = 20.7 \qquad \text{(E15.4.16)}$$

$$\tau = 1/(\zeta\omega_n) = 0.5 \text{ s} \qquad \text{(E15.4.17)}$$

$$t_s = 4\tau = 2.0 \text{ s} \qquad \text{(E15.4.18)}$$

The results are summarized in Table E15.4.1, which shows that the reduced-order model is a good approximation. The frequency response of the two systems can also be compared to again confirm the validity of the approximation.

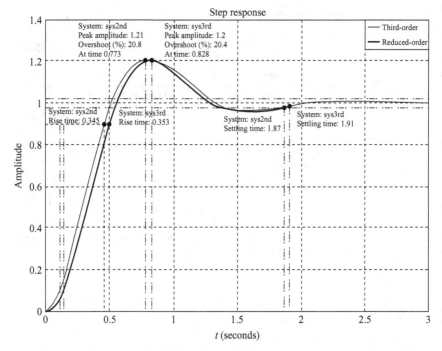

FIGURE E15.4.4 MATLAB plots of time domain response of voice coil actuator.

TABLE E15.4.1 Comparison of Responses of Systems Models of Actuator

Time Domain Specification	Third-order System Simulation Results	Reduced-order Simulation Results	Reduced Order Using Equation 15.4.2
Percent Overshoot	20.4%	20.8%	20.7%
t_p (s)	0.828	0.773	0.785
t_s (s)	1.91	1.87	2.0

Example 15.5 Frequency Response of Linear Voice Coil Actuator   The frequency response of the voice coil actuator of Example 15.4 is to be found using MATLAB. Its Bode function is to be used to plot the response versus frequency for both the second-order and third-order models.

Solution   The *m*-file commands are as follows.

```
L = 0.02; R = 0.5; K = 1.6; Kf = 0.8;
M = 0.08; B = 0.25;
num1 = Kf/(L*M); % third-order system numerator
den1 = [1 (B/M)+(R/L) (K/M)+(B/M)*(R/L) (K/M)*(R/L)];
                  % third-order system den
```

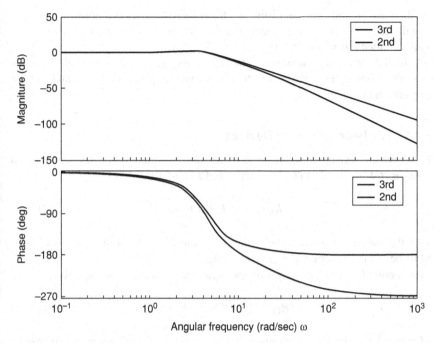

FIGURE E15.5.1   MATLAB bode plots of frequency domain response of voice coil actuator. At high frequencies the third-order model gives lower curves than the second-order model.

```
num2=num1/(R/L);        % reduced-order system numerator
den2 = [1 B/M K/M];     % reduced-order system denominator
sys3rd = tf(num1, den1) % third-order system
sys2nd = tf(num2, den2) % reduced-order system
disp('Example 15.4')
bode(sys3rd, 'b', sys2nd, 'r')
```

which obtain the Bode magnitude and phase plots shown above in Figure E15.5.1. Note that they agree closely at low frequencies but not at high frequencies. Note also that the peak magnitude occurs slightly above $\omega = 4$ rad/s. This is in reasonable agreement with the resonance formula $(K/M)^{1/2}$ of (14.21) that obtains $\omega = 4.47$ rad/s.

15.4   INCLUDING EDDY CURRENT DIFFUSION USING A RESISTOR

As discussed in Chapters 8 and 9, actuators and sensors made of solid steel can have significant eddy currents and associated diffusion time delays. One example is the Bessho DC axisymmetric plunger actuator, which is made of solid steel. Also, Example 9.2 has shown that planar devices also exhibit transient eddy currents. To include transient eddy current diffusion effects in systems models, one method is to solve the eddy current finite-element model at each time step along with the systems solution, called *cosimulation* [10].

A simpler way to include eddy current effects in systems models is to add a resistor as described below. The resistor will first be derived for planar problems and then for axisymmetric problems.

15.4.1 Resistor for Planar Devices

The equivalent resistor for eddy current diffusion in a planar slab has been derived for linear *B–H* material of conductivity σ and a one-turn coil as [5]:

$$R_{EL1} = \pi^2 h_z/(\sigma\, h_y\, w) \tag{15.14}$$

where the material depth into the plane is h_z, its width in direction of diffusion is w, and its height in the remaining direction is h_y.

For a coil of *N* turns, the equation for linear eddy resistance in one slab is:

$$R_{EL} = (\pi N)^2 h_z/(\sigma\, h_y\, w) \tag{15.15}$$

Chapter 9 has shown that nonlinear *B–H* curves cause the diffusion time to change. To include nonlinear *B–H* effects, the eddy resistor is multiplied by the ratio of time constants:

$$R_E = (\tau_{mL}/\tau_{mN})(\pi N)^2 h_z/(\sigma\, h_y\, w) \tag{15.16}$$

where τ_{mL} is the diffusion time of Chapter 9 for linear *B–H* and τ_{mN} is the diffusion time of Chapter 9 for nonlinear *B–H*.

Example 15.6 R_E for Inductor of Example 9.2   The eddy current resistor is to be found for the inductor of Example 9.2, shown in Figure E9.2.1, at a current of 0.5 A.

Solution   The width $w = 0.02$ m, and the diffusion time ratio is $(\tau_{ml}/\tau_{mn}) = (54.6/18.9)$. Also, the total number of turns $N = (26.4)(2)(0.05/0.002) = 1320$. Also, the height $h_y = 0.05$ m and the depth $h_z = 0.1$ m. In addition, since the inductor has two legs (slabs), its power loss is doubled, and thus the eddy resistor of (15.14) must be divided by two, giving $R_E = 1462\ \Omega$.

Example 15.7 Simplorer Model of Inductor of Example 9.2 with Input Step Current and R_E   The inductor of Example 9.2, shown in Figure E9.2.1, is to be modeled in Simplorer with an eddy current resistor added to account for linear magnetic diffusion. Its transient performance is to be computed when a step current of 0.5 A is input.

Solution   From Example 15.6, the eddy current resistor is 1462 Ω. The magnetostatic flux linkage versus inductance is found by multiple Maxwell finite-element solutions of the model of Example 9.2 over a wide range of currents. The resulting Simplorer model is shown in Figure E15.7.1.

The input step current and computed output flux versus time are shown in Figure E15.7.2. Note that the flux is delayed somewhat less than the nonlinear diffusion time of 18.9 ms of Example 9.2. Since the diffusion time is the time for the center flux to rise, and most of the flux is outside the center of the leg, the Simplorer results are reasonable.

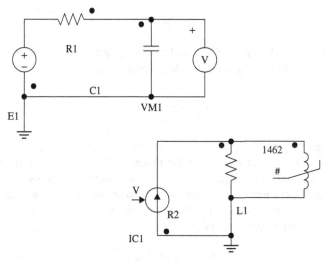

FIGURE E15.7.1   Simplorer model of inductor made of rectangular solid steel.

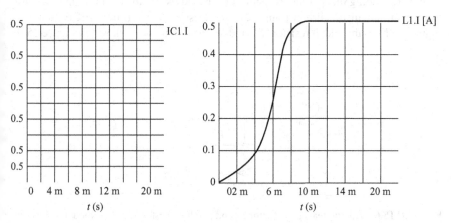

FIGURE E15.7.2   Current and flux obtained by Simplorer for Figure E15.7.1.

15.4.2   Resistor for Axisymmetric Devices

To derive the equivalent resistor for axisymmetric devices, begin with the constant permeability case. The classical formula for the resistance seen by current flowing peripherally on an arc around an axis is [3]:

$$R_{EL1} = \frac{4\pi}{2\sigma h \ln(b/a)} \tag{15.17}$$

where h is the axial length, b is outer radius, and a is inner radius. For a cylindrical armature and/or stator bore, the effective inner radius a varies with time according to the magnetic diffusion concept of Chapter 9, that is:

$$a = be^{-t/\tau_m} \tag{15.18}$$

Substituting (15.18) into (15.17) and referring resistance to the N turn stator winding gives the linear time-varying eddy current resistor [3]:

$$R_{EL} = 4\pi \frac{(N/k)^2}{2\sigma ht} \tag{15.19}$$

where the coupling coefficient k is not unity because the peripheral arc acts as a shorted turn and does not link all of the magnetic flux lines. Since both the flux and the eddy current decay together below the steel surface, k equals $\frac{1}{2}$. Also, time-varying resistors are not available in most versions of SPICE and other electric circuit software. Often the ratio (τ_m/t) is approximately unity at closing time, and thus (15.19) becomes approximately:

$$R_{EL} = 8\pi N^2/(\sigma h) \tag{15.20}$$

Besides adding this parallel resistor, it is also advisable to add a leakage inductor to prevent unrealistic current jumps. For the Bessho actuator, the SPICE model of Figure 15.1 has the added resistor R_{EL} and leakage inductor L_1 shown in Figure 15.6.

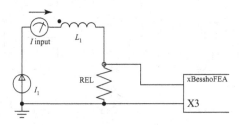

FIGURE 15.6   Adding leakage inductor L_1 and parallel resistor to SPICE model to account for eddy current diffusion.

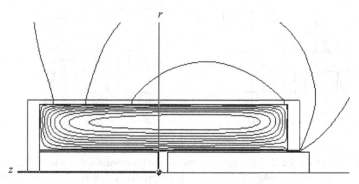

FIGURE 15.7 Computer display of flux line plot for Bessho actuator with flux excluded from steel, used to obtain value of leakage inductance L_1.

For the Bessho axial height of 0.27 m, $N = 3300$ and $\sigma = 1.7\text{E}6$ S/m, $R_{EL} = 596\ \Omega$. The leakage inductor L_1 is for flux that does not link the induced eddy currents. Thus L_1 is found by magnetostatic finite-element analysis of the actuator with its steel permeability set to the small value 0.1% of air. The resulting flux plot is shown in Figure 15.7 and the corresponding inductance L_1 is 65 mH.

To include nonlinear B–H effects, as shown earlier the resistance must be multiplied by the ratio of nonlinear to linear diffusion times. Thus the nonlinear eddy resistor is [11]:

$$R_{EN} = \left(\frac{\tau_{mL}}{\tau_{mN}}\right) \frac{8\pi N^2}{\sigma h} \tag{15.21}$$

where the nonlinear diffusion time τ_{mN} varies with applied field intensity H_o as described in Chapter 9. One must keep in mind, however, that the diffusion time calculations of Chapter 9 were for 1D problems, whereas airgaps and other 2D geometric features of real magnetic devices will cause changes in diffusion. Thus (15.21) is approximate, even when actuators are closed with "zero" airgap. No gap is exactly zero, but instead equals approximately 36 μm for polished steel surfaces [12].

For the Bessho actuator, Example 9.3 has found that $H_o = 6600$ A/m for the nominal current of 0.5 A. Using the actual Bessho steel B–H curves (not the step B–H curve of Example 9.3), the nonlinear diffusion time τ_{mN} computed by Maxwell finite-element analysis is 42 ms. Then the corresponding eddy resistor using (15.21) with $(\tau_{mL}/\tau_{mN}) = (93\ \text{ms}/42\ \text{ms})$ is $R_{EN} = 1320\ \Omega$ [11]. The resulting Simplorer output position for the same input current as Figure 15.5 is shown in Figure 15.8. The closing time is now 101 ms, increased by 4% from Figure 15.5, as expected from Table 14.1. Hence in this case the eddy currents cause only a 4% increase in operation time, but larger increases occur in other cases, as will be seen later in this chapter.

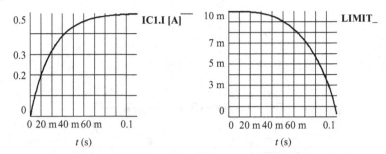

FIGURE 15.8   Computed Simplorer input and output for Bessho actuator including nonlinear 1320-Ω eddy resistor and 6-kg moving mass. The input on the left is the current (in amperes) waveform, and the output on the right is the airgap (in meters) as a function of time (in seconds).

15.5   MAGNETIC ACTUATOR SYSTEMS FOR 2D PLANAR MOTION

Besides the linear and rotary magnetic actuators discussed above and in previous chapters, magnetic actuators are increasingly being developed for 2D motion over a plane with optional motion in other degrees of freedom. Applications include pick-and-place machines for industry, inspection systems, and wafer scanners. In some cases magnetic bearings (levitators) are used [13] that are variations of the bearings discussed in Chapter 7. Alternative bearings include air bearings.

Planar actuators are typically designed to develop Lorentz force in either of the following two ways.

- A moving coil armature travels over a stator made with stationary permanent magnets. Like the voice coil actuator of Chapter 7, this type requires flexible connections to the moving coil.
- An armature made with permanent magnets travels over a stator made with current-carrying coils which do not require flexible connections. Thus the armature is said to be *contactless*.

Designs of planar magnetic actuators vary depending on stroke requirements. As mentioned in Chapter 7, commutation may be required for long strokes.

Figure 15.9 shows a recently developed planar actuator with a permanent magnet armature [14, 15]. It also has a manipulator energized by *contactless energy transfer* (CET) via inductive magnetic coupling. The control of the manipulator is by wireless link. The planar actuator uses commutation for long strokes in the xy plane, while the manipulator does the picking and placing of objects.

Designers of planar actuators face many challenges. The first challenge is the design of the magnetic suspension and propulsion that should be capable of levitating the heavy platform. Next, the planar actuator should have an unlimited long-stroke movement capability in the xy plane and short stroke in the other degree of freedom. In addition, the CET must be integrated in the system.

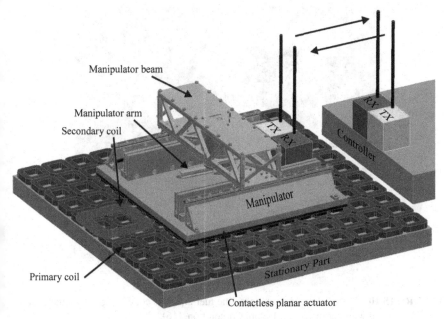

FIGURE 15.9   Contactless planar actuator with manipulator [14].

The heavy payload of the planar actuator requires large magnets and the use of a Halbach magnet array. Recall that Halbach magnets were discussed in Chapter 5. The coil topology is mainly determined by the design requirement to combine both magnetic actuation and CET by means of the same coil set. Rectangular coils are more efficient in generating forces than square or round coils. However, rectangular coils can only produce forces in two directions, so a pattern of rectangular coils in two orientations 90° apart is necessary to create forces in all three directions. Despite higher power loss to lift a certain mass, round coils are used in this system due to their suitability for CET. The final design of the planar actuator consists of a Halbach magnet array with 10 × 10 poles, with a pole pitch of 40 mm. The round coils are made of Litz wire to reduce power losses as mentioned in Chapter 12.

Sensors to control the planar actuator include both lasers and magnetic eddy current proximity sensors as discussed in Chapter 11. Preliminary results showed [14] that the planar actuator performed basic actuator functions, but had problems with its CET and long-stroke movements. More work is needed to perfect planar actuators.

15.6   OPTIMIZING MAGNETIC ACTUATOR SYSTEMS

Optimization of magnetic actuators as they operate electromechanical systems is achieved nowadays using finite-element software coupled to systems software. The

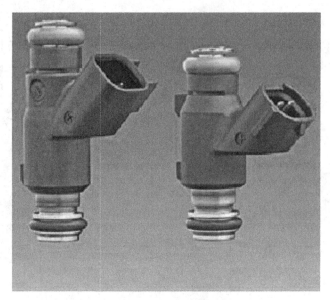

FIGURE 15.10 Two Delphi Multec® gasoline fuel injectors containing 3D magnetic actuators. Published with permission of Delphi Corporation. [16].

systems optimized here are a fuel injection system for automotive engines followed by a hydraulic valve actuated by a plunger solenoid.

15.6.1 3D Analysis of Fuel Injector System

Figure 15.10 shows two automotive fuel injectors manufactured by Delphi Corporation. They contain magnetic actuators similar to that shown in Figure 15.11. Note that one-half of the magnetic actuator is shown, which must be modeled in 3D to account for geometry details.

Some of the Maxwell software setup for the fuel injector system model is shown in Figure 15.12. Note that both electrical and mechanical design variables are described. The empirically obtained load force equation involves the injector parameters and is specified by inputting code (where $*$ symbolizes multiply and $**2$ symbolizes 2nd power) [16]:

$$F_{\text{load}} = (-(2 + k * \text{Position}) - \text{weight} - (\pi * \text{FuelP} * \text{sealdia} * *2/$$

$$(4 * (1 + \text{PressRatio} * (\text{Position/maxstroke}) * *2)))) \qquad (15.22)$$

Also part of the Maxwell input is the schematic circuit shown in Figure 15.13. As in Example 15.3, a diode is used in the switching circuit.

The computed current, armature speed, and flux versus time are plotted in Figure 15.14. Note that the speed increases until the armature closes the gap at time approximately equal to 0.7 ms.

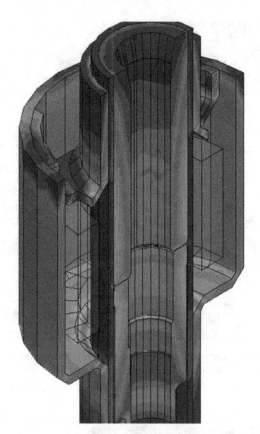

FIGURE 15.11   One-half of 3D magnetic actuator for fuel injector of Figure 15.10. Published with permission of Delphi Corporation and Aɴsʏs, Inc. [16].

Next the moving mass was increased by 50% and the optimum design was sought by varying the wire gauge and number of turns while maintaining the coil resistance. This was accomplished using many parametric analyses on several computers simultaneously, a method called *distributed processing*. Alternative optimization methods include *genetic optimization*, which imitates the evolutionary process of biological natural selection [17].

Parametric results were obtained with and without conductivity in the solid steel armature and stator. Figure 15.15 graphs the current versus time plotted without conductivity, that is, with no eddy currents and zero diffusion time. Figure 15.16 graphs the current versus time computed with actual finite-steel conductivity. Note that the dip in the current waveform indicates closing time, which is made longer (as expected) by the diffusion time in the conducting steel. Note also the variation with wire size, which helps the designer choose the wire size and coil turns. The faster the closing time, the more accurate is the fuel injection, which reduces hydrocarbon emissions by the automotive gasoline engine.

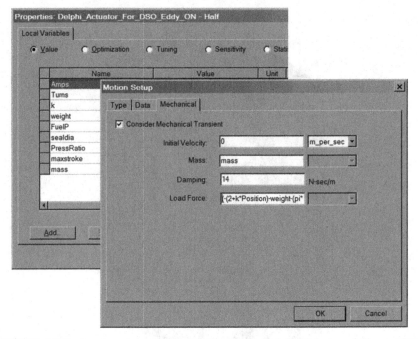

FIGURE 15.12 Input parameters for optimized design of fuel injector of Figure 15.11. Published with permission of Delphi Corporation and ANSYS, Inc. [16].

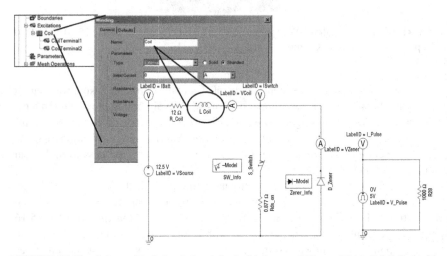

FIGURE 15.13 Schematic of circuit input for fuel injector of Figure 15.11. Published with permission of Delphi Corporation and ANSYS, Inc. [16].

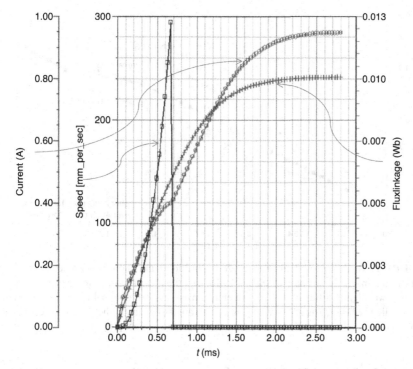

FIGURE 15.14   Computed results for current, speed, and flux for half model of fuel injector of Figure 15.11. Note that the armature closes in approximately 0.7 ms. Published with permission of Delphi Corporation and ANSYS, Inc. [16].

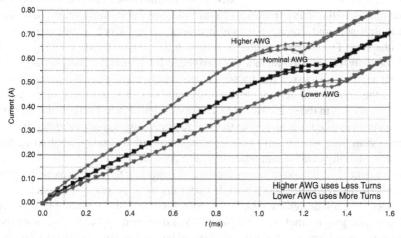

FIGURE 15.15   Computed current waveforms for zero conductivity in the magnetic actuator of Figure 15.11. The dips occur at closing time. Each curve pair has two slightly different curves for nominal mass and half nominal mass, the half mass closing slightly faster. The AWG wire gauge was varied to maintain the same coil resistance. Published with permission of Delphi Corporation and ANSYS, Inc. [16].

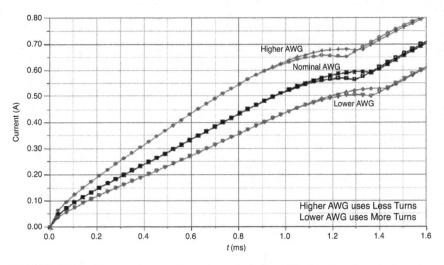

FIGURE 15.16   Computed current waveforms including diffusion time for actual conductivity in the magnetic actuator of Figure 15.10. The dips occur at closing time. Each curve pair has two slightly different curves for nominal mass and half nominal mass, the half mass closing slightly faster. The AWG wire gauge was varied to maintain the same coil resistance. Published with permission of Delphi Corporation and Ansys, Inc. [16].

15.6.2   Analysis of 3D Solenoid for Valve Actuation

Engineers designing solenoid actuators for applications such as hydraulic valves need to optimize the entire system. The system design method used on a TRW Automotive application is outlined in Figure 15.17 [18]. The lower portion shows the schematic of a typical solenoid driving a typical hydraulic system (such as discussed in the next chapter). The design method begins with 2D axisymmetric magnetic finite-element analysis, but then includes 3D magnetic finite-element analysis as well as thermal analysis (discussed in Chapter 12) and finally uses optimization software to achieve the best system design. Optimum solenoid design meets force versus stroke requirements, fits in the space available, and operates within a specified closing time. The solenoid design must also meet thermal requirements and be integrated into the entire system function. Systems optimized include antilock brakes, traction control, electronic stability control, and adaptive cruise control.

The solenoid to be optimized is shown in Figure 15.18. An approximate initial model is the axisymmetric 2D model, but later the more accurate 3D model will be optimized. The coil design equations are incorporated as design variables in Maxwell, with the wire gauge input by the designer. Based on coil space, the software then easily calculates the number of turns and the coil resistance. For all of the various wire gauges, parametric finite-element analyses compute the performance over a range of coil currents and armature positions. The parametric ranges can also include armature and stator shapes, material changes, etc. Computed performance includes the static magnetic force versus current. Figure 15.19 shows that the computed force curve

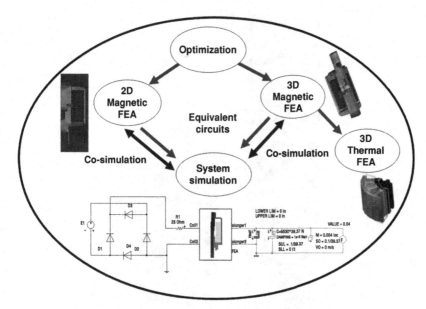

FIGURE 15.17   Design method for solenoid actuators driving hydraulic systems. Published with permission of TRW Automotive and Ansys, Inc. [18].

compares very well with the measured force curve, which varies as shown due to friction.

Next the dynamic performance of the solenoid, including eddy currents and magnetic diffusion, is computed. Figure 15.20 shows the dynamic (or transient) drive circuit model of the actuator, which is driven by a voltage source and switches controlled by subcircuits. The computed waveforms of current and position are shown in Figure 15.21. Note that the voltage is initially switched to a high value to move the armature quickly, and then the current is reduced electronically between upper and lower limits to maintain closure, ending with switch-off and the armature moving back to its initial zero position.

After computing the above performance of the base solenoid electromechanical model, optimization of the solenoid system was carried out in the following steps. First, the shape of the actuator pole was varied to meet the mechanical spring curve. The spring curve is made by two springs as shown in Figure 15.22, which also shows typical magnetic pull curves for different pole shapes. The intersection of the magnetic curve and the spring curve is the operating position of the armature. The magnetic pull curves are of course affected by the magnetic circuit, including the details of the armature and stator shapes and the airgap. The dimensional variables in the airgap region are shown in Figure 15.23, all of which were input parametric variables in the Maxwell 3D finite-element analysis.

As the optimization software ran through the design variables of Figure 15.23, it was first programmed to minimize the *cost function* that meets the spring curve of Figure 15.22 at the critical maximum force points. The pole shape optimized design,

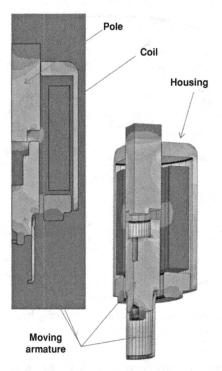

FIGURE 15.18 Typical solenoid to be optimized for use in a hydraulic system. A 2D axisymmetric model and a 3D model are shown. Published with permission of TRW Automotive and ANSYS, Inc. [18].

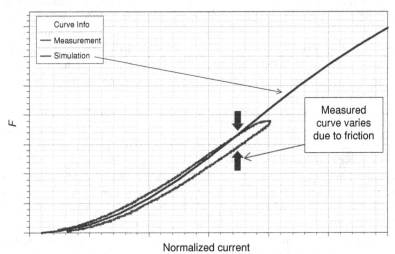

FIGURE 15.19 Computed and measured curves of static magnetic force versus coil current for solenoid of Figure 15.18. Published with permission of TRW Automotive and ANSYS, Inc. [18].

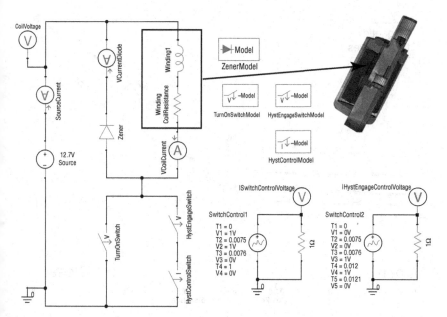

FIGURE 15.20 Circuit schematic of transient Maxwell model of solenoid of Figure 15.18. Published with permission of TRW Automotive and ANSYS, Inc. [18].

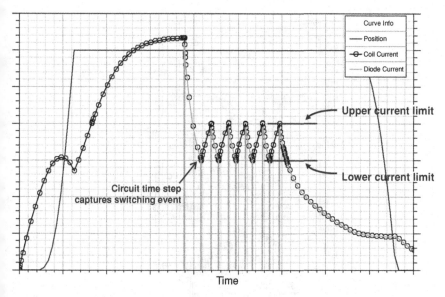

FIGURE 15.21 Computed coil current and position versus time for solenoid model of Figure 15.18. Published with permission of TRW Automotive and ANSYS, Inc. [18].

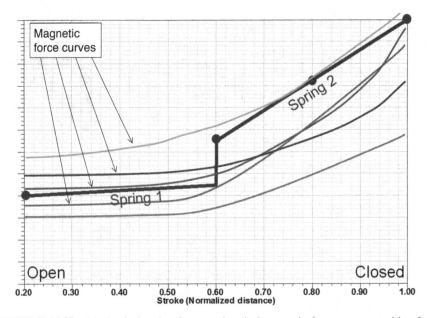

FIGURE 15.22 Mechanical spring forces and typical magnetic forces versus position for solenoid model of Figure 15.18. Published with permission of TRW Automotive and ANSYS, Inc. [18].

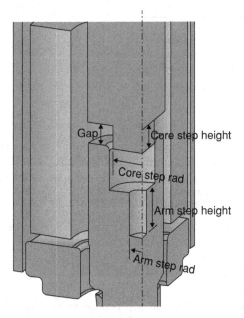

FIGURE 15.23 Dimensional variables for 3D model of solenoid of Figure 15.18. Published with permission of TRW Automotive and ANSYS, Inc. [18].

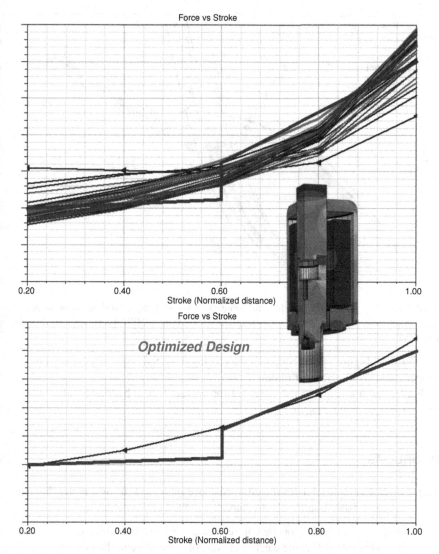

FIGURE 15.24 Graphs of force versus position for 3D model of solenoid of Figure 15.18. The upper graph has many curves for many design iterations. The lower graph shows the final magnetic pull curve and the mechanical spring curve. Published with permission of TRW Automotive and Ansys, Inc. [18].

achieved after about 50 design iterations, is shown in Figure 15.24 and meets the force curve.

The next optimization was for the cost function set to minimize the closing time of the solenoid in its electromechanical system. Again the 3D Maxwell transient model was used along with the related Optimetrics software, but the input variable this time was the number of coil turns. The families of position and current versus time curves

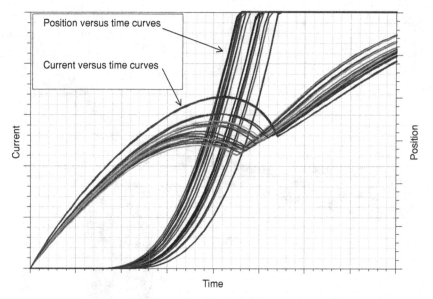

FIGURE 15.25   Curves of position and coil current versus time for various numbers of coil turns in solenoid of Figure 15.24. Published with permission of TRW Automotive and ANSYS, Inc. [18].

are shown in Figure 15.25. Note that the closing time can be greatly reduced by the proper choice of the number of coil turns.

Further optimization of this solenoid for operation of a hydraulic valve will be presented in the next chapter.

PROBLEMS

15.1   Redo Example 15.1 for the coil resistance increased (using smaller diameter wire) by 50%. Use the electric circuit software of your choice.

15.2   Redo Example 15.1 for the coil resistance increased (using smaller diameter wire) by 100%. Use the electric circuit software of your choice.

15.3   Redo Example 15.2 with $V_{speed} = 11t^2$.

15.4   Redo Example 15.2 with $V_{speed} = 9t^2$.

15.5   Redo VHDL–AMS Example 15.3 with the solenoid constant $k = 391.5$.

15.6   Redo VHDL–AMS Example 15.3 with the solenoid constant $k = 300$.

15.7   Redo MATLAB Example 15.4 with mass reduced to 0.06 kg.

15.8   Redo MATLAB Example 15.4 with damping reduced to 0.25 N s/m.

15.9   Redo MATLAB Example 15.5 with mass reduced to 0.06 kg.

REFERENCES

1. Kielkowski R. *Inside SPICE*. McGraw-Hill, Inc.; 1994.

2. McDermott TE, Zhou P, Gilmore J, Cendes ZJ. Electromechanical system simulation with models generated from finite element solutions. *IEEE Trans Magn* 1997;33:1682–1685.

3. Brauer JR, Chen QM. Alternative dynamic electromechanical models of magnetic actuators containing eddy currents. *IEEE Trans Magn* 2000;36:1333–1336.

4. *Simplorer V7 VHDL–AMS Tutorial*. Pittsburgh, PA: ANSYS, Inc.; 2004.

5. Woodson HH, Melcher JR. *Electromechanical Dynamics*. New York: John Wiley & Sons; 1968.

6. Hanselman D, Littlefield B. *Mastering MATLAB® 6*. Upper Saddle River, NJ: Prentice Hall; 2001.

7. Strum RD, Kirk DE. *Contemporary Linear Systems Using MATLAB®*. Pacific Grove, CA: Brooks/Cole; 2000.

8. Dorf RC, Bishop RH. *Modern Control Systems*, 9th ed. Upper Saddle River, NJ: Prentice Hall; 2001.

9. Saadat H. *Control Systems Laboratory Manual*. Milwaukee, WI: Milwaukee School of Engineering; 2003.

10. Zhou P, Lin D, Fu WN, Ionescu B, Cendes ZJ. A general co-simulation approach for coupled field-circuit problems. *IEEE Trans Magn* 2006;42:1051–1054.

11. Brauer JR, Mayergoyz ID. Finite element computation of nonlinear magnetic diffusion and its effects when coupled to electrical, mechanical, and hydraulic systems. *IEEE Trans Magn* 2004;40:537–540.

12. Roters HC. *Electromagnetic Devices*. New York: John Wiley & Sons; 1941.

13. Jansen JW, van Lierop CMM, Lomonova EA, Vandenput AJA. Magnetically levitated planar actuators with moving magnets. *IEEE Trans Indust Appl* 2008;44:1108–1115.

14. deBoeij J, Lomonova E, Duarte J. Contactless planar actuator with manipulator: a motion system without cables and physical contact between the mover and the fixed world. *IEEE Trans Indust Appl* 2009;45:1930–1938.

15. Lomonova E. Advanced actuation systems—state of the art: fundamental and applied research. *Proceedings of the International Conference on Electrical Machines and Systems*, October 10–13, 2010, pp 13–24.

16. Solveson M, Steward D, Mastro N. Design optimization of fast-acting actuators including eddy current effects and diffusion. Presentation at *ANSYS Inspiring Engineering Workshop*, Southfield, MI, October, 2007.

17. Pellikka M, Ahola V, Söderlund L, Uusitalo J-P. Genetic optimization of a fast solenoid actuator for a digital hydraulic valve. *Int J Fluid Power* 2011;12:49–56.

18. Solveson M, Christini M, Collins D. Actuator design: Meeting your customer's requirements for success. Presentation at *ANSYS First-pass System Success Workshop*, Southfield, MI, October, 2008.

CHAPTER 16

Coupled Electrohydraulic Analysis Using Systems Models

Airplanes, automobiles, tractors, ships, robots, and other vehicles commonly contain hydraulic actuation systems such as brakes. Besides such *mobile hydraulic* systems, hydraulics is also common in stationary applications such as factories and elevators. In many cases, especially with computer control, the input to the hydraulic system is electrical. Such coupled systems are called *electrohydraulic systems*. The usual coupling devices are magnetic actuators and sensors.

16.1 COMPARING HYDRAULICS AND MAGNETICS

Hydraulic actuation systems have certain advantages over magnetic actuators [1]. As discussed in Chapter 5, magnetic pressures on steel surfaces are given by:

$$P_{\mathrm{mag}} = B^2 / (2\mu_o) - w_{\mathrm{co}} \qquad (16.1)$$

where coenergy w_{co} appears as in (5.17) and (5.13). The magnetic pressure P_{mag} of a typical steel was graphed in Figure 5.7 and shown to be less than 20 bar, even for magnetic flux densities B as large as 2.4 T. Recall that one bar equals 1.E5 Pa, which approximates the standard atmospheric pressure at sea level.

In comparison, hydraulic pressures of hundreds of bar are commonly used. Since force is pressure times area, the higher hydraulic pressures can yield smaller, lighter, and less expensive actuators than magnetic actuators. High hydraulic pressures produce high forces.

Another big advantage of hydraulic actuators is that magnetic actuator armatures often must be made of steel, a heavy material that reduces acceleration. Since acceleration must be high for fast actuation, armatures made of lighter material such as aluminum are advantageous. Hydraulic armatures do not carry any magnetic flux and need not be made of steel, and thus for a given force can produce faster acceleration.

Pure hydraulic systems, however, have the disadvantage of requiring a mechanical input, as shown in Figure 16.1. A typical example of Figure 16.1 is the automobile

Magnetic Actuators and Sensors, Second Edition. John R. Brauer.
© 2014 The Institute of Electrical and Electronics Engineers, Inc. Published 2014 by John Wiley & Sons, Inc.

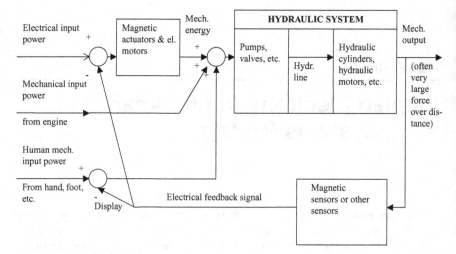

FIGURE 16.1 Block diagram showing energy conversions in a general system that includes a hydraulic system. The circles represent summing junctions. As shown, magnetic actuators require input electrical energy, which is not necessarily required by magnetic sensors with permanent magnets. The sensors may illuminate a display, such as vehicle speed, which is read by the human operator and thus used as one form of feedback control.

brake. The required large braking force can be produced by hydraulic cylinders, but in early days the only input was the human foot for the so-called manual brakes. Later, power brakes used the mechanical power of the gasoline engine to power a pump to provide additional hydraulic braking power. Nowadays, electric motors are often used to provide the power, and as discussed in Chapter 10, magnetic actuators and sensors are used for braking feedback control.

The most common way of utilizing electric power as well as optional computer control is through the magnetically operated hydraulic valve, which will be discussed later in this chapter. Thus coupled electrohydraulic systems are becoming more and more popular.

In converting energy from mechanical to hydraulic and back again, some energy is inevitably lost. Yet if the force output of Figure 16.1 is not large enough, a second hydraulic system can be added as a second hydraulic *stage*. While such multistage systems have been common in the past to produce large forces in large aircraft and tractors, to increase efficiency the trend nowadays is toward one stage that uses electrohydraulics.

16.2 HYDRAULIC BASICS AND ELECTRICAL ANALOGIES

As discussed in the preceding chapter, analogies are helpful in modeling various types of physical systems. Table 16.1 compares various physical system relationships, with its last column containing hydraulic relationships [2].

TABLE 16.1   Physical System Relationships

Parameter	Electrical	Mechanical (Linear)	Mechanical (Rotary)	Hydraulic
Inductance	L	M (mass)	J (inertia)	I (fluid inertia)
Units	henrys	kg	N m s^2	N s^2/m^5
Equation	$V = L\,di/dt$	$F = M\,dv/dt$	$T = J\,d\omega/dt$	$p = I\,dQ/dt$
Energy	$W = Li^2/2$	$W = Mv^2/2$	$W = J\omega^2/2$	$W = IQ^2/2$
Capacitance	C	K (spring)	K (spring)	C (capacitance)
Units	farads	N/m	N m/rad	m^3/Pa
Equation	$V = (1/C)\int i\,dt$	$F = Kx$	$T = K\theta$	$p = (1/C)\int q\,dt$
Energy	$W = CV^2/2$	$W = Kx^2/2$	$W = K\theta^2/2$	$W = Cp^2/2$
Resistance[a]	R	B (damper)	B (damper)	Orifice
Units	Ω (ohms)	N/m/s	N m s	(m^3/s)/N/m^2)$^{1/2}$
Equation	$V = IR$	$F = Bv$	$T = B\omega$	$Q = Kvp^{1/2}$
Power	$P = V^2/R$	$P = Bv^2$	$P = B\omega^2$	$P = pQ$
Effort	V (voltage)	F (force)	T (torque)	p (pressure)
Units	V	N	N-m	Pa $(= N/m^2)$
Equation	$\int V\,dt = LI$	$\int F\,dt = mv$	$\int T\,dt = J\omega$	$\int p\,dt = QI$
Flow	I (current)	v (velocity)	ω (angular velocity)	Q (flow rate)
Units	amperes	m/s	rad/s	m^3/s
Equation	$\int I\,dt = q$	$\int v\,dt = x$	$\int \omega\,dt = \theta$	$\int Q\,dt = q$
Power	$P = VI$	$P = Fv$	$P = T\omega$	$P = pQ$

[a]Turbulent resistance.

Just as standard symbols are used worldwide for electric circuit elements such as resistors, inductors, and capacitors, hydraulic circuits have standard symbols. Like wires in electric circuits, hydraulics lines and hoses are represented by lines.

Figure 16.2 shows several common hydraulic ISO standard symbols. The tank shown in Figure 16.2 is generally at zero pressure, similar to ground in electric circuits. As indicated in Table 16.1, pressure P in hydraulic circuits is analogous to voltage, and fluid flow rate Q is analogous to current. The pressure source is usually a pump. Some types of pumps (such as in artificial hearts) require reciprocating motion and are thus operated by reciprocating magnetic actuators. More often, however,

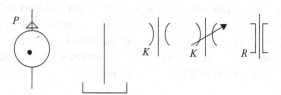

FIGURE 16.2   Symbols are shown above for several hydraulic components. From left to right they are: pressure source (e.g., a pressure-compensated pump), hydraulic line and tank, common orifice (turbulent flow with square law relation in Table 16.1), control valve (variable turbulent orifice), and linear (laminar) orifice.

pumps are powered by rotary machines such as internal combustion engines or electric motors.

Analogous to electrical capacitance is hydraulic capacitance, obtained by storage devices such as accumulators. While hydraulic capacitance is common, hydraulic inductance is usually negligible.

Analogous to the electrical resistor is the hydraulic orifice, a narrowing of a hydraulic line. For a linear orifice:

$$P = QR \qquad (16.2)$$

The more common nonlinear orifice obeys the relation in Table 16.1:

$$P = (Q/K)^2 \qquad (16.3)$$

The nonlinear orifice contains turbulent flow, while linear orifices contain laminar flow [3–5]. As shown in Figure 16.2, the two orifice types have differing symbols. Also shown in Figure 16.2, an orifice with variable R or K acts as a *control valve*. The relation between P and Q is often found by measurement, but also can be computed using fluid finite-element software [6] or other computational fluid dynamics (CFD) software. As will be examined later in this chapter, the P–Q relation for hydraulic valves is important, because valves commonly serve in hydraulic actuators. Hydraulic units are listed in Appendix A.

The most common hydraulic fluid is oil, but other fluids such as water are sometimes used. Air is another popular fluid, but systems using it are called *pneumatic* systems. Because air is much more compressible than hydraulic liquids, thermodynamic effects can no longer be ignored. Thus pneumatic systems are usually more difficult to analyze than hydraulic systems. Also, pneumatic pressures are seldom as high as common hydraulic pressures, and thus pneumatics are not covered in this book. Both hydraulics and pneumatics are examples of *fluid power systems*.

16.3 MODELING HYDRAULIC CIRCUITS IN SPICE

The varieties of systems software examined in the preceding chapter can be applied to hydraulic and electrohydraulic systems. As that chapter discussed, systems software is often preferred to pure finite-element software. While the coupled finite elements in Chapter 14 solved electric circuits coupled to magnetics and then coupled to structural mechanics, adding coupling to hydraulic finite elements is very difficult. In fact, no papers can be found that attempt coupled electrohydraulic finite-element analysis, although a few couple electromagnetic finite elements with those for acoustics [7] or fluid flow [8].

Because one of the most common systems programs used by electrical engineers is SPICE (in commercial and free versions), it is here used to model hydraulic circuits. The electric circuits in SPICE can be used to model hydraulic circuits by the analogies of Table 16.1. For linear orifices, the ordinary constant resistor suffices. However, for the nonlinear P–Q relation (16.3) of turbulent orifices, the electric circuit requires a resistor with a nonlinear V–I relation. Several versions of SPICE feature such a nonlinear resistor.

Example 16.1   SPICE Model of Hydraulic Circuit with Linear (Laminar) Orifices   A hydraulic circuit containing two laminar orifices in parallel hydraulic lines is shown in Figure E16.1.1. The orifice hydraulic resistances are $R_1 = 1$ MPa s/m^3 and $R_2 = 2$ MPa s/m^3. They are both supplied by a pump of constant pressure $P_S = 5$ kPa. Find the flow rates in both orifices.

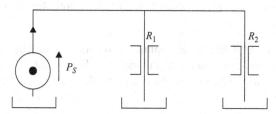

FIGURE E16.1.1   Hydraulic circuit with two parallel laminar orifices.

Solution   The analogous SPICE electric circuit is shown in Figure E16.1.2. Since there are no capacitances and no inductances (neither hydraulic nor electric), the flows are DC. The computed values agree exactly with Ohm's law, being 5 mA in R_1 and 2.5 mA in R_2. Thus the flow rates are 5E–3 m^3/s and 2.5E–3 m^3/s, respectively.

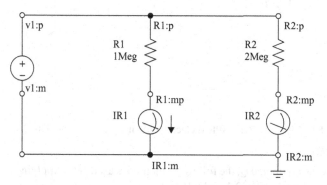

FIGURE E16.1.2   SPICE electric circuit model in Figure E16.1.1.

Example 16.2   SPICE Model of Hydraulic Circuit with Linear and Nonlinear (Turbulent) Orifices   A hydraulic circuit containing a linear laminar orifice in series with a turbulent orifice is shown in Figure E16.2.1 [9, 10]. The turbulent orifice has the constant K in its square law P–Q relation of Table 16.1. The laminar orifice has hydraulic resistance R. The given values are $K = 1$, $R = 1$, and source pressure $P_S = 1$ Pa. Find the DC flow rate Q.

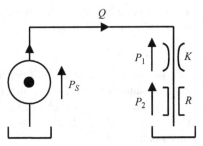

FIGURE E16.2.1   Hydraulic circuit with a turbulent orifice in series with a laminar orifice. Reprinted with permission from SAE Paper 2000-01-2633 © 2000 SAE International.

Solution   Maxwell SPICE includes nonlinear resistors, indicated by the crooked line superimposed on the upper resistor in the model in Figure E16.2.2. The computed current is 0.618 A, which means that the computed flow rate $Q = 0.618$ m^3/s.

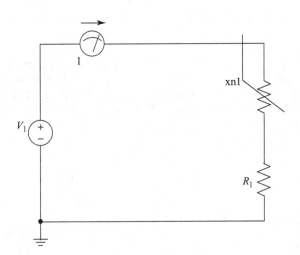

FIGURE E16.2.2   SPICE electric circuit model in Figure E16.2.1.

To verify the computed Q, the following method is used [11]. Applying Kirchhoff's voltage law to Figures E16.2.1 and E16.2.2,

$$P_S = P_1 + P_2 \qquad (E16.2.1)$$

in which from (16.2) and (16.3):

$$P_1 = \frac{Q^2}{K^2}, \quad P_2 = QR \qquad (E16.2.2)$$

Substituting (E16.2.2) into (E16.2.1) gives:

$$P_S = \frac{Q^2}{K^2} + RQ \tag{E16.2.3}$$

where Q is the unknown. Putting (E16.2.3) in quadratic form:

$$Q^2 + RK^2 Q - K^2 P_S = 0 \tag{E16.2.4}$$

Solving using the quadratic formula gives two roots [12]:

$$Q = -\frac{RK^2}{2} \pm \frac{1}{2}\sqrt{R^2 K^4 + 4K^2 P_S} \tag{E16.2.5}$$

Substituting the given unity values of R, K, and P_S gives $Q = +0.618$ and $Q = -1.618$. The negative value is impossible, and thus $Q = +0.618$ m^3/s, agreeing exactly with the SPICE computation.

Example 16.3 SPICE Model of Switched Hydraulic Circuit with Hydraulic Capacitance The hydraulic circuit shown in Figure E16.3.1 contains a manual valve "switch" that is thrown at time zero. Before time zero, all of the flow from the constant flow source Q is shunted through the valve to the tank "ground." At time zero, the valve sends the flow through the hydraulic components to the right, consisting of a laminar orifice with hydraulic capacitance. The values are $Q = 10$ m^3/s, $R = 100$ Pa s/m^3, and $C = 100\text{E--}6$ m^3/Pa. To be found is the hydraulic pressure as a function of time.

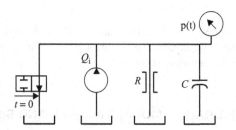

FIGURE E16.3.1 Hydraulic circuit with capacitance C, laminar resistance R, constant flow source Q_1 that is switched from tank at time zero. Reprinted with permission from SAE Paper 2000-01-2633 © 2000 SAE International.

Solution The analogous SPICE electric circuit model is shown in Figure E16.3.2. It consists of a current source of 10 A that is switched on at time zero, and a parallel 100-Ω resistor and 100-μF capacitor. Since the RC time constant is 0.01 s, the rise time of the pressure should be 0.01 s. The SPICE output voltage versus time is shown

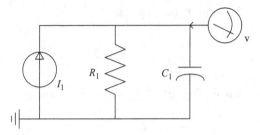

FIGURE E16.3.2 SPICE electric circuit model in Figure E16.3.1.

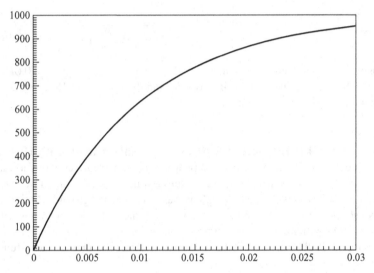

FIGURE E16.3.3 Voltage versus time (in seconds) computed by model in Figure E16.3.2.

in Figure E16.3.3. As expected, its time constant is 0.01 s. It approaches the final value of 1000 V, which equals the expected 10 A times 100 Ω. Thus the final steady-state pressure of the hydraulic circuit is 1000 Pa.

16.4 ELECTROHYDRAULIC MODELS IN SPICE AND SIMPLORER

SPICE hydraulic models can be combined with the SPICE electrical/magnetic/ mechanical models of the preceding chapter to make electrohydraulic models. If desired, other electric circuit software such as Simplorer can be used instead of SPICE.

The simplest electrohydraulic models assume ideal proportional hydraulic *control valves*. Such valves produce fluid flow proportional to valve position regardless of pressure. An example familiar to everyone is a faucet that has flow rate directly proportional to its position x. If its position is zero, there is no flow, and the flow rate is always approximately proportional to the position. In magnetic actuators, oftentimes the flow is directly through the "air" gap, and thus such actuators are often called *wet solenoids*. A dry alternative is to have the magnetic actuator armature push a rod through a physically separate proportional valve.

Such proportionality between position and flow rate is readily modeled using the *dependent sources* available in SPICE and Simplorer. SPICE includes the following four types of dependent sources identified by four letters.

- *E*, voltage-controlled voltage source with proportionality constant = (source voltage)/(controlling voltage at any node).

- *F*, current-controlled current source with proportionality constant = (source current)/(controlling current in any branch).

- *G*, voltage-controlled current source with proportionality constant = (source current)/(controlling voltage at any node).

- *H*, current-controlled voltage source with proportionality constant = (source voltage)/(controlling current in any branch).

Use of these dependent sources is shown in the SPICE electrohydraulic model in Figure 16.3 [10]. It is based on the Bessho magnetic actuator Maxwell SPICE model in Figure 15.1. Added to it are two dependent sources and other elements to model hydraulics. The hydraulic circuit is in the lower left corner of Figure 16.3, and consists of a linear orifice with $R = 1$ and the dependent source G_1. The flow source G_1 is set to 1000 m^3/(s m) and is dependent on the magnetic actuator airgap x. The second dependent source is G_2 added to the mechanical circuit on the right. G_2 is set to 1 N/Pa and is a flow force proportional to the flow in the hydraulic valve. The assumed proportionality of sources G_1 and G_2 is usually only approximately obeyed in actual hydraulic valves; the next section will discuss more realistic valve and hydraulic system models.

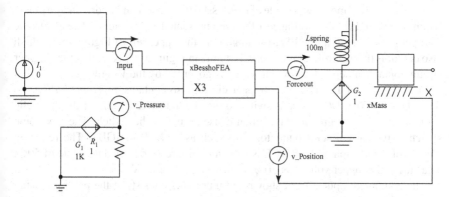

FIGURE 16.3   Maxwell SPICE model of simple electrohydraulic system with Bessho actuator driving a proportional control valve. Reprinted with permission from SAE Paper 2000-01-2633 © 2000 SAE International.

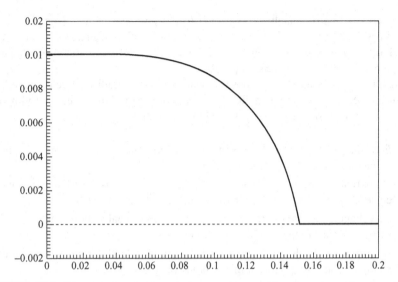

FIGURE 16.4 Computed magnetic actuator armature position in m (y axis) versus time in s for electrohydraulic system of Figure 16.3. Reprinted with permission from SAE Paper 2000-01-2633 © 2000 SAE International.

The computed Bessho actuator armature position versus time for Figure 16.3 is plotted [10] in Figure 16.4. Compared to the plot of Figure 15.3 (curve 1), the closure time has increased from 99 to 150 ms. The hydraulic force has caused the armature to move more slowly.

The next SPICE model has the linear orifice of Figure 16.3 replaced by a nonlinear orifice with $K = 3$. The computed armature closure time increases to 160 ms. Next, a hydraulic capacitance $C = 0.1$ is added in parallel with the nonlinear orifice, and the computed closure time changes to 106 ms. The capacitance reduces the hydraulic force acting on the valve, thereby increasing the speed of armature motion.

Another SPICE model was made of the Bessho magnetic actuator driven by a pulse-width modulated (PWM) voltage of the type presented in Chapter 7. The electronic chopper circuit creating the PWM voltage is in the upper left of Figure 16.5, which also contains the electrohydraulic SPICE model in Figure 16.3. Note that the chopper circuit contains a dependent switch which is governed by the V_3 voltage source that has a frequency of 100 Hz. The chopper circuit also contains a *"flyback" diode* so that when the 50-V DC source is switched off, the inductive current of the magnetic actuator can flow to ground rather than cause arcing in the switch. The electronic switch is typically a semiconductor chip such as an SCR or IGBT. The resulting PWM voltage computed by SPICE is shown in Figure 16.6 for a duty cycle of 50%. Thus the DC chopper voltage is 50% of the input 50, or 25 V.

The computed input current shown in Figure 16.7 consists of the previous rising exponential with a dip caused by the moving armature, but also it has a ripple superimposed. The ripple is at the 100-Hz chopper frequency. The resulting magnetic

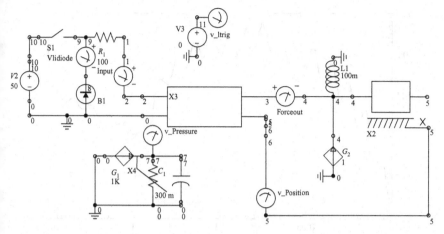

FIGURE 16.5   Chopper-driven PWM magnetic actuator in an electrohydraulic system modeled in SPICE. The chopper circuit is in the upper left, driving the Bessho actuator represented by the block-labeled X_3.

force is shown in Figure 16.8 and also contains a ripple. The computed magnetic actuator armature airgap versus time is shown in Figure 16.9; note that closure takes about 140 ms, similar to Figure 16.4. The computed hydraulic pressure and flow rate are shown in Figures 16.10 and 16.11, respectively. Since the flow rate is assumed proportional to armature position (which is 10 mm minus the airgap), the flow rate increases with time [12].

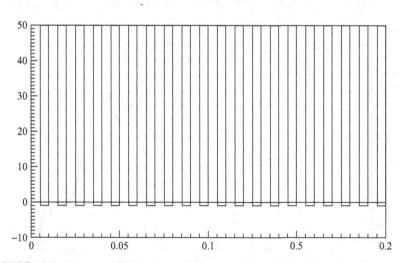

FIGURE 16.6   Computed PWM voltage (V along y axis) versus time (s) produced by chopper of Figure 16.5.

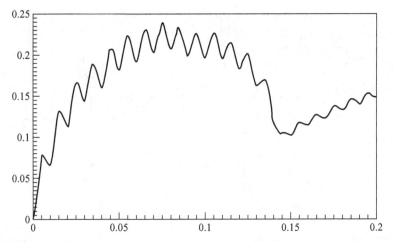

FIGURE 16.7 Computed magnetic actuator current (A on y axis) versus time (s) for Figure 16.5.

A hydraulics library is available for Simplorer. One of its hydraulic elements, the variable orifice, is used here to model a control valve operated by the Bessho magnetic actuator. The variable orifice is labeled as *RHYD*1 in the Simplorer model in Figure 16.12, which also contains a Simplorer fixed orifice with $R = 1E9$ Pa s/m^3. To achieve faster performance than the previous electrohydraulic systems of Figures 16.4–16.11, the 0.5-A current is turned on instantly (with 0 rather than 20 ms rise time) and the moving mass is reduced from 6 to 1.4 kg. The resulting armature position versus time computed by Simplorer is shown in Figure 16.13 assuming no

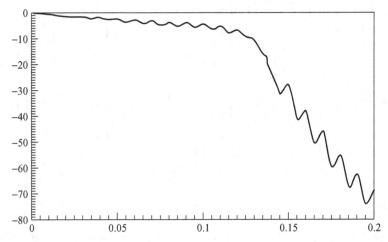

FIGURE 16.8 Computed magnetic force (N on y axis) versus time (s) for Figure 16.5.

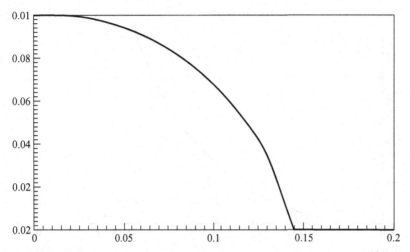

FIGURE 16.9   Computed magnetic armature position (m on *y* axis) versus time (s) for Figure 16.5.

flow forces. As expected, movement is now much faster, closing in only 34 ms. The pressure source of 1 MPa (10 bar) then produces the flow rate also graphed in Figure 16.13 for wet flow through the airgap that ceases upon its closure.

The Simplorer hydraulic orifice also allows flow forces to be included. For a flow force factor of 10 on the variable orifice, the computed results are graphed in Figure 16.14. Compared with Figure 16.13, the armature now closes in 31.5 ms, which is 2.5 ms faster.

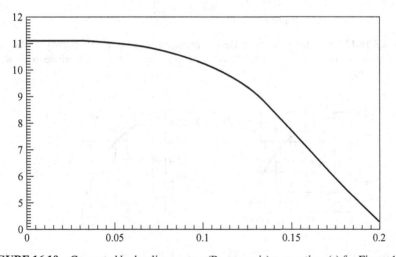

FIGURE 16.10   Computed hydraulic pressure (Pa on *y* axis) versus time (s) for Figure 16.5.

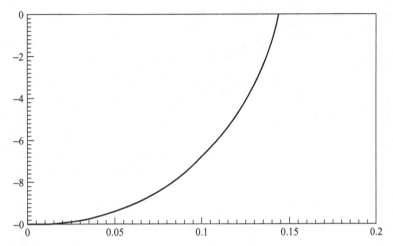

FIGURE 16.11   Computed hydraulic flow rate (m³/s on *y* axis) versus time (s) for Figure 16.5.

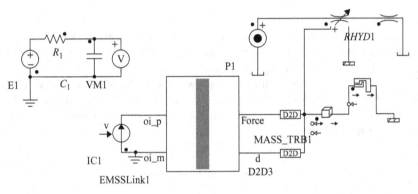

FIGURE 16.12   Simplorer model of Bessho magnetic actuator in an electrohydraulic system containing the variable orifice *RHYD*1. To its right is the fixed orifice with $R = $ 1E–9 m³/(s Pa).

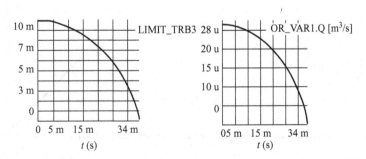

FIGURE 16.13   Computed results for Figure 16.12 with no flow forces. The left graph is armature airgap (m) versus time (s). The right graph is the flow rate (m³/s) versus time (s).

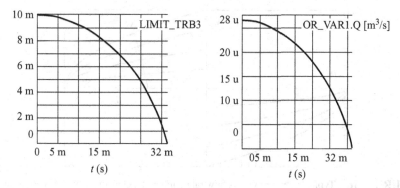

FIGURE 16.14   Computed results for Figure 16.12 with valve *RHYD1* flow force factor of 10. The left graph is armature airgap (m) versus time (s). The right graph is the flow rate (m³/s) versus time (s).

16.5   HYDRAULIC VALVES AND CYLINDERS IN SYSTEMS MODELS

16.5.1   Valves and Cylinders

The most common hydraulic actuator is the hydraulic cylinder, which contains a piston operated by hydraulic pressure. The cross-section of a hydraulic cylinder placed in a typical electrohydraulic system is shown in Figure 16.15. The differential pressure P between the two sides of the piston of area A creates the hydraulic force. Hydraulic cylinders are easily visible on many tractors, dozers, power shovels, dump trucks, garbage trucks, snowplows, vertical lift bridges, etc.

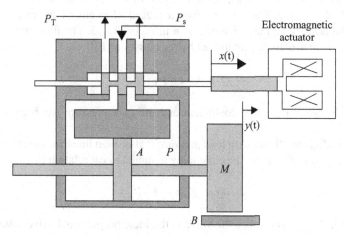

FIGURE 16.15   Electrohydraulic system containing a magnetic actuator driving a control valve that operates a hydraulic cylinder with moving mass *M*.

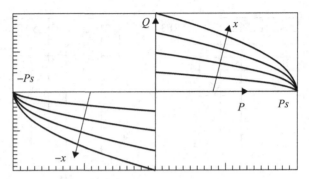

FIGURE 16.16   Typical pressure versus flow rate curves of a control valve. P is the pressure to the load and Q is the flow rate to the load. The valve spool position x is zero when the valve is closed and Q is zero.

If the cylinder position is y as shown in Figure 16.15, then the flow rate (in cubic meters per second) must obey:

$$Q = A\,dy/dt \qquad (16.4)$$

The electrohydraulic system of Figure 16.15 also shows the cross section of a typical control valve [1] driven by a magnetic actuator. The armature of the magnetic actuator is directly connected to the *valve spool*, the movable part of the valve that is shaped like a spool for thread. The spool in Figure 16.15 has three *lands*, which are large diameter cylindrically shaped regions. The valve in Figure 16.15 is called a *four-way valve* because it has a total of four flow paths [13, 14]. Its two hydraulic inputs are tank pressure P_T and supply pressure P_S.

The input–output characteristics of the valve are those shown in the typical family of curves in Figure 16.16. Each curve is a plot of output (load) pressure P versus flow rate to the load Q. The valve position x is zero when closed and can be moved in either the positive and negative direction in Figure 16.15. The flow rate curves of Figure 16.16 are usually approximated by the equation [13]:

$$Q = K_V x \sqrt{P_S - P} \qquad (16.5)$$

where K_V is called the valve coefficient and the tank pressure has been assumed zero.

Since (16.5) is nonlinear with load pressure P, it is often linearized to allow easier analysis. Over a small range, the commonly assumed linear relation is:

$$Q = K_x x - K_p P \qquad (16.6)$$

Note that if K_p is zero, then the valve is the ideal proportional valve assumed in the preceding section. This section, however, is more realistic in allowing K_p to be nonzero in the following derivations.

16.5.2   Use in SPICE Systems Models

The above valve and cylinder can be modeled in SPICE based on the linearized valve equation (16.6), the hydraulic cylinder equation (16.4), and Newton's laws. For the cylinder load mass M and damping B shown in Figure 16.15, the force balance equation is:

$$PA = B(dy/dt) + M(d^2y/dt^2) \tag{16.7}$$

To model (16.4), (16.6), and (16.7) in SPICE, they can be written as three simultaneous equations in Q, P, and y:

$$(1)Q + (K_p)P + (0)y = K_x x \tag{16.8}$$

$$(1)Q + (0)P - A\,dy/dt = 0 \tag{16.9}$$

$$(-B/A)Q + (A)P - M\,d^2y/dt^2 = 0 \tag{16.10}$$

The above three simultaneous differential equations are here simulated using three circuit meshes in SPICE. The three meshes appear in the bottom of the SPICE schematic of Figure 16.17. The upper portion of Figure 16.17 is the same as the previous SPICE electrical/magnetic/mechanical model of the Bessho magnetic actuator of Figure 15.1. As before, the block-labeled X_3 contains results of parametric magnetic finite-element analyses of the Bessho magnetic actuator.

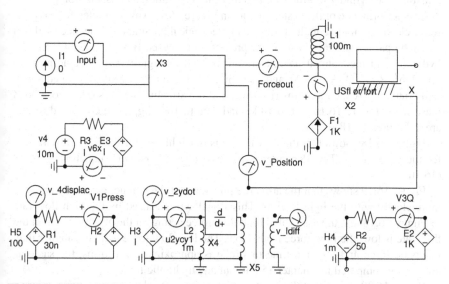

FIGURE 16.17   SPICE schematic of electrohydraulic system of Figure 16.15. The upper right armature mass is 6 kg. The upper left current source rises to 0.5 A with a time constant of 20 ms.

The lower left circuit mesh in Figure 16.17 represents (16.8). P is its mesh current to be computed. Its resistor is K_p. Its two dependent sources are related to the other two terms of (16.8). The left-hand-dependent H (current-controlled voltage) source is K_x times the armature position x. Its right–hand-dependent H (current-controlled voltage) source is equal to flow rate Q. Thus according to Kirchhoff's voltage law, this circuit mesh simulates (16.8).

The middle circuit mesh in the bottom of Figure 16.17 represents (16.9). Its mesh current is the cylinder position y to be computed. The inductor is of value A, and thus its voltage is $A(dy/dt)$. This voltage equals that of the dependent H (current-controlled voltage) source equal to Q according to Kirchhoff's voltage law. A differentiator is added to produce an additional voltage $A(d^2y/dt^2)$, which is transformed by an added ideal transformer to (d^2y/dt^2) for use elsewhere.

The lower right circuit mesh in Figure 16.17 represents (16.10). Its mesh current is the flow rate Q to be computed. The resistor is of value B/A. The left-hand-dependent H (current-controlled voltage) source is set to A times the pressure P. Its right-hand-dependent E (voltage-controlled voltage) source is set to $M(d^2y/dt^2)$, where (d^2y/dt^2) is found from the lower middle mesh.

To model the flow forces, the dependent F (current-controlled current) source is placed in the mechanical circuit in the upper right-hand corner of Figure 16.17. The flow force acting on the magnetic armature and valve spool is generally a nonlinear function of flow rate Q and position x, but is here modeled (as commonly done [14]) as a force proportional to Q directed to oppose the valve opening.

Thus the SPICE model in Figure 16.17 couples the solutions of electric currents, magnetic forces, flux linkages, hydraulic pressures and flow rates, and mechanical armature and cylinder motions, all together into one simultaneous solution [1].

The performance of the system model in Figure 16.17 was computed for a typical set of electrical, magnetic, hydraulic, and mechanical parameters. The electrical and magnetic parameters are those of the previously analyzed Bessho solenoid, and the hydraulic and mechanical parameters are labeled in Figure 16.17, including $A = 1.\text{E}{-}3 \text{ m}^2$, $B = 50A$, $K_x = 100$, and $K_p = 30.\text{E}{-}9$. The input current is assumed to rise to its rated 0.5 A with a 20 ms rise time, and the armature mass is 6 kg. For a load mass M (of Figure 16.16) set to 10 kg and then to 1000 kg, some computed results are in Figures 16.18–16.21.

Figure 16.18 shows that the flow rate Q is much higher for the light load M than for the heavy load. This dependence agrees with the governing equations (16.8)–(16.10).

Figure 16.19 shows that the armature position x of the magnetic actuator changes more slowly with the light load M. This slowdown is caused by the high flow rate of Figure 16.18 and the associated high flow force acting on the armature. When the flow rate is low, the flow force is also low. The low flow force is much smaller than magnetic force and thus the time to stroke is approximately 100 ms, the same as previously computed in Chapter 15 without any hydraulic force.

Figure 16.20 shows that the position y of the mass M changes much more slowly with the heavy mass than with the light mass, as expected from Newton's law of (16.7).

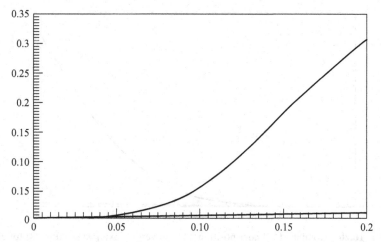

FIGURE 16.18 Computed Q (m^3/s on y axis) versus time (s) for electrohydraulic model in Figure 16.17. The upper curve is for $M = 10$ kg and the lower curve is for $M = 1000$ kg.

Finally, Figure 16.21 shows that the differential pressure P for both light and heavy values of M peaks at about 330.E5 Pa $= 330$ bar, a much higher pressure than magnetic pressures of less than 20 bar. Multiplying by $A = 1.$E-3 m^2, the peak hydraulic force is a respectable 33 kN.

The electrohydraulic system can next be modeled when chopper driven. The chopper is shown in the upper left corner of the SPICE model in Figure 16.22, and is similar to that of Figure 16.5. Note that the DC supply voltage is now 42 V, a standard voltage proposed for automobiles. The duty cycle is again 50%, and thus the output of

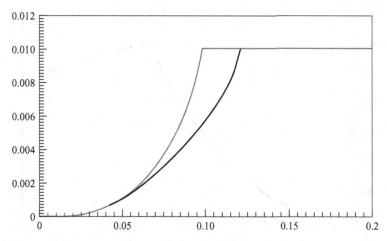

FIGURE 16.19 Computed magnetic armature position x (m along vertical axis) versus time (s) for electrohydraulic model in Figure 16.17. The right curve is for $M = 10$ kg and the left curve is for $M = 1000$ kg.

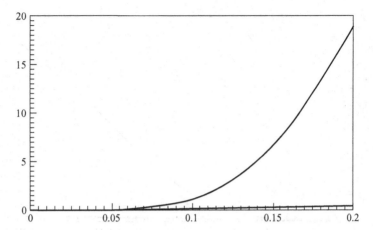

FIGURE 16.20   Computed cylinder position y (m on vertical axis) versus time (s) for electrohydraulic model in Figure 16.17. The upper curve is for $M = 10$ kg and the lower curve is for $M = 1000$ kg.

the chopper has 21 V DC. Figure 16.22 shows the computed current waveforms in the coil of the magnetic actuator. For both load mass values, the currents of Figure 16.22 show the 100-Hz chop rate superimposed on an LR circuit rise time, as expected. The currents of Figure 16.23 also exhibit the expected dip due to back voltage induced by motion that stops when the armature reaches its full 10-mm stroke. Figure 16.24 shows the computed armature position x. The armature moves its full 10 mm in 0.16 s for $M = 1000$ kg, but takes 0.19 s for $M = 10$ kg. The slowdown is caused by the high flow rate for the light mass, similar to the case without the chopper drive. Curves of

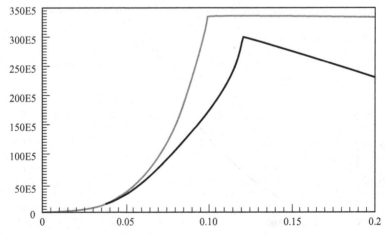

FIGURE 16.21   Computed differential pressure P (Pa on vertical axis) versus time (s) for electrohydraulic model in Figure 16.17. The lower curve is for $M = 10$ kg and the upper curve is for $M = 1000$ kg.

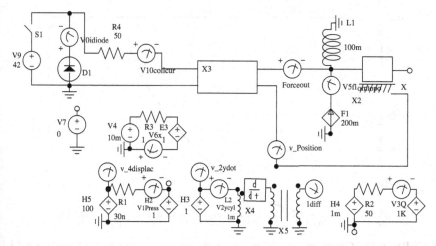

FIGURE 16.22 SPICE schematic of Figure 16.17 with added chopper drive in upper left corner. Reprinted with permission from SAE Paper OH 2002-01-1346 © 2002 SAE International and NFPA.

pressure and flow have also been obtained [15] that are qualitatively similar to those shown for the system without chopper drive.

While no magnetic sensor is shown in the electrohydraulic systems of Figures 16.15 and 16.22, oftentimes magnetic sensors are used. To determine the cylinder position, the magnetostrictive and linear variable differential transformer (LVDT) sensors of Chapter 11 are commonly employed, especially in feedback control systems. Also, to measure hydraulic flow rates, steel paddle wheels are occasionally placed

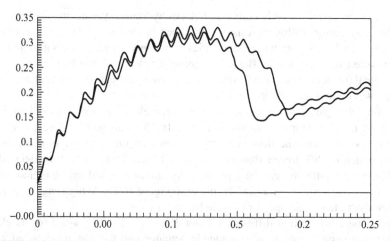

FIGURE 16.23 Computed coil current (A on y axis) versus time (s) for model in Figure 16.22. For $M = 10$ kg the current dips at 0.19 s, while for $M = 1000$ kg the current dips at 0.16 s. Reprinted with permission from SAE Paper OH 2002-01-1346 © 2002 SAE International and NFPA.

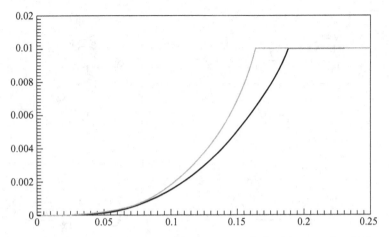

FIGURE 16.24 Computed armature position x (m on vertical axis) versus time (s) for model in Figure 16.22. The left curve is for $M = 1000$ kg, while the right curve is for $M = 10$ kg. Reprinted with permission from SAE Paper OH 2002-01-1346 © 2002 SAE International and NFPA.

in hydraulic lines; the passing steel paddles behave similarly to the gear teeth of the Hall sensors of Chapter 10. Such *flowmeters* can also use the inductive pickup coils discussed in Chapter 11.

16.6 MAGNETIC DIFFUSION RESISTOR IN ELECTROHYDRAULIC MODELS

To include the effects of eddy currents in electrohydraulic systems models, one can use the eddy current diffusion resistor introduced in Chapter 15 [16]. As described in Section 15.4, this parallel resistor varies in value with magnetic saturation. Its values were calculated for the Bessho magnetic actuator and will be used here. For the nominal 0.5-A current in the Bessho actuator, the nonlinear eddy current diffusion resistor was computed to be $R_{EN} = 1320\ \Omega$.

Adding this parallel resistor to the Simplorer electrohydraulic model in Figure 16.12, the new output is shown in Figure 16.25. With no flow force, both the armature position and the flow rate graphs show that the armature now closes in 40 ms, which is 18% longer than the 34 ms of Figure 16.13. This time delay due to eddy current diffusion can be a problem in automotive fuel injectors and other electrohydraulic systems with models similar to Figure 16.12. At high engine speeds, fuel injectors often must open or close in less than 1 ms.

The same eddy current diffusion resistor $R_{EN} = 1320\ \Omega$ is added to the electrohydraulic system with Bessho magnetic actuator and the hydraulic cylinder of Figure 16.15. The new SPICE model is shown in Figure 16.26, which has a step input current of 0.5 A and cylinder load mass of 1000 kg [17]. It has a lighter armature mass of 1.4 kg than the previous model in Figure 16.17, as well as a higher $K_x = 3000$.

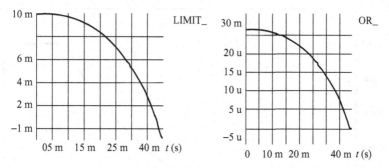

FIGURE 16.25   Computed results for Simplorer model in Figure 16.12 with added parallel resistor of 1320 Ω and with no flow forces. The left graph is armature airgap (m) versus time (s). The right graph is the flow rate (m³/s) versus time (s).

Computed results are shown with and without the 1320-Ω eddy resistor in Figures 16.27–16.31. Figure 16.27 plots armature position and Figure 16.28 shows armature magnetic force. Note that the step current and lighter armature mass cause the airgap to close much faster than 100 ms, as expected. Hydraulic pressure and flow rate are shown in Figures 16.29 and 16.30, respectively. Multiplying the peak pressure by cylinder area $A = 1.E{-}3 \text{ m}^2$, the peak hydraulic force is a respectable 30 kN. Finally, Figure 16.31 plots the cylinder load position versus time. All of the computed results show that eddy currents produce significant delays in the electrohydraulic system response [17].

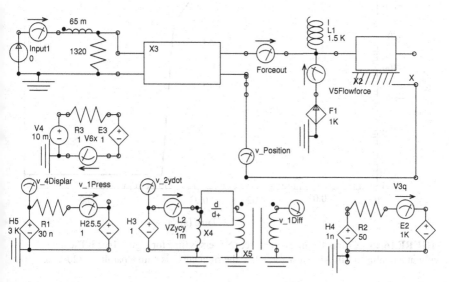

FIGURE 16.26   SPICE model of electrohydraulic system with hydraulic cylinder of Figure 16.15 with added diffusion resistor of 1320 Ω. The upper right armature mass is 1.4 kg. The upper left current source steps instantly to 0.5 A.

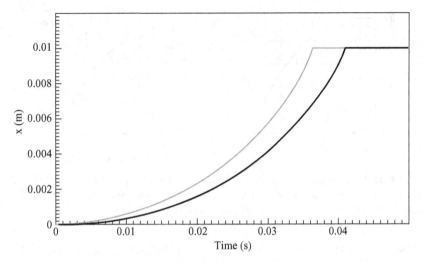

FIGURE 16.27   Computed armature position x (m) versus time for Figure 16.26. The left curve is without a nonlinear diffusion resistor, while the right curve is with a parallel 1320-Ω resistor.

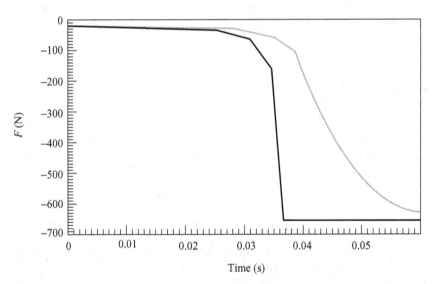

FIGURE 16.28   Computed magnetic force (N) versus time for Figure 16.26. The left curve is without a nonlinear diffusion resistor, while the right curve is with a parallel 1320-Ω resistor.

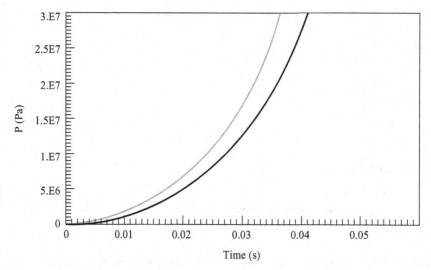

FIGURE 16.29   Computed hydraulic pressure (Pa) versus time for Figure 16.26. The left curve is without a nonlinear diffusion resistor, while the right curve is with a parallel 1320-Ω resistor.

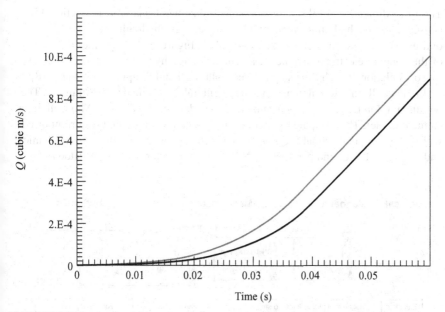

FIGURE 16.30   Computed flow rate (m³/s) versus time for Figure 16.26. The left curve is without a nonlinear diffusion resistor, while the right curve is with a parallel 1320-Ω resistor.

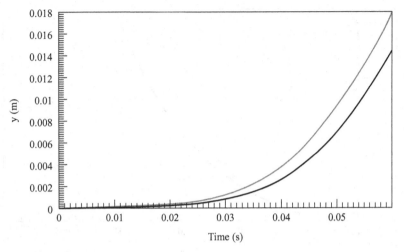

FIGURE 16.31   Computed cylinder position y versus time for Figure 16.26. The left curve is without a nonlinear diffusion resistor, while the right curve is with a parallel 1320-Ω resistor.

16.7   OPTIMIZATION OF AN ELECTROHYDRAULIC SYSTEM

The three dimensional (3D) magnetic solenoid actuator analyzed in Section 15.6.2 is used to drive a hydraulic valve, and the entire electrohydraulic system needs to be optimized. The system is shown schematically in Figure 16.32. Here the cosimulation of the coupled electrical, magnetic, mechanical, and hydraulic systems is achieved by cosimulation at each time step between Maxwell and Simplorer software [18].

In Maxwell, an external transient to transient link is specified with Simplorer. Then within Simplorer, a finite-element subcircuit is added and linked to the Maxwell finite-element model. The computed system performance includes all eddy current effects within the two dimensional (2D) or 3D magnetic actuator, as well as all mechanical and hydraulic forces and parameters. No diffusion resistor needs to be added.

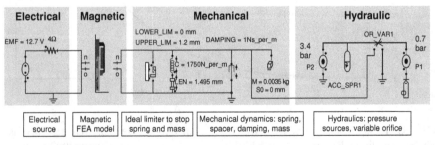

FIGURE 16.32   Schematic of time-by-step cosimulation of electrical, magnetic, mechanical, and hydraulic systems for a solenoid operating a hydraulic valve. Published with permission of TRW Automotive and ANSYS, Inc.

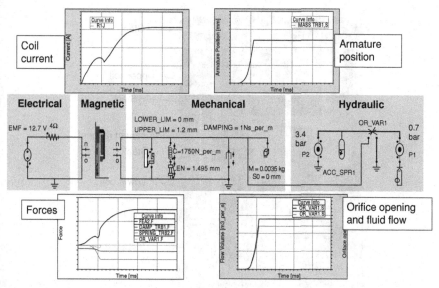

FIGURE 16.33   Computed results for the electrohydraulic system of Fig. 16.32, showing waveforms of current, position, various forces, and hydraulic flow rate. Published with permission of TRW Automotive and ANSYS, Inc.

The computed electrohydraulic system results are summarized in Figure 16.33. The computed coil current waveform shows the expected dip at closing and then rises to its final DC value. The armature position versus time display verifies the closing time of the current waveform. The force curves displayed include magnetic, mechanical, and hydraulic forces. Finally, the flow versus time curve shows the hydraulic flow rate as the valve orifice closes. The engineer can determine the effects of any electrical, magnetic, mechanical, or hydraulic design change on the system performance and thereby optimize the electrohydraulic system performance.

16.8   MAGNETIC ACTUATORS FOR DIGITAL HYDRAULICS

Instead of the valves discussed so far in this chapter, which can be called analog or servo valves, one can use more recently developed *digital valves*. Digital valves are less prone to faults (such as due to hydraulic fluid impurities) and can be less complex and thus less expensive [19].

Digital hydraulics includes digital valves, pumps, and cylinders. Digital valves have on/off or binary operation and are often placed in parallel in valve packages with each valve sized differently (such as 1, 2, 4, 8 etc. for binary control).

Digital hydraulics can be safer and more energy efficient than conventional analog hydraulics. For example, digital hydraulics can theoretically unload a railcar full of logs or coal with zero net power consumption.

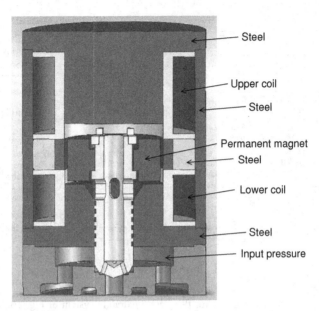

FIGURE 16.34 Digital hydraulic hammer valve with radially magnetized permanent magnet armature. Reprinted by permission from *Int J Fluid Power* [20].

The on/off valve is the key to digital hydraulics, and is most commonly operated by a magnetic solenoid actuator. Since many such on/off valves are usually required, it is necessary to reduce the size and power consumption of each solenoid actuator and its valve. A solenoid valve design developed for digital hydraulics is shown in Figure 16.34. It has an upper coil, a lower coil, and a radially magnetized permanent magnet armature, all forming a variation of the design of Figure 7.4. Called a hammer actuator, the hydraulic fluid flow around the valve input is designed to be compatible with the magnetic circuit design. This magnetic actuator is called *bistable* because it has two stable positions which are reversed by a short current pulse [20, 21]. The computed magnetic flux density in the actuator is shown in Figure 16.35 for both no coil current and for a current pulse in a coil. A typical hammer valve size is roughly that of a common 9-V battery [20].

For a typical voltage pulse swinging between +24 V and −24 V, the current waveform at a typical drive frequency is plotted in Figure 16.36. Note the good agreement between computed and measured current waveforms. For a higher drive frequency of 200 Hz (the maximum), Figure 16.37 plots both the voltage and pressure difference waveforms. The measured response time of the hammer valve was about 2.5 ms [20]. For a supply pressure of 210 bar, Figure 16.38 plots the measured pressure difference versus flow rate curve.

After prototyping the single hammer valve, 16 such valves were incorporated into a digital valve package [19,21]. With all valves in parallel, the package has 17 possible discrete flow states. There is a large number of possible valve combinations for most of the 17 flow states. Future valve packages containing more valves would enable more precise flow control.

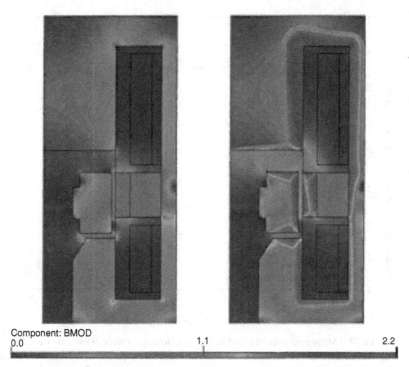

Component: BMOD
0.0   1.1   2.2

FIGURE 16.35   Flux densities (T) computed in actuator of Figure 16.34. The left display is for no coil current with the valve closed, while a coil is fed an opening current pulse in the right display (note the visible skin effects). Reprinted by permission from *Int J Fluid Power* [20].

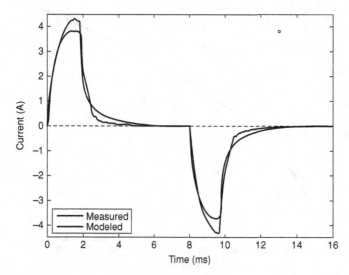

FIGURE 16.36   Comparison of computer-modeled and measured currents in the hammer valve of Figures 16.34 and 16.35 operating a typical hydraulic circuit, driven by 24 V switched at 62.5 Hz. The measured curve has slightly greater peak amplitude than the modeled curve. Reprinted by permission from *Int J Fluid Power* [20].

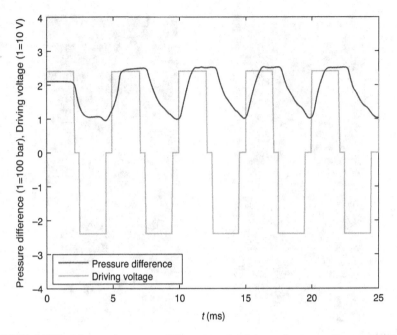

FIGURE 16.37   Measured pressure difference and driving voltage waveforms at 200 Hz switching rate in hammer valve of Figure 16.35. Reprinted by permission from *Int J Fluid Power* [20].

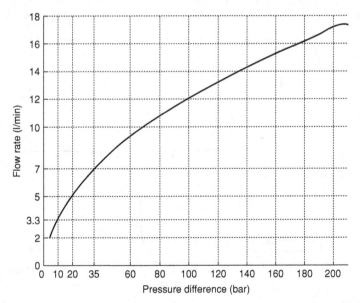

FIGURE 16.38   Measured pressure difference versus flow rate (in liters per minute) for hammer valve of Figure 16.35 with 210 bar input supply pressure and hydraulic oil temperature of 30°C. Reprinted by permission from *Int J Fluid Power* [20].

It remains to be seen whether digital hydraulics will become as commonplace as digital electronics. A recent application of digital hydraulics is flow control in petroleum production, particularly for waterflood injector wells [22].

PROBLEMS

16.1   Use SPICE to solve Example 16.1 with $R_1 = 2$ MPa s/m^3 and $R_2 = 4$ MPa s/m^3. Compare your results with simple electric circuit theory.

16.2   Use SPICE to solve Example 16.1 with $R_1 = 3$ MPa s/m^3 and $R_2 = 5$ MPa s/m^3. Compare your results with simple electric circuit theory.

16.3   Redo Example 16.2 using both SPICE and the quadratic formula for $K = 1$, $R = 1$, and source pressure $P_S = 10$ Pa.

16.4   Redo Example 16.2 using both SPICE and the quadratic formula for $K = 2$, $R = 1$, and source pressure $P_S = 10$ Pa.

16.5   Redo Example 16.2 using both SPICE and the quadratic formula for $K = 1$, $R = 2$, and source pressure $P_S = 10$ Pa.

16.6   Redo Example 16.2 using both SPICE and the quadratic formula for $K = 2$, $R = 2$, and source pressure $P_S = 10$ Pa.

16.7   Use SPICE to solve Example 16.3 with $Q = 10$ m^3/s, $R = 50$ Pa s/m^3, and $C = 100$E-6 m^3/Pa. Find the time constant of your results and compare it with RC.

16.8   Use SPICE to solve Example 16.3 with $Q = 10$ m^3/s, $R = 50$ Pa s/m^3, and $C = 50$E-6 m^3/Pa. Find the time constant of your results and compare it with RC.

REFERENCES

1. Brauer JR, Lumkes Jr. JH. Coupled model of a magnetically-actuated valve controlling a hydraulic cylinder and load. *IEEE Trans Magn* 2002;38:917–920.

2. Brauer JR. Electromagnetic analogies for modeling hydraulics. *Int Compumag Soc Newsl* 2000;7:3–5.

3. White FM. *Fluid Mechanics*, 2nd ed. New York: McGraw-Hill Book Co.; 1986.

4. Slater JG, Wanke T, Bitant J. *Introduction to Hydraulics Seminar Manual*. Milwaukee, WI: Milwaukee School of Engineering Fluid Power Institute; 2005.

5. Giles RV, Evett JB, Liu C. *Fluid Mechanics and Hydraulics*, 3rd ed. New York: McGraw-Hill Book Co.; 1994.

6. Brauer JR (ed.). *What Every Engineer Should Know About Finite Element Analysis*, 2nd ed. New York: Marcel Dekker, Inc.; 1993.

7. Rausch M, Gebhardt M, Kaltenbacher M, Landes H. Computer-aided design of clinical magnetic resonance imaging scanners by coupled magnetomechanical-acoustic modeling. *IEEE Trans Magn* 2005;41:72–81.

8. Tallbaeck GR, Lavers JD, Erraki A, Beitelman LS. Influence of model parameters on 3-D turbulent flow in an electromagnetic stirring system for continuous billet casting. *IEEE Trans Magn* 2004;40:597–600.

9. Brauer JR, Lumkes JH Jr., Slater JG. Coupled electromagnetic and hydraulic devices modeled by finite elements and circuits, *Digests of IEEE Conference on Electromagnetic Field Computation*, Milwaukee, WI, June 2000.

10. Brauer JR, Lumkes Jr. JH. Electrohydraulic systems simulations containing electromagnetic finite element models of magnetic actuators, *SAE Off-Highway & Powerplant Congress*, Milwaukee, WI, Sept. 2000, paper 2000-01-2633.

11. Johnson JL. *Design of Electrohydraulic Systems for Industrial Motion Control*. Cleveland, OH: Parker Corp.; 1991.

12. Brauer JR. Model of a Chopper-driven magnetic actuator in an electrohydraulic system, *Proceedings of the IEEE International Electric Machines and Drives Conference*, Cambridge, MA, June 2001.

13. Lumkes Jr. JH. *Control Strategies for Dynamic Systems*. New York: Marcel Dekker, Inc.; 2002.

14. Merritt HE. *Hydraulic Control Systems*. New York: John Wiley & Sons, 1967.

15. Brauer JR, Lumkes Jr. JH, Lin D. Modeling an electronically-controlled magnetic actuator operating a hydraulic valve and cylinder, *Proceedings of the International Fluid Power Expo/SAE/National Fluid Power Conference*, Las Vegas, NV, March 2002.

16. Brauer JR. Magnetic actuator models including prediction of nonlinear eddy current effects and coupling to hydraulics and mechanics, *Proceedings of Congresso Brasileiro de Eletromagnetismo*, Gramado, Brazil, November 2002.

17. Brauer JR, Mayergoyz ID. Finite element computation of nonlinear magnetic diffusion and its effects when coupled to electrical, mechanical, and hydraulic systems. *IEEE Trans Magn* 2004;40:537–540.

18. Solveson M, Christini M, Collins D. Actuator design: meeting your customer's requirements for success, Presentation at *ANSYS First-Pass System Success Workshop*, Southfield, MI, October 2008.

19. Uusitalo JP, A novel digital hydraulic valve package: a fast and small multiphysics design, Ph.D. thesis, Tampere University, Finland (2010).

20. Uusitalo JP, Ahola V, Söderlund L, Linjama M, Juhola M, Kettunen L. Novel bistable hammer valve for digital hydraulics. *Int J Fluid Power* 2010;11:35–44.

21. Uusitalo JP, Söderlund L, Kettunen L, Linjama M, Vilenius M. Dynamic analysis of a bistable actuator for digital hydraulics. *IET Science, Meas & Techn* 2009;3:235–243.

22. Konopscynski M. *A Sign of Good Things to Come?*, Oilfield Technology, May 2012.

▄▄▄▄ APPENDIX A

Symbols, Dimensions, and Units

Base dimensions and their SI unit symbols

M = mass (kg), L = length (m), T = time (s), Q = charge (C),

τ = temperature (K or °C)

TABLE A.1 Electromagnetics

Parameter and its Symbol	Dimensions	Unit Name	Unit Symbol
Charge Q	Q	coulombs	C
Electric field intensity **E**	$MLT^{-2}Q^{-1}$	volts/m	V/m
Electric flux density **D**	QL^{-2}	coulombs/m^2	C/m^2
Electric scalar potential ϕ_v	$ML^2T^{-2}Q^{-1}$	volts	V
Current I	QT^{-1}	amperes	A
Current density **J**	$QL^{-2}T^{-1}$	amperes/m^2	A/m^2
Conductivity σ	$M^{-1}L^{-2}TQ^2$	siemens/m	S/m
Resistance R	$ML^2T^{-1}Q^{-2}$	ohms	Ω
Permittivity ε	$M^{-1}L^{-3}T^2Q^{-2}$	farads/m	F/m
Capacitance C	$M^{-1}L^{-2}T^2Q^{-2}$	farads	F
Magnetic field intensity **H**	$L^{-1}T^{-1}Q$	amperes/m	A/m
Magnetic flux density **B**	$MT^{-1}Q^{-1}$	webers/m^2 = teslas	Wb/m^2 = T
Magnetic vector potential **A**	$MLT^{-1}Q^{-1}$	webers/m	Wb/m
Magnetization **M**	$L^{-1}T^{-1}Q$	amperes/m	A/m
Permeability μ	MLQ^{-2}	henrys/m	H/m
Inductance L	ML^2Q^{-2}	henrys	H
Flux ϕ	$ML^2T^{-1}Q^{-1}$	webers	Wb
Reluctance $\mathcal{R}$	$M^{-1}L^{-2}Q^2$	amperes/weber	A/Wb

Magnetic Actuators and Sensors, Second Edition. John R. Brauer.
© 2014 The Institute of Electrical and Electronics Engineers, Inc. Published 2014 by John Wiley & Sons, Inc.

TABLE A.2   Mechanics

Parameter and its Symbol	Dimensions	Unit Name	Unit Symbol
Mass M	M	kilograms	kg
Length l	L	meters	m
Time t	T	seconds	s
Velocity $\mathbf{V}$	LT^{-1}	meters/second	m/s
Force $\mathbf{F}$	MLT^{-2}	newtons	N
Pressure p	$ML^{-1}T^{-2}$	newtons/m^2 = pascals	N/m^2 = Pa
Density ρ	ML^{-3}	kilogram/m^3	kg/m^3
Energy or work W	ML^2T^{-2}	newton meters = joules	N m = J
Power P	ML^2T^{-3}	watts	W
Stiffness K	MT^{-2}	kilogram/second^2	kg/s^2
Damping B	MT^{-1}	kilogram/second	kg/s
Modulus of elasticity E	$ML^{-1}T^{-2}$	newtons/m^2 = pascals	N/m^2 = Pa

TABLE A.3   Hydraulics

Parameter and its Symbol	Dimensions	Unit Name	Unit Symbol
Pressure p	$ML^{-1}T^{-2}$	newtons/m^2 = pascals = 1.E–5 bar	N/m^2 = Pa = 1.E-5 bar
Flow rate Q	L^3T^{-1}	m^3/s = 1000 liters/s	m^3/s = 1000 L/s
Laminar orifice resistance R	$M^{-2}L^{-1}T^{-1}$	pascal s/m^3	Pa s/m^3
Turbulent orifice coefficient K	$M^{-2}L^9T^3$	m^7/(N^{2}s)	m^7/(N^{2}s)
Hydraulic capacitance C	$M^{-1}L^4T^2$	m^3/pascal	m^3/Pa

TABLE A.4   Heat

Parameter and its Symbol	Dimensions	Unit Name	Unit Symbol
Temperature T	τ	kelvin = 273 + degree celsius	K = 273 + °C
Quantity of heat energy W	ML^2T^{-2}	newton meters = joules	N m = J
Heat flow or heat flux Q	ML^2T^{-3}	watts	W
Heat flux density q	MT^{-3}	watts/m^2	W/m^2
Thermal conductivity k	$MLT^{-3}\tau^{-1}$	watts/(m °C)	W/(m °C)
Film coefficient h	$MT^{-3}\tau^{-1}$	watts/(°C m^2)	W/(°C m^2)

APPENDIX B

Nonlinear *B–H* Curves

TABLE B.1   Constants for $\mu = B/H = [1/(k_1\,e^{k_2\,B^2} + k_3)] + \mu_o$ furnished by Mark A. Juds with permission of Eaton Corporation. Note that these are all approximate and highly dependent on manufacturing methods, etc.

Ferromagnetic Material	k_1	k_2	k_3
1010 annealed—US Steel	4.847	1.908	227.3
1010 cold rolled—US Steel	36.62	1.331	534.9
1020 annealed—US Steel	4.770	2.055	302.2
1020 cold rolled—US Steel	14.23	1.699	806.5
1030 annealed—US Steel	50.00	1.371	645.3
1030 cold rolled—US Steel	40.00	1.416	1212.
Armco® H0	0.00001500	4.650	11.08
Armco M6	0.006819	3.195	10.59
Armco M15	0.8795	2.666	94.27
Armco M19	2.150	2.477	83.03
Armco M22	2.214	2.412	90.38
Armco M36	1.683	2.432	103.5
Armco M45	3.500	2.148	124.8
Armco M47	0.1247	3.335	70.83
Carpenter® Hiperco® 15	5.137	1.700	389.8
Carpenter Hiperco 50 A	0.0009388	2.816	49.92
Carpenter HiPerm® 49 annealed	0.001857	6.265	8.250
Carpenter HiPerm 49 mill processed	53.86	0.9941	149.3
Carpenter HiMu® 80	0.0002031	26.73	1.858
Carpenter HiMu 800	0.00002000	31.32	2.758
Carpenter silicon core iron B-FM	1.192	2.812	174.5
Cast iron gray—Metals HDBK	4093	0.9865	142.1
Cast iron nodular—Metals HDBK	55.04	2.324	1630
Magnetics® SqPermalloy	0.04500	9.091	6.344

(*continued*)

Magnetic Actuators and Sensors, Second Edition. John R. Brauer.
© 2014 The Institute of Electrical and Electronics Engineers, Inc. Published 2014 by John Wiley & Sons, Inc.

TABLE B.1 (*Continued*)

Ferromagnetic Material	k_1	k_2	k_3
Magnetics Supermalloy®	0.1073	9.176	3.046
Magnetics Supermendur®	0.03715	0.9109	11.12
Metglas® 2605S-2	0.00003232	5.632	2.164
Metglas 2605S-3A	0.0007339	5.178	1.588
Metglas 2605SM	0.003031	6.138	1.171
Metglas 2826MB	0.00002981	17.82	0.9301
Micrometals 26	2000	1.092	889.4
Micrometals 52	2002	1.023	1901
Stainless steel 416—Metals HDBK	11.92	2.749	1036
Stainless steel 430F $H_{RB} = 78$—Carpenter	0.01186	7.701	418.2
Stainless steel 430F $H_{RB} = 87$—Carpenter	0.1384	6.117	708.9
Stainless steel 430FR 9.53 mm Dia–Carpenter	0.02981	6.880	407.4
Stainless steel 430FR 15.9 mm Dia–Carpenter	0.01375	8.428	723.5

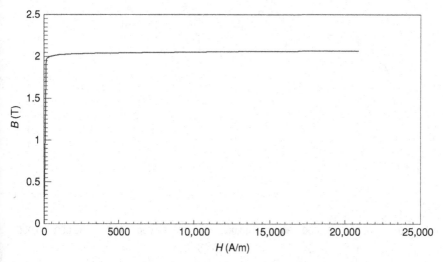

FIGURE B.1   "Step" *B–H* curve suitable for finite-element analysis, in which the slope below 1.93 T is 10,000 times the permeability of free space and the slope above 2 T approaches the permeability of free space. The data points are in Table B.2.1.

TABLE B.2.1   Data points for "step" *B–H* curve of Figure B.1.

B (teslas)	H (A/m)
0	0
1.93	153.5
1.94	155.1
1.95	158.3
1.96	164.7
1.97	177.4
1.98	202.9
1.99	304.9
2.00	406.9
2.01	610.9
2.02	1009
2.03	1805
2.04	3396
2.05	6578
2.06	12,942
2.07	20,897

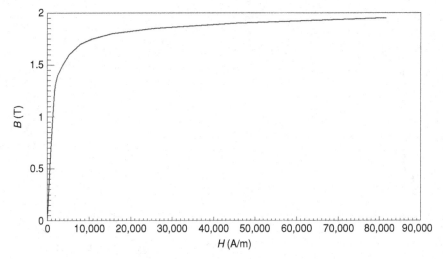

FIGURE B.2   Real *B–H* curve used for finite-element analysis of cylindrical plunger and stopper of the Bessho actuator. The relative permeability for $B < 0.55$ T is $(0.55/695)/12.57E–7 = 630$. The data points are in Table B.2.2.

TABLE B.2.2   Data points for "real" *B–H* curve of Figure B.2.

B (teslas)	H (A/m)
0	0
0.55	695
1.0	1350
1.15	1600
1.25	1800
1.3	1950
1.35	2200
1.4	2600
1.5	3800
1.6	5300
1.7	8000
1.75	10,652
1.8	15,624
1.85	25,568
1.9	45,457
1.95	81,618
1.96026	89,780
2.15513	244,854

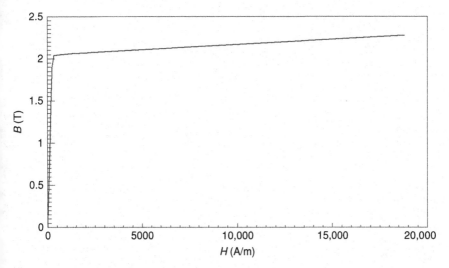

FIGURE B.3   Ramp *B–H* curve used for finite-element analysis. The relative permeability of the initial ramp is 630, and of the final ramp is 1. See Table B.2.3 for data points.

**TABLE B.2.3   Data points for
ramp *B–H* curve of Figure B.3.**

B (teslas)	*H* (A/m)
0	0
1.0	1263
1.5	1894
1.9	2399
2.04	3396
2.05	6578
2.06	12,942
2.07	20,897
2.07263	22,987
2.12253	62,691
2.13041	68,960
2.28009	188,073

Final Answers for Odd-Numbered Problems

(2.1) $\nabla T = 10\mathbf{u}_x + 60y^2\mathbf{u}_y + 30\mathbf{u}_z\,°/\mathrm{m}$

(2.3) $\nabla\!\cdot\!\mathbf{A} = 32(1)^3 + 10 = 42$, $\nabla \times \mathbf{A} = (-11)\mathbf{u}_x + (-4)\mathbf{u}_y + (-15)\mathbf{u}_z$

(2.5) $\mathbf{J} = -12\mathbf{u}_x - 4\mathbf{u}_y - 279\mathbf{u}_z\ \mathrm{A/m^2}$

(2.7) (a) $\mathbf{B} = (1.2\mathrm{E}\text{–}3\ \mathbf{u}_x + 1.6\mathrm{E}\text{-}3\mathbf{u}_y)\mathrm{T}$
   (b) $\mathbf{B} = (3\mathbf{u}_x + 4\mathbf{u}_y)\mathrm{T}$
   (c) $\mathbf{B} = (0.93\mathbf{u}_x + 1.24\mathbf{u}_y)\mathrm{T}$

(2.9) $V = -4524\cos(2\pi 60t)\ \mathrm{V}$, $I = -1131\cos(2\pi 60t)\mathrm{A}$,

$E = -514\cos(2\pi 60t)\ \mathrm{V/m}$, $J = -18.2\mathrm{E}9\cos(2\pi 60t)\ \mathrm{A/m^2}$

(2.11) $\mathbf{E} = -4\mathbf{u}_y - 75.4x\cos(2\pi 60t)\mathbf{u}_z\ \mathrm{V/m}$

(2.13) $\mathbf{J}_{\mathrm{disp}} = 12.07\mathrm{E}\text{–}6\cos(2\pi 50t)\mathbf{u}_y\ \mathrm{A/m^2}$,

$I_C = 42.25\mathrm{E}\text{–}9\cos 314.16t\ \mathrm{A}$

(3.1) $\phi = 10.93\mathrm{E} - 4\ \mathrm{Wb}$, $B = 0.102\ \mathrm{T}$

(3.3) $B_{\mathrm{align}} = 0.6285\ \mathrm{T}$, $B_{\mathrm{misalign}} = 0.06285\ \mathrm{T}$

(4.1)

$$\begin{pmatrix} 1352 & 1750 & 2148 & 0 & 0 \\ 1750 & 2307 & 2864 & 0 & 0 \\ 2148 & 2864 & (3580+2.704) & 3.500 & 4.296 \\ 0 & 0 & 3.500 & 4.614 & 5.728 \\ 0 & 0 & 4.296 & 5.728 & 7.160 \end{pmatrix} \begin{bmatrix} A_1 \\ A_2 \\ A_3 \\ A_4 \\ A_5 \end{bmatrix} = \begin{bmatrix} 1.333 \\ 1.333 \\ 1.333 \\ 0 \\ 0 \end{bmatrix}$$

(4.3) Energy stored $= 137.28\ \mathrm{J}$ with energy error $= 0.41\%$

Magnetic Actuators and Sensors, Second Edition. John R. Brauer.
© 2014 The Institute of Electrical and Electronics Engineers, Inc. Published 2014 by John Wiley & Sons, Inc.

(5.1) $\mathbf{B} = -1\mathbf{u}_y$

(5.3) $\mathbf{B} = 2y\mathbf{u}_x$

(5.5) $P_{mag} = 480,984$ Pa

(5.7) $F_{mag} = 62.03$ N

(5.9) $\mathbf{F} = 80\mathbf{u}_z$ N/m

(5.11) The computed force on the lower magnet is 3.43 N upward.

(5.13) The computed force on the lower magnet is 68.45 N upward.

(5.15) This proof uses the differential volume of a cylinder.

(6.1) The finite-element value of flux for 0.1 m depth is 0.00314 Wb for one turn. The reluctance method obtains flux = 0.00125 Wb.

(6.3) Proof. Evaluation gives magnetic force = −113 N, agreeing exactly with corrected result of Example 6.2.

(6.5) $L_{11} = L_{22} = 6.06E{-}8$ H and $L_{12} = L_{21} = 1.799E{-}8$ H, all for one turn.

(6.7) $L = 12.477E{-}4$ H

(6.9) $Z = (-565.5 + j377)$ Ω. Note: for realistic positive R, given λ should have negative imaginary part.

(6.11) For 20 turns in the lower coil, the matrices for one turn have $L_{11} = L_{22} = 8.90E{-}8$ H, $L_{12} = L_{21} = 1.785E{-}8$ H, $R_{11} = R_{22} = 8.9E{-}8$ Ω, and $R_{12} = R_{21} = 6.23E{-}8$ Ω.

(7.1) **(a)** force on the left gap = 529 N in the −y direction for 1 m depth.
 (b) same force on right gap, so total force = 1058 N/m depth.
 (c) 1431 N/m depth.
 (d) 1431.6 N/m depth
 (e) plot shows force as high as 2200 N.

(7.3) **(a)** force = 1244 N for 1 m depth.
 (b) force = 1533 N for 1 m depth (for 1600 A).
 (c) force = 1566 N for 1 m depth (for 1600 A).

(7.5) 15.1 N

(7.7) Plot shows total force as high as 1000 N.

(8.1) Matrix for one turn has $L_{11} = L_{22} = 6.058E{-}8$ H, $L_{12} = L_{21} = 1.795E{-}8$ H, $R_{11} = R_{22} = 5.6E{-}9$ Ω and $R_{12} = R_{21} = 4.05E{-}9$ Ω

(8.3) (a) 64.97 m, (b) 8.53 mm, (c) 325 μm

(8.5) $L_p = 1.7$ mH, $R_p = 9.612$ Ω

(8.7) Force has time-average value of 8.566 N and an "AC fluctuation" of 8.528 N

(8.9) Power loss $= 0.198$ W/m depth

(9.1) $a = 3613$ m/s^2, for $s = 5.$E-3 m $t = 1.664$ ms

(9.3) (a) 581.4 ms, (b) energy $= 0.658$ J, (c) 581.4 ms

(9.5) (a) 0.0581 ms, (b) energy $= 0.653$ J, (c) 0.0581 ms

(9.7) Finite-element diffusion times are 28 and 14 ms

(10.1) (a) $\sigma_o = 6.4$ S/m, (b)

$$\begin{pmatrix} \sigma_{xx} & \sigma_{xy} \\ \sigma_{yx} & \sigma_{yy} \end{pmatrix} = \begin{pmatrix} 6.1761 & -1.1759 \\ 1.1759 & 6.1761 \end{pmatrix}$$

(10.3) Power loss $= 5$ W without Al, 5.224 W with Al

(10.5) $V_y = k_H J_x d_y (0.03143 + 0.2828 \sin n_T \theta)$V

(11.1) $V = -\Omega N n_T (283$E$-6) \cos(n_T \Omega t)$V

(11.3) (a) $Z(0.002) = 228$E$-5 + j664$E-5 Ω, $Z(0.004) = 2055$E-5 $+ j757$E-5 Ω

(b) $Z(0.002) = 318.7$E$-5 + j1206$E-5 Ω, $Z(0.004) = 280$E-5 $+ j1408$E-5 Ω.

(c) $Z(0.002) = 489$E$-5 + j2215$E-5 Ω, $Z(0.004) = 426$E-5 $+ j2652$E-5 Ω.

(11.4) 15.55 V

(12.1) $N = 826, I = 1.211$ A

(12.3) $N = 1077, I = 0.929$ A

(12.5) $R = 0.5726$E-4 Ω/m, J plot is as high as almost 350,000 A/m^2

(12.8) $P = 0.365$ W, J plot shows differing densities.

(12.9) Maximum temperature is 48.1°C.

(12.11) Maximum temperature is 71.3°C.

(13.1) (a) plot with skin depth $= 0.6$ mm
(b) plot shows aperture has 0.038 T inside and 0.020 T outside
(c) plot with skin depth $= 0.08$ mm
(d) outside location has 104 mT

(13.3) Characteristic impedance $= 67.7$ Ω

(13.5) Characteristic impedance $= 65.55$ Ω

(14.1) Reluctance $\mathcal{R} = 95.47$, $|I_1| = 738.5$ A, $|I_2| = 184.6$ A

(15.1) Plot shows $I(t = 1) = 0.79987$ A

(15.3) The current minimum at 0.8 s is now 0.721 A

(15.5) The maximum (negative) force is now 51.22 N

(15.7) Comparison of responses of actuator models with mass 0.06 kg

Time domain Specification	Third-order system Simulation results	Reduced-order Simulation results	Reduced-order Using (E15.4.2)
P.O.	14.7%	15.0%	15.04%
t_p, s	0.745	0.704	0.7104
t_s, s	1.56	1.52	1.5

(15.9) The peak magnitude is now at a higher frequency, at approximately $\omega = 5.164$ rad/s $= (1.6/.06)^{1/2}$. At $\omega = 1000$ rad/s, the second-order magnitude is approximately -90 dB and the third-order magnitude is approximately -120 dB. At the same frequency, the second-order phase angle is close to $-180°$ and the third-order phase angle is close to $-270°$. At lower frequencies the two curves become quite close.

(16.1) The flow rates are 2.5E–3 m^3/s in R_1 and 1.25E–3 m^3/s in R_2

(16.3) $Q = 2.70$ m^3/s

(16.5) $Q = 2.31$ m^3/s

(16.7) At $t = 30$ ms, the pressure is 499.755 Pa

INDEX

Magnetic Actuators and Sensors, Second Edition. John R. Brauer.
© 2014 The Institute of Electrical and Electronics Engineers, Inc. Published 2014 by John Wiley & Sons, Inc.

Printed in the United States
By Bookmasters

Printed in the United States
By Bookmasters